Lecture Notes in Computer Sci

Founding Editors

Gerhard Goos
Juris Hartmanis

The series Lecture Notes in Computer Science (LNCS), including its subseries Lecture Notes in Artificial Intelligence (LNAI) and Lecture Notes in Bioinformatics (LNBI), has established itself as a medium for the publication of new developments in computer science and information technology research, teaching, and education.

LNCS enjoys close cooperation with the computer science R & D community, the series counts many renowned academics among its volume editors and paper authors, and collaborates with prestigious societies. Its mission is to serve this international community by providing an invaluable service, mainly focused on the publication of conference and workshop proceedings and postproceedings. LNCS commenced publication in 1973.

Gordon Waiter · Tryphon Lambrou ·
Georgios Leontidis · Nir Oren · Teresa Morris ·
Sharon Gordon
Editors

Medical Image Understanding and Analysis

27th Annual Conference, MIUA 2023
Aberdeen, UK, July 19–21, 2023
Proceedings

Springer

Editors
Gordon Waiter ⓘD
University of Aberdeen
Aberdeen, UK

Tryphon Lambrou ⓘD
University of Aberdeen
Aberdeen, UK

Georgios Leontidis ⓘD
University of Aberdeen
Aberdeen, UK

Nir Oren ⓘD
University of Aberdeen
Aberdeen, UK

Teresa Morris
University of Aberdeen
Aberdeen, UK

Sharon Gordon ⓘD
University of Aberdeen
Aberdeen, UK

ISSN 0302-9743   ISSN 1611-3349 (electronic)
Lecture Notes in Computer Science
ISBN 978-3-031-48592-3   ISBN 978-3-031-48593-0 (eBook)
https://doi.org/10.1007/978-3-031-48593-0

This Springer imprint is published by the registered company Springer Nature Switzerland AG
The registered company address is: Gewerbestrasse 11, 6330 Cham, Switzerland

Paper in this product is recyclable.

Preface

We are very pleased to present the proceedings of the 27th Conference on Medical Image Understanding and Analysis (MIUA 2023), a UK-based international conference for the communication of image processing and analysis research and its application to biomedical imaging and biomedicine. The conference was held at the University of Aberdeen, UK, during July 19–21, 2023, and featured presentations from the authors of all accepted papers.

This year's edition was co-chaired by Gordon Waiter, Tryphon Lambrou, Georgios Leontidis, Nir Oren, Teresa Morris, and Sharon Gordon at the Suttie Centre, University of Aberdeen.

The organizers were academic members from: the Aberdeen Biomedical Imaging Centre (https://www.abdn.ac.uk/ims/research/abic/); the Aberdeen Centre for Health Data Science (https://www.abdn.ac.uk/achds/); and the School of Natural and Computing Sciences (https://www.abdn.ac.uk/ncs/).

The conference was organized with sponsorship received from Platinum Sponsor: SINAPSE (Scottish Imaging Network: A Platform for Scientific Excellence) (http://www.sinapse.ac.uk/), Gold Sponsor: Canon Medical Research Europe (https://research.eu.medical.canon/), Silver Sponsors: MathWorks (http://www.mathworks.com/), Philips (https://www.philips.co.uk/healthcare), Aidence (https://www.aidence.com/), and The British Machine Vision Association (https://britishmachinevisionassociation.github.io/). The conference proceedings were published in partnership with Springer (https://www.springer.com).

The diverse range of topics covered in these proceedings reflects the growth in development and application of biomedical imaging. The conference proceedings feature the most recent work in the fields of: (i) Image interpretation, (ii) Radiomics, Predictive models and Quantitative imaging, (iii) Image classification, (iv) Biomarker detection.

This year's edition of MIUA received a large number of high-quality submissions, making the review process particularly competitive. In total, 81 (42 full papers) submissions were submitted to the Conference Management Toolkit (CMT), and after an initial quality check, the papers were sent out for the peer-review process, completed by the Scientific Review Committee consisting of 83 reviewers. To keep the quality of the reviews consistent with the previous editions of MIUA, the majority of the reviewers were selected from (i) a pool of previous MIUA conference reviewers and (ii) authors and co-authors of papers presented at past MIUA conferences.

All submissions were subject to double-blind review by at least three members of the Program Committee and meta-reviewed by at least one member of the Scientific Review Committee. Based on their recommendations, a ranking was created and the best 24 papers (out of 42, i.e. 57%) were accepted as full papers for presentation at the conference. Furthermore, the papers included in the proceedings were revised by the authors following feedback received from the reviewers.

Submissions were received from authors at 80 different institutes from 13 countries across five continents, including China (2), Egypt (9), France (3), Germany (5), India (3), Mexico (1), Morocco (2), Nepal (1), Norway (1), South Korea (1), Spain (1), the United Arab Emirates (3), the UK (40), and the USA (7). Papers were accepted from a total of 350 authors, with an average of ~4.5 co-authors per paper.

We thank all members of the MIUA 2023 Organizing, Steering, Program, and Scientific Review Committees. In particular, we sincerely thank all who contributed greatly to the success of MIUA 2023: the authors for submitting their work, the reviewers for insightful comments improving the quality of the proceedings, the sponsors for financial support, and all participants in this year's in-person MIUA conference.

We also thank our exceptional keynote speakers Greg Slabaugh (Professor of Computer Vision and AI at the Queen Mary University of London), Michael Kampffmeyer (Associate Professor and Head of the Machine Learning Group at the University of Tromsø (UiT), The Arctic University of Norway), Tim Cootes (Professor of Computer Vision at the University of Manchester), and Lesley Anderson (Chair in Health Data Science at the University of Aberdeen) for sharing their success, knowledge, and experiences.

July 2023

Gordon Waiter
Tryphon Lambrou
Georgios Leontidis
Nir Oren
Teresa Morris
Sharon Gordon

Organization

General Chair

Gordon Waiter   University of Aberdeen, UK

Program Committee Chairs

Tryphon Lambrou   University of Aberdeen, UK
Georgios Leontidis   University of Aberdeen, UK
Nir Oren   University of Aberdeen, UK
Teresa Morris   University of Aberdeen, UK
Sharon Gordon   University of Aberdeen, UK

MIUA Steering Committee

Ke Chen   University of Liverpool, UK
Tryphon Lambrou   University of Aberdeen, UK
Sasan Mahmoodi   University of Southampton, UK
Stephen McKenna   University of Dundee, UK
Mark Nixon   University of Southampton, UK
Nasir Rajpoot   University of Warwick, UK
Constantino Carlos   City, University of London, UK
    Reyes-Aldasoro
Maria del C. Valdes-Hernandez   University of Edinburgh, UK
Bryan M. Williams   Lancaster University, UK
Xianghua Xie   Swansea University, UK
Xujiong Ye   University of Lincoln, UK
Yalin Zheng   University of Liverpool, UK
Reyer Zwiggelaar   Aberystwyth University, UK
Guang Yang   Imperial College London, UK
Angelica Aviles-Rivero   University of Cambridge, UK
Michael Roberts   University of Cambridge, UK
Carola-Bibiane   Schönlieb University of Cambridge, UK
Bartłomiej W. Papiez   University of Oxford, UK
Ana I. L. Namburete   University of Oxford, UK

Mohammad Yaqub Mohamed bin Zayed	University of Artificial Intelligence, United Arab Emirates, and University of Oxford, UK
J. Alison Noble	University of Oxford, UK

Additional Reviewers

Adrian Galdran
Amaya Gallagher-Syed
Anna Breger
Aras Asaad
Carlos F. Moreno-Garcia
Chaoyan Huang
Cheng Xue
Chenyu You
Clarisse F. de Vries
Cohen Ethan
Constantino Reyes-Aldasoro
Cristian Linte
Damian J. J. Farnell
David Rodriguez Gonzalez
Elena Loli Piccolomini
Fuying Wang
Gilberto Ochoa-Ruiz
Gordon Waiter
Henning Muller
Hok Shing Wong
Jaidip M. Jagtap
Jialu Li
Jinah Park
Jorge Novo
Juan I. Arribas

Lei Qi
Lihao Liu
Ming Li
Nicholas Senn
Nicolas Basty
Omnia Alwazzan
Pablo G. Tahoces
Qiangqiang Gu
Rakkrit Duangsoithong
Rihuan Ke
Sarah Lee
Shan Raza
Timothy Cootes
Tryphon Lambrou
Vasiliki Mallikourti
Weiqin Zhao
Weiwei Zong
Xiaodan Xing
Xiaohui Zhang
Xujiong Ye
Yang Nan
Yashbir Singh
Yijun Yang
Yingying Fang
Zakaria Belhachmi

Contents

Image Classification

Image Interpretation

Segmentation of White Matter Hyperintensities and Ischaemic Stroke Lesions in Structural MRI

Jesse Phitidis[1,2]([✉]) [ID], Alison Q. O'Neil[1,2] [ID], Stewart Wiseman[1] [ID], David Alexander Dickie[3] [ID], Eleni Sakka[1] [ID], Agniete Kampaite[1] [ID], William Whiteley[1,4,5] [ID], Miguel O. Bernabeu[1] [ID], Beatrice Alex[1] [ID], Joanna M. Wardlaw[1] [ID], and Maria Valdés Hernández[1] [ID]

[1] University of Edinburgh, Edinburgh, UK
J.Phitidis@ed.ac.uk
[2] Canon Medical Research Europe, Edinburgh, UK
[3] University of Glasgow, Glasgow, UK
[4] University of Oxford, Oxford, UK
[5] McMaster University, Hamilton, Canada

Abstract. White matter hyperintensities (WMH) and ischaemic stroke lesions are frequently seen on brain magnetic resonance images (MRI) of people with cerebral small vessel disease (SVD). Segmentation and differentiation of these lesions is important in diagnosis, prognosis and management, but this is challenging to automate because they have similar appearance. In this study, we used MRI scans from four cohorts of people with sporadic SVD with both WMH and ischaemic stroke, acquired under diverse imaging protocols, totalling 297 individuals. We compared two state-of-the-art medical image segmentation frameworks and investigated the data characteristics affecting the performance. We found that nnU-Net and an ensemble of two Auto3DSeg-trained models outperform ($p < 0.05$ for WMH and stroke Dice Coefficients (DSC)) a model previously proposed for this task, achieving mean DSC of 0.613 and 0.615 respectively compared with 0.512. The segmentation performance was better when the stroke lesions were subcortical than when they were cortical, with a mean DSC of 0.635 compared with 0.580, for nnU-Net. We found that including only a small number ($n = 12$) of scans with lower contrast and lower resolution to the high quality training data ($n = 150$) resulted in improvements ($p < 0.05$) in the WMH DSC and 95th percentile of the Hausdorff Distance (95% HD) for stroke on the 61 lower quality scans in the test set. In conclusion, cortical stroke lesions are more likely to be confounded with WMH due to their size and shape, despite their different locations, and including only a few would-be out-of-distribution examples in the training data can result in statistically significant performance gains.

Keywords: Cerebral small vessel disease · Machine learning · Medical image segmentation

Supported by Medical Research Scotland and Canon Medical Research Europe.

G. Waiter et al. (Eds.): MIUA 2023, LNCS 14122, pp. 3–17, 2024.
https://doi.org/10.1007/978-3-031-48593-0_1

1   Introduction

Cerebral small vessel disease (SVD) is common with ageing and is associated with increased risk of dementia and stroke (both ischaemic and haemorrhagic) [4]. While the underlying pathogenesis is not fully understood, established imaging markers visible on magnetic resonance imaging (MRI) include white matter hyperintensities (WMH) of presumed vascular origin, subcortical infarcts, lacunes of presumed vascular origin, enlarged perivascular spaces, cerebral microbleeds, and brain atrophy. Automated segmentation of these markers is desirable for volumetric and regional quantification in research studies, as well as for automated reporting in clinical environments and patient prognosis and management [3]. In this paper, we focused on segmenting WMH of presumed vascular origin and ischaemic stroke lesions (Fig. 1).

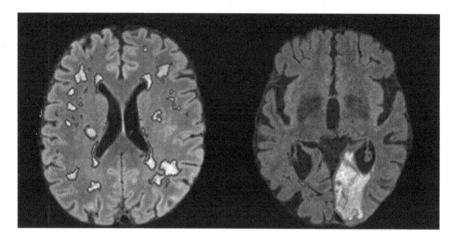

Fig. 1. Sample FLAIR MRI images and ground truth showing WMH (blue) and stroke lesions (red). The left image shows a subcortical stroke with similar size and intensity to surrounding WMH. The right image shows a large cortical stroke lesion. (Color figure online)

1.1   Clinical Segmentation Task

WMH are regions of abnormally high signal intensity within the white matter in T2-weighted MRI sequences. WMH can vary in volume, brain region location, symmetry, intensity distribution, and border definition depending on the disease. WMH of presumed vascular origin are usually symmetrically distributed within the periventricular and deep white matter, corpus striatum and thalami. There is a high inter-rater and inter-algorithmic variability for WMH, which imposes an upper bound on segmentation model performance when evaluating across multiple datasets.

Ischaemic stroke lesions are infarcts visible in diagnostic imaging following an ischaemic stroke. Diffusion weighted imaging (DWI) is most sensitive to these lesions in the acute phase, although only in approximately 70% of cases [15]. Stroke lesions appear as hyperintense on T2-weighted sequences. In the long term, they occasionally disappear and when subcortical, can cavitate to become lacunes. Cortical strokes originate in the large vessels of the cortex, while subcortical strokes originate in the deeper small vessels. The visible lesion resulting from a cortical stroke often extends into the subcortical space.

Since WMH and ischaemic stroke lesions are both bright in appearance, differentiation can be challenging. Segmentation of both WMH of presumed vascular origin and ischaemic stroke lesions, separately, has been the subject of public challenges [10,12,13], which have garnered plenty of attention from the medical image analysis community. However, joint segmentation of both features, which are frequently seen together, has only been investigated by a few works, perhaps due to the lack of data available with both types of annotations.

1.2   Imaging Protocols and Processing

Different works have explored the use of different imaging sequences for the automatic detection of different types of lesions. Wang et al. [20] used T1-weighted and FLAIR images to segment WMH, cortical, and subcortical ischaemic strokes lesions. Their method requires prior brain-tissue segmentation of the T1-weighted scan and follows this by fitting Gaussian mixture models and utilising binary morphological operations. The method is slow compared to deep learning based segmentation of these lesions. Duan et al. [6] used four independent vanilla 2D U-Nets to segment four SVD features: WMH, subcortical infarcts, lacunes, and cerebral microbleeds. T1-weighted and FLAIR sequences were used for both the WMH and lacune segmentation networks, and T2*-weighted for the microbleed segmentation network. The input to the infarct segmentation network was the DWI sequence only, and this is sensitive to acute and sub-acute ischaemic stroke lesions in only approximately 70% of cases [15]. Thus, DWI-negative acute lesions would not be identified, and neither would chronic stroke lesions. LLRHNet, a model proposed by Liu et al. [14] uses DWI and FLAIR as input in a 2D U-Net network which is modified to incorporate long range features via the addition of a global branch utilising a transformer layer. While this enables the model to detect chronic and DWI-negative acute lesions, it limits its applicability in a clinical setting, since DWI is not always acquired along with FLAIR (especially in the absence of acute stroke symptoms). To the best of our knowledge only the work of Guerrero et al. [7] addresses segmentation of WMH and stroke lesions in all their stages, without the requirement for DWI. They use T1-weighted and FLAIR as input to a U-Net style architecture, which they term uResNet [7]. Their dataset contained scans acquired under three different protocols (protocols 2, 3 and 4 in Table 1). However, they obtained a Dice coefficient (DSC) of only 0.4 for stroke.

The choice of processing method is also a topic of debate. There is mounting evidence that utilising 3D volumes improves brain image segmentation algorithms [2]. In previous work [7], it has been argued that since slice thicknesses are often large in 2D FLAIR, 3D convolutions may be a poor choice, because aggregating information across disparate slices (e.g. 15 mm separation with a slice thickness of 5 mm and a $3 \times 3 \times 3$ kernel) may have limited biological basis. While this is true, we hypothesise that in the case of mixed resolution data (as is often the case when combining multiple datasets), the potential benefits of 3D convolutions outweigh the costs, since the network should learn to be robust to resampled low resolution patches, while taking advantage of the additional information for true high resolution patches.

Real-world data may vary in quality and resolution. We would like any models to be robust to distribution shift when exposed to lower quality, lower resolution, data at inference time. It was shown by Dorent et al. [5], in a different setting, that including only a small number of would-be out-of-distribution training samples can lead to robust generalisation. Investigation of how this finding translates to our task is important, due to the scarcity of data with both ischaemic stroke lesion and WMH masks.

1.3 Medical Image Segmentation Frameworks

Most deep learning-based segmentation methods have an underlying U-Net structure, i.e. encoder-decoder with skip connections. While many additions such as attention mechanisms and residual connections have been observed to improve performance in specific instances, no additions consistently improve performance across tasks in medical image segmentation challenge submissions [11]. Recent challenge leaderboards have shown that freely available generic frameworks can give state-of-the-art results for biomedical segmentation problems, notably nnU-Net [11] and Auto3DSeg[1] from MONAI [16].

nnU-Net encompasses preprocessing, network configuration, training and inference. It won 39/53 medical image segmentation challenges it competed in in 2019 and has remained state-of-the-art.

Auto3DSeg is a framework developed by MONAI which, like nnU-Net, takes care of the preprocessing, training and inference. This framework does not configure a model, but instead utilises multiple highly performing models from the literature and international challenges and trains them as suggested in their associated publications (with a few differences, made clear in the documentation). The different algorithms can then be easily ensembled at test time. The default bundle of algorithms contains DiNTS [9], SegResNet (2D and 3D) [17] and Swin UNETR [8].

In this work, we apply and compare nnU-Net and Auto3DSeg to understand their efficacy for this challenging segmentation problem.

1.4 Contributions

In this paper, we make the following contributions:

---

[1] https://monai.io/apps/auto3dseg.

- Comparison of two easy-to-use and state-of-the-art end-to-end segmentation frameworks on the task of simultaneous segmentation of WMH and stroke lesions (from acute and sub-acute to *chronic*) on FLAIR MRI.
- Comparison of the performance of 2D versus 3D models on mixed resolution data on this task.
- Analysis of the performance of models with respect to different imaging characteristics (different imaging protocols, pathology locations, and pathology volumes).
- Investigation of the domain adaptation ability of models when trained with only a small proportion of data from lower image quality domains (supplementary material:) https://github.com/Jesse-Phitidis/WMHandISLseg.

2   Data

Ethical approval and patients' or next-of-kin consent was obtained as required for the primary studies of mild ageing and non-disabling stroke that provided data to this study [3, 21–23](mean age $\pm$ std $= 67 \pm 11$; 64% male; 20% diabetes; 70% hypertension; 54% current/ex smokers). The data used in this work comprises structural MRI acquired under four different imaging protocols, with expert-delineated masks of stroke lesions (203 acute/sub-acute; 8 chronic; 58 both), and semi-automatically derived WMH masks (corrected by an expert). We used only the FLAIR sequence, since previous work [7] observed that the addition of the T1-weighted sequence did not improve model performance. The variation in acquisition parameters (Table 1) results in significant differences in resolution and image quality, contrast and appearance. While in Protocol 1, 3D images are acquired at 3T with isotropic $1\,\text{mm}^3$ voxels, the other protocols have 2D acquisitions in the axial plane at 1.5T and with much lower spatial resolution.

To explore the applicability of Dorent et al.'s [5] finding to our task, we split our data (Table 2) so that 150/162 training cases are from protocol 1, with the

Table 1. FLAIR MRI acquisition protocols from the four data sources used in this work. Voxel spacings are stated as originally acquired and differ slightly from the images accessed in this work due to prior processing necessary for the original studies.

Parameters	Protocol 1	Protocol 2	Protocol 3	Protocol 4
Acquisition	3D	2D	2D	2D
Field strength (T)	3	1.5	1.5	1.5
TR/TE/TI (ms)	5000/388/1100	9000/140/2200	9002/147/2200	9000/140/2200
Slice thickness (mm)	1	5	5	4
Inter-slice gap (mm)	0	1	1.5	0
No. slices	192	28	20	40
Voxel spacing (mm)	$1 \times 1 \times 1$	$0.47 \times 0.47 \times 6$	$0.94 \times 0.94 \times 6.5$	$1 \times 1 \times 4$

Table 2. Train-test data split showing the distribution of stroke subtypes within the different datasets.

	Stroke subtype	Protocol 1	Protocol 2	Protocol 3	Protocol 4	Total
Train	Cortical	67	2	2	3	74
	Subcortical	66	1	2	2	71
	Both	1	0	0	0	1
	No stroke	16	0	0	0	16
	Total	150	3	4	5	162
Test	Cortical	36	16	2	3	57
	Subcortical	30	32	0	0	62
	Both	1	8	0	0	9
	No stroke	7	0	0	0	7
	Total	74	56	2	3	135

remaining 12 scans coming from the other protocols. We will refer to protocols 2, 3 and 4 data as "edge-of-distribution" (analysis in supplementary material: https://github.com/Jesse-Phitidis/WMHandISLseg). We also incorporated a similar number of cortical and subcortical stroke lesions in the training data so as to avoid bias and enable an analysis of the trained model performance on each of these stroke locations.

3 Methods

We used two well-known segmentation frameworks, nnU-Net and Auto3DSeg, in their default configurations. We compared these methods against each other and against uResNet (the only method we know of applied specifically to this task) which was faithfully re-implemented and trained according to the original paper.

3.1 Deep Learning Models

Below we describe our nnU-Net configuration, each of the algorithms trained when Auto3DSeg is run, and uResNet. All models except uResNet are trained with 5-fold cross-validation and the resulting 5 models are ensembled. The models perform multi-class segmentation (background, WMH, and stroke), since this is less computationally expensive than training two binary models.

nnU-Net. We used the set of fixed parameters, rule-based parameters and empirical parameters nnU-Net authors identified from numerous experiments on ten datasets [11]. The fixed parameters include the loss function, learning rate, data augmentation and optimiser; the rule-based parameters include the patch and batch size, target spacing and resampling strategy and are determined by the dataset characteristics and available GPU memory; the empirical parameters configure the postprocessing and ensemble selection based on the validation data. Three models can be configured - the 2D model, the 3D model and a 3D cascade which uses the output of a 3D low resolution model as additional input to the 3D high resolution model and should be used when the hardware is insufficient to fit patch sizes resulting in a large enough receptive field into memory. Models are trained with deep supervision. The configured 3D model used in this work had a batch size of 2, patch size of $112 \times 160 \times 128$, and input images were resampled to $0.94 \times 0.94 \times 0.90$ mm. No "non-largest component suppression" postprocessing was applied, since multiple instances of each class can be present.

SegResNet. This model is a U-Net style architecture with residual blocks replacing all convolutional blocks and deep supervision. Trained with Auto3DSeg, it was the winning method in the head and neck tumour segmentation challenge (HECKTOR) 2022 [18,19]. The Auto3DSeg framework implementation we use does not include the variational autoencoder regularisation module used in the original publication and the configured 3D model uses a batch size of 2, patch size of $224 \times 224 \times 144$, and voxel spacing of $0.94 \times 0.94 \times 0.90$ mm. The 2D model uses the same batch size and voxel spacing (first two dimensions), but uses a patch size of $256 \times 256 \times 3$, meaning that although 2D convolutions are used, one slice on each side of the target slice is visible to the network.

Swin UNETR. This network is specifically designed to enable the modelling of long-range dependencies, thus addressing a common limitation of convolutional neural networks (CNNs). The network forms the familiar U-shaped encoder-decoder structure with skip connections, however the encoder blocks are transformer blocks utilising the multi-head self-attention mechanism which allows for each output voxel to gain global context relating to the input patch. This method is currently second on the leaderboard in the ongoing medical segmentation decathlon (MSD) challenge [1]. The Auto3DSeg implementation configured uses a batch size of 3, patch size of $96 \times 96 \times 96$ and again, resamples the images to $0.94 \times 0.94 \times 0.90$ mm resolution. Notably, while the patch size is smaller than nnU-Net and SegResNet, the transformer blocks offset this by allowing for a full sized receptive field.

DiNTS. DiNTS is a self configuring network. The basic structure ensures that multi-scale features enter the topology search space and are processed and upsampled upon leaving it. The search space is a lattice of interconnecting feature nodes. For each connection, the choice of a skip connection, a 3D convolution, or a pseudo 3D convolution in one of three possible dimensions is optimised over. The method achieved the top ranking on the MSD leaderboard at the time of release. The Auto3DSeg implementation configured uses a batch size of 2, patch size of $96 \times 96 \times 96$, and $0.94 \times 0.94 \times 0.90$ mm resolution.

uResNet. This network is a 2D U-Net style architecture, with residual blocks in place of the usual convolutional blocks. It was proposed specifically for the task at hand. It uses a batch size of 128, patch size of 64×64 and resolution of 1×1 mm, without resampling in the axial direction. Unusually for 2D models, the authors opt for a small patch size to ease the class imbalance problem, through selective sampling. They additionally introduce a scheme whereby once the patch is sampled, it is randomly translated by up to half of the patch size, so that the target class is not always in the centre. The method outperformed state-of-the-art methods at the time.

3.2   Evaluation Metrics

Our evaluation metrics are precision, recall, Dice similarity coefficient (DSC), and the 95th percentile of the Hausdorff Distance (HD).

When a class is absent from the ground truth, precision, recall and DSC score zero or are undefined, depending on the implementation. For this reason, in the evaluation, we do not average over cases which do not contain positive ground truth labels. The test data contains only seven examples without stroke (and none without WMH), all from protocol 1. These seven cases are assessed using only the proportion of false positives.

4   Results

4.1   Segmentation Models Comparison

Table 3 shows the results, with the distribution of DSC and 95% HD shown in Fig. 2. The segmentation of the stroke lesions was more challenging than that of WMH, as evidenced by the lower DSC, higher 95% HD and increased variability between images. nnU-Net performed best in terms of precision for both classes, but was less successful in terms of recall - a very important metric for medical applications, where the consequences of false negatives are generally worse than false positives.

In addition to the ensembling linked to cross-validation folds, Auto3DSeg offers an option to further ensemble the top N models according to the cross validation, where N can be chosen by the user. Here we selected the top four (Ensemble 4) and the top two (Ensemble 2). Ensemble 2 uses DiNTS and Seg-ResNet 3D and had the highest mean DSC across both classes (0.615), with nnU-Net in second place (0.613). nnU-Net showed the lowest mean 95% HD (20.2) with Ensemble 2 in third place (21.4) behind SegResNet 3D (21.0). nnU-Net and Ensemble 2 took joint first place in making only one false positive out of the seven cases without stroke. We hence consider nnU-Net and Ensemble 2 to be the two best performing methods tested here and performed further analysis with them. Notably, both of these general segmentation methods significantly outperformed the task-specific uResNet (Wilcoxon signed-rank test, p< 0.05 for WMH and stroke DSC).

The two worst performing models were SegResNet 2D and uResNet (which is also 2D). Specifically, SegResNet 2D performed worse than SegResNet 3D; this supports the hypothesis that in mixed resolution datasets, 3D convolutions are beneficial.

Table 3. Mean evaluation metrics across all volumes with the relevant lesion. False positives (FP) are out of 7 non-stroke cases. 95% HD in mm. Best result in bold.

Method	DSC		Precision		Recall		95% HD		FP
	WMH	Stroke	WMH	Stroke	WMH	Stroke	WMH	Stroke	Stroke
nnU-Net	**0.7469**	0.4795	**0.7745**	**0.6520**	0.7506	0.4278	10.16	**30.21**	1
DiNTS	0.7196	**0.4974**	0.7169	0.5674	0.7554	**0.5078**	9.739	43.11	2
SegResNet 2D	0.6932	0.3840	0.7066	0.4204	0.7161	0.4214	10.83	57.67	7
SegResNet 3D	0.7391	0.4770	0.7381	0.6165	**0.7709**	0.4369	**8.962**	32.98	4
Swin UNETR	0.7259	0.4519	0.7403	0.5622	0.7428	0.4404	10.18	43.11	5
Ensemble 4	0.7386	0.4764	0.7561	0.6488	0.7525	0.4265	10.19	34.98	1
Ensemble 2	0.7385	0.4914	0.7433	0.6503	0.7648	0.4472	9.723	33.14	1
uResNet	0.6840	0.3394	0.7201	0.3316	0.6814	0.4608	9.914	69.18	7

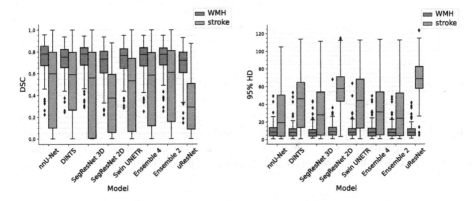

Fig. 2. Distribution of metrics. Whiskers extend to 1.5 the interquartile range (IQR). 95% HD in mm.

4.2 Performance Analysis

Volumetric Analysis. In some research studies, the volume of a pathology is used as a summary measure, and so it is important to identify how well this is predicted by the models. Figure 3 shows the linear fit between predicted and ground truth volumes. At the default threshold values both models (nnU-Net and Ensemble 2) had a tendency to underestimate the WMH volume in cases with higher volumes. For stroke volumes the same pattern of underestimation can be observed. Figure 4 also shows this trend and highlights the similarity between the models, suggesting that they were similarly biased. The tendency to underestimate agrees with the results of Guerrero et al. [7].

We also investigated the effect of the true volumes on the evaluation metrics. Table 7 in the supplementary material makes clear the confounding nature of the two pathologies. An increase in the volume of WMH with no change in the stroke volume generally results in worse segmentation of the stroke lesion, and vice versa.

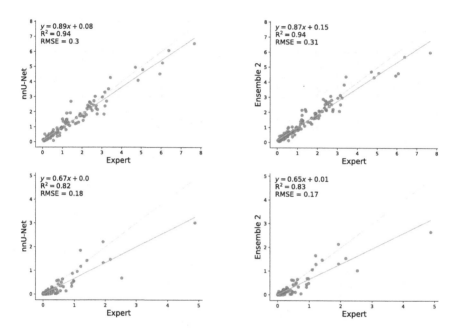

Fig. 3. Linear relationship between predicted and ground truth WMH (top) and stroke lesion (bottom) volumes with volumes expressed as percentage of intracranial volume (ICV). Line of best fit in orange with $y = x$ in gray. (Color figure online)

Cortical Vs Subcortical Ischaemic Stroke Lesions. For this analysis, we used only the test cases with either cortical or subcortical strokes.

WMH of presumed vascular origin are a subcortical feature. Given the similarity in appearance of WMH and stroke lesions, we would expect the segmentation performance to improve when the stroke is cortical in nature and thus less likely to be confounded with nearby WMH. On the contrary, Table 4 shows the opposite, with a statistically significant (Mann-Whitney U, $p < 0.05$) decrease in precision for WMH (Table 8 in the supplementary material). Table 5 confirms that, counterintuitively, for both models a higher proportion of false positive WMH voxels are false negative stroke voxels in cases with cortical strokes than in cases with subcortical strokes. We hypothesise that this effect is due to cortical strokes being larger and/or less uniform than subcortical strokes, making

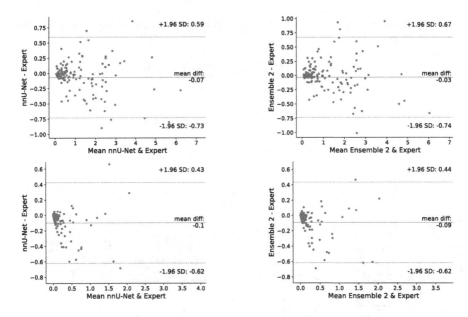

Fig. 4. Bland-Altman plots of predicted and ground truth WMH (top) and stroke lesion (bottom) volumes with volumes expressed as percentage of intracranial volume (ICV).

them more similar in appearance to WMH. We can see this through the image statistics. Figure 5 shows that cortical strokes have more volume (mean 0.41% ICV/6414 mm^3) than subcortical strokes (mean 0.19% ICV/2903 mm^3). Despite this larger volume, they have higher area to volume ratio (mean 0.536 for cortical and 0.533 for subcortical). If a volume increases in size, while maintaining the same shape, its area to volume ratio will always decrease, for example, doubling the volume of a sphere results in an approximately 20% reduction in its area to volume ratio. This shows that cortical strokes are not only larger, but are also "thinner", visually giving them similar characteristics to WMH. Figure 6a shows two cases where cortical strokes are mistaken for WMH by nnU-Net.

Table 4. Comparison of test performance on cases with cortical and subcortical stroke lesions. 95% HD in mm.

Data	Method	DSC		Precision		Recall		95% HD	
		WMH	Stroke	WMH	Stroke	WMH	Stroke	WMH	Stroke
Cortical	nnU-Net	0.7159	0.4447	0.7366	0.6141	0.7375	0.3892	11.33	33.41
	Ensemble 2	0.7082	0.4525	0.7059	0.6114	0.7523	0.4096	10.76	38.64
Subcortical	nnU-Net	0.7630	0.5079	0.8028	0.6601	0.7466	0.4644	9.552	24.57
	Ensemble 2	0.7583	0.5376	0.7771	0.6746	0.7584	0.4962	9.436	26.40

Table 5. Percentage of false positive voxels which are false negative voxels for the other pathology.

Data	Method	WMH	Stroke
Cortical	nnU-Net	46.8	51.3
	Ensemble 2	33.2	41.9
Subcotical	nnU-Net	28.7	55.8
	Ensemble 2	16.0	52.9

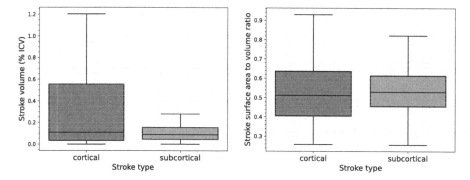

Fig. 5. Plots showing (left) volume and (right) surface area to volume ratio, for cortical and subcortical stroke lesion cases.

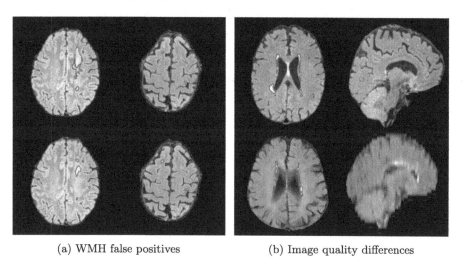

(a) WMH false positives   (b) Image quality differences

Fig. 6. (a) Cortical stroke lesions (red) incorrectly classified as WMH (blue) by nnU-net (top: ground truth, bottom: prediction). The left lesion appears similar to WMH due to its large size, while the right lesion appears similar to WMH because of its irregular shape and scattered appearance. (b) High quality protocol 1 scan (top) and lower quality protocol 4 scan (bottom). The left images show the difference in contrast. The right images show the difference in axial resolution. (Color figure online)

5   Conclusion

In this work, we tested the level of performance of two current state-of-the-art segmentation frameworks in WMH and stroke lesion segmentation from FLAIR MRI, in a heterogeneous multi-domain sample. We also investigated the pathology characteristic conditional performance to identify areas that must be improved on in the future.

nnU-Net outperformed the task specific uResNet method, as did Auto3DSeg with appropriate ensemble selection. The two 2D networks performed worse than all 3D networks, suggesting that for mixed resolution data, aggregation of depth-wise information is not as undesirable as previously assumed.

The strong linear relationship between the predicted and true WMH volumes shows that for studies concerned with the direction of changes in volume, these frameworks can perform well - especially for cases with mild to moderate WMH burden. Their simplicity of training and deployment is an additional benefit to the clinical researchers conducting such studies, who may not be working within a multidisciplinary team.

In contrast to our intuition, the precision of WMH segmentation was higher when the stroke was subcortical. Our analysis suggests that, unlike humans who are able to use their knowledge that WMH do not appear in cortex, the CNNs rely more on intensity and shape and frequently misclassify larger and less uniform cortical stroke lesions as WMH. We propose that future work consider some ways of improving the WMH precision: (i) perform rough segmentation of the cortex at inference and convert all WMH predictions overlapping this region to stroke predictions; (ii) incorporate cortex as an additional label in the training to force the models to learn relevant features; (iii) register images to an atlas and incorporate the corresponding atlas patch as conditioning in the model.

Acknowledgements. We would like to acknowledge Medical Research Scotland and Canon Medical Research Europe for providing funding for this work. Data collection and preparation, ground truth generation and supervision received funding from the Row Fogo Charitable Trust (BRO-D. FID3668413), MRC UK (G0701120, G1001245, MR/M013111/1 and MR/R024065/1), Dementia Research Institute, and Wellcome Trust (WT088134/Z/09/A).

References

1. Antonelli, M., et al.: The medical segmentation decathlon. Nat. Commun. **13**(1), 4128 (2022). https://doi.org/10.1038/s41467-022-30695-9
2. Avesta, A., et al.: Comparing 3D, 2.5D, and 2D approaches to brain image auto-segmentation. Bioengineering **10**(2) (2023). https://doi.org/10.3390/bioengineering10020181
3. Clancy, U., et al.: Rationale and design of a longitudinal study of cerebral small vessel diseases, clinical and imaging outcomes in patients presenting with mild ischaemic stroke: mild stroke study 3. Eur. Stroke J. **6**(1), 81–88 (2021). https://doi.org/10.1177/2396987320929617

4. Debette, S., et al.: Clinical significance of magnetic resonance imaging markers of vascular brain injury: a systematic review and meta-analysis. JAMA Neurol. **76**(1), 81–94 (2019). https://doi.org/10.1001/jamaneurol.2018.3122

5. Dorent, R., et al.: Learning joint segmentation of tissues and brain lesions from task-specific hetero-modal domain-shifted datasets. Med. Image Anal. **67**, 101862 (2021). https://doi.org/10.1016/j.media.2020.101862

6. Duan, Y., et al.: Primary categorizing and masking cerebral small vessel disease based on "deep learning system". Front. Neuroinform. **14** (2020)

7. Guerrero, R., et al.: White matter hyperintensity and stroke lesion segmentation and differentiation using convolutional neural networks. NeuroImage: Clin. **17**, 918–934 (2018). https://doi.org/10.1016/j.nicl.2017.12.022

8. Hatamizadeh, A., et al.: Swin UNETR: swin transformers for semantic segmentation of brain tumors in MRI images. In: Crimi, A., Bakas, S. (eds.) Brainlesion: Glioma, Multiple Sclerosis, Stroke and Traumatic Brain Injuries, pp. 272–284. Springer, Cham (2022). https://doi.org/10.1007/978-3-031-08999-2_22

9. He, Y., et al.: Dints: differentiable neural network topology search for 3d medical image segmentation. In: 2021 IEEE/CVF Conference on Computer Vision and Pattern Recognition (CVPR), pp. 5837–5846 (2021). https://doi.org/10.1109/CVPR46437.2021.00578

10. Hernandez Petzsche, M.R., et al.: ISLES 2022: a multi-center magnetic resonance imaging stroke lesion segmentation dataset. Scientific Data **9**(1), 762 (2022). https://doi.org/10.1038/s41597-022-01875-5

11. Isensee, F., et al.: nnU-Net: a self-configuring method for deep learning-based biomedical image segmentation. Nat. Methods **18**(2), 203–211 (2021). https://doi.org/10.1038/s41592-020-01008-z

12. Kuijf, H.J., et al.: Standardized assessment of automatic segmentation of white matter hyperintensities and results of the WMH segmentation challenge. IEEE Trans. Med. Imaging **38**(11), 2556–2568 (2019). https://doi.org/10.1109/TMI.2019.2905770

13. Liew, S.L., Lo, B.P., Donnelly, M.T.: A large, curated, open-source stroke neuroimaging dataset to improve lesion segmentation algorithms. Scientific Data **9**(1), 320 (2022). https://doi.org/10.1038/s41597-022-01401-7

14. Liu, L., et al.: LLRHNet: multiple lesions segmentation using local-long range features. Front. Neuroinform. **16** (2022)

15. Makin, S.D., et al.: Clinically confirmed stroke with negative diffusion-weighted imaging magnetic resonance imaging. Stroke **46**(11), 3142–3148 (2015). https://doi.org/10.1161/STROKEAHA.115.010665

16. MONAI: Medical open network for AI (2022). https://doi.org/10.5281/zenodo.7459814

17. Myronenko, A.: 3D MRI brain tumor segmentation using autoencoder regularization. In: Crimi, A., Bakas, S., Kuijf, H., Keyvan, F., Reyes, M., van Walsum, T. (eds.) BrainLes 2018. LNCS, vol. 11384, pp. 311–320. Springer, Cham (2019). https://doi.org/10.1007/978-3-030-11726-9_28

18. Myronenko, A., Siddiquee, M.M.R., Yang, D., He, Y., Xu, D.: Automated head and neck tumor segmentation from 3D PET/CT HECKTOR 2022 challenge report. In: Andrearczyk, V., Oreiller, V., Hatt, M., Depeursinge, A. (eds.) Head and Neck Tumor Segmentation and Outcome Prediction, pp. 31–37. Springer, Cham (2023). https://doi.org/10.1007/978-3-031-27420-6_2

19. Oreiller, V., et al.: Head and neck tumor segmentation in PET/CT: the HECKTOR challenge. Med. Image Anal. **77**, 102336 (2022). https://doi.org/10.1016/j.media.2021.102336

20. Wang, Y., et al.: Multi-stage segmentation of white matter hyperintensity, cortical and lacunar infarcts. Neuroimage **60**(4), 2379–2388 (2012). https://doi.org/10.1016/j.neuroimage.2012.02.034

21. Wardlaw, J.M., et al.: Brain aging, cognition in youth and old age and vascular disease in the lothian birth cohort 1936: rationale, design and methodology of the imaging protocol. Int. J. Stroke **6**(6), 547–559 (2011). https://doi.org/10.1111/j.1747-4949.2011.00683.x

22. Wardlaw, J.M., et al.: Lacunar stroke is associated with diffuse blood-brain barrier dysfunction. Ann. Neurol. **65**(2), 194–202 (2009). https://doi.org/10.1002/ana.21549

23. Wardlaw, J.M., et al.: Blood-brain barrier failure as a core mechanism in cerebral small vessel disease and dementia: evidence from a cohort study. Alzheimer's & Dementia **13**(6), 634–643 (2017). https://doi.org/10.1016/j.jalz.2016.09.006

A Deep Learning Based Approach to Semantic Segmentation of Lung Tumour Areas in Gross Pathology Images

Matthew Gil[1,2(✉)] , Craig Dick[2], Stephen Harrow[3], Paul Murray[1] ,
Gabriel Reines March[1,2] , and Stephen Marshall[1]

[1] University of Strathclyde, Glasgow, UK
matthew.gil@strath.ac.uk
[2] NHS Greater Glasgow and Clyde, Glasgow, UK
[3] NHS Lothian, Edinburgh, UK

Abstract. Gross pathology photography of surgically resected specimens is an often overlooked modality for the study of medical images that
can provide and document useful information about a tumour before it is
distorted by slicing. A method for the automatic segmentation of tumour
areas in this modality could provide a useful tool for both pathologists
and researchers. We propose the first deep learning based methodology
for the automatic segmentation of tumour areas in gross pathological
images of lung cancer specimens. The semantic segmentation models
applied are Deeplabv3+ with both a MobileNet and Resnet50 backbone
as well as UNet, all models were trained and tested with both a DICE and
cross entropy loss function. Also included is a pre and post-processing
pipeline for the input images and output segmentations respectively. The
final model is formed of an ensemble of all the trained networks which
produced a tumour pixel-wise accuracy of 69.7% (96.8% global accuracy)
and tumour area IoU score of 0.616. This work on this novel application
highlights the challenges with implementing a semantic segmentation
model in this domain that have not been previously documented.

Keywords: NSCLC · Semantic Segmentation · Gross Pathology
Photography

1 Introduction

Pathology photography can be a useful tool for documenting ground truth
anatomy before it has been distorted by the slicing processes that are used for
whole slide imaging (WSI). Segmentation of regions in pathology photographs
can therefore provide ground truth for the shape of an area, or volume if three
dimensions are considered, of a particular anatomical region. Additionally, current pathological assessment of tumour size, which is a strong predictor of patient
outcomes [14], is generally made by measuring the gross length of the tumour

G. Waiter et al. (Eds.): MIUA 2023, LNCS 14122, pp. 18–32, 2024.
https://doi.org/10.1007/978-3-031-48593-0_2

across its largest dimension by hand with a ruler which often has to be reevaluated at the time of microscopic assessment [26]. Automatic segmentation of gross tumour area would provide a more reliable method of estimating the tumour volume and cellular load which are the metrics that are being estimated by gross measurements with a ruler. Additionally, if a method of automatic segmentation of singular tumours is successful it could then be expanded to be used to pick up other more subtle nodules that could be easily missed by the naked eye but may have been seen in radiology images and if used in real-time this would allow the pathologist to sample these nodules at the time of dissection. An automatic segmentation method for non-small cell lung cancer (NSCLC) tumours in gross pathology photographs, therefore, has both clinical and research applications. The work in this study aims to produce and test a methodology for the automatic semantic segmentation of lung tumours in pathology photographs of specimens that have been surgically removed from patients with NSCLC.

The procedures generally used to capture gross pathology photographs have been described in [17]. Best practices include placing the pathology specimens on a background that provides a good contrast between the specimen and background. The specimen should be well-lit with lighting located to the sides of the specimen as overhead lighting is more likely to cause reflections that may obscure anatomy. Excess moisture should also be removed from the surface of the specimen as this may obscure the underlying anatomy through the liquids opacity or the increased reflections this may cause. The specimen should also be well framed, in focus and the imaging plane should be the same as the slicing plane. The International Association for the Study of Lung Cancer (IASLC) recommends pathology photography as a standard part of pathology processing for NSCLC specimen processing after neoadjuvant therapy [27].

Gross pathology photography has been applied in some studies to provide the information necessary to transform WSI so that the geometry of the images more accurately represents what would have been observed in-vivo. This has often been for the application of registering images from the PET and pathology modalities [12,16]. Gross pathology photography has been used as an important feature in many studies where regions are generally segmented by experienced pathologists. These studies include investigations into the mechanical properties of tissues [21], ablation treatment monitoring [29] and histologically diagnosed cardiac sarcoidosis [15]. A semi-automatic vector quantisation based pathology segmentation approach has been applied to segment regions of fibrosis in gross photographs to determine the overall prevalence of fibrotic tissue in lymph nodes [10]. Hyperspectral image based tumour segmentation has also been applied for application in real-time tissue classification during laparoscopic surgery [2].

An area where gross pathology photography has been applied more extensively than lung the lung cancer domain is skin lesion photography. There are similarities between lung lesion photography and skin lesion photography that make the greater catalogue of previous work on skin lesion segmentation relevant to this study. One such example is the work by Y. Yuan et al. who produced a fully connected (FC) convolutional neural network (CNN) based approach for skin lesion segmentation with a Jaccard distance based loss function with their highest performing method consisting of an ensemble of six separate FC CNNs [31]. Q. Ha et al. detail their work on skin lesion segmentation that achieved 1st place in the 2020 SIIM-ISIC melanoma classification challenge [6]. Their method involved using an ensemble-based model that averages the average pixel prediction scores of models using various versions of EfficientNet, SE-ResNeXt and ResNeSt as the network backbone. Also included was a thorough image augmentation pipeline. Additionally, the ISIC skin lesion segmentation challenge [4] has run every year from 2016 until 2020 so there is a large back-catalogue of skin lesion segmentation methods all trained and tested on a standardised dataset. A detailed review of the skin lesion segmentation literature can be found in [7] which summarises 356 publications on skin lesion segmentation and 238 on skin lesion classification published between 2011 and 2022.

At the time of writing, the authors are not aware of any deep learning based semantic segmentation of gross pathology photographs of surgically removed tumours for any region of anatomy, this is supported by a google scholar search on 23/02/2023 using the key words "gross", "pathology", "deep", "learning" and "segmentation". Therefore the main contribution of this work is establishing the feasibility of this particular application of deep learning.

1.1 Loss Functions

The choice of loss function when training a deep learning based semantic segmentation model can have a large impact on the performance of the model. This is especially true for problems with unbalanced datasets where a model may greatly focus on increasing the accuracy of the class with the most instances causing the accuracy of segmentation of the underrepresented class to be low. For this particular study and more generally in many oncology based segmentation problems, the tumour class is underrepresented compared to the background class but would be considered a more important class to accurately segment. For this reason, the choice of loss function is important in this study.

Cross entropy is a distribution-based metric used for solving optimisation problems that has seen much use in deep learning applications [20]. Balanced cross-entropy (BCE) loss was introduced to improve performance when training models on unbalanced datasets that have more examples of a particular class than others by adding a weighting to the contribution each class to the loss function [30]. For binary classification problems, the balanced cross-entropy loss is expressed by Eq. 1 [25].

$$L_{BCE}(y, \hat{y}) = -\frac{1}{N} \sum_{n=1}^{N} (\beta y_n log(p_n) + (1 - \beta)(1 - y_n)log(1 - p_n)) \qquad (1)$$

where L_{BCE} is the balanced cross-entropy loss, N is the total number of individual pixels n. y_n is a ground truth pixel value, p_n is a predicted pixel probability outcome and β is a factor used to apply a weighting to the classes.

The DICE score was first proposed for use as a loss function for two class segmentation problems by Milletari et al. [13]. This was expanded to the generalised DICE loss function for multi-class classification problems by C. H. Sudre et al. [25]. As the problem presented in this study is a two-class classification problem, only the standard DICE loss function is required. This is described in Eq. 2 where L_{DICE} is the DICE loss [25].

$$L_{DICE}(y_n, p_n) = 1 - \frac{\sum_{n=1}^{N} 2y_n p_n}{\sum_{n=1}^{N} y_n + \sum_{n=1}^{N} p_n} \tag{2}$$

2 Methods

Unless mentioned otherwise, any computational steps applied in the following sections were applied using MATLAB ver R2022b (The MathWorks, Inc.).

2.1 Datasets

The data for this study comes from two separate datasets. Both datasets consist of photographs of lung specimens that have been surgically resected from patients with NSCLC. The specimens were all sliced as they would be for WSI and photographs were taken of the slices. Manual segmentations of the tumour boundary were produced by an expert pathologist using the software ImageJ [24].

The first dataset, referred to from now on as dataset-A, contains gross pathology photographs that were sliced and segmented using non-standard approaches. The lung specimens from this dataset consist of entire lung lobes that were inflated with agar in an attempt to create a better correlation between the shape of the ex-vivo specimens with their in-vivo shape. The lung specimens were then suspended in agar, so that the three-dimensional information was preserved through the slicing process, and sliced at 5 mm intervals with photographs being taken after every slice was removed. During the collection of this dataset, care was taken in the lighting of the samples as well as partially drying the samples so that minimal reflection and maximum tissue contrast could be produced. An example of a photograph and its corresponding manual segmentation can be seen in Fig. 1. More information on the data gathering and pathology processing procedures for dataset-A can be found in [18].

The second dataset, which will be referred to as dataset-B, consists of pathology specimens that were sliced and photographed freehand with only the standard pathology lab lighting used to light the specimens. In this dataset, less care was taken to remove excessive moisture and reflections on the samples meaning these photographs are of poorer quality than dataset-A. The number of patients and images in each dataset is summarised in Table 1.

Examples of gross pathology photographs from four separate patients are shown in Fig. 2. This shows the variability in the tumours and some of the different features that can be seen. For example Fig. 2 (b) shows a tumour with

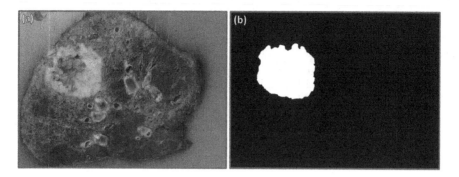

Fig. 1. An example of (**a**) a gross pathology photograph and (**b**) its corresponding manual tumour segmentation from dataset-A.

Table 1. Pathology photograph dataset information.

Dataset	Number of Patients	Number of Images
Dataset-A	9	64
Dataset-B	6	52
Total	15	116

a large necrotic core whereas Fig. 2 (d) shows a tumour with no necrotic regions. Figure 2 (a), (b) and (d) all display regions with increased red or pink colour, this is due to the tumour reducing the quality of fixation in these regions. Figure 2 (c) shows an example where the tumour is displaying poorer contrast to the healthy tissue than the other examples, likely due to this example being an adenocarcinoma tumour.

2.2   Image Pre-processing and Data Augmentation

Before the pathology images and labels were used in training some pre-processing steps were applied.

Non-tissue Background Removal. Many of the images contain a large level of background area compared to the area of lung tissue. This was reduced by manually cropping the images down to a rectangular shape closely bounding the lung tissue. The aim of this step was to reduce the computational load of training the models by decreasing the image sizes and to get the CNN to focus more on areas of the lung specimen.

In all of the pathology photographs, the non-tissue background is well distinguished from the tissue regions of the image. This allows for the application of non-learning based segmentation techniques to create a mask that removes the background regions. As the pathology samples in dataset-A and dataset-B were prepared using different methodologies, the background regions in both datasets

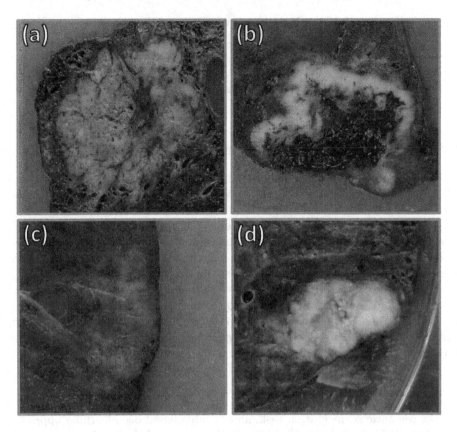

Fig. 2. Examples of tumour regions in gross photographs for four separate patients.

are reasonably different. Dataset-A contains only regions of agar, these regions are of a similar colour to much of the tissue regions. The tissue has also been inflated with agar causing there to be regions of agar within the outer tissue boundary. This means a colour-based approach to background segmentation is not appropriate. Here we use spectral residual saliency detection approach as described in [11]. This was applied in MATLAB using methods adapted from [23].

Dataset-B contains samples either taken on a pathology slicing board or the perforated metal pathology workstation. There are also often separate objects such as rulers contained within the images. These images are less suited to the spectral residual approach used for dataset-A but work well with a colour-based approach due to the clear colour contrast between the tissue and non-tissue regions of the image. For dataset-B a k-means clustering approach [1] is applied for colour segmentation where in this case $k = 2$. The parameters used in the k-means clustering algorithm are shown in Table 2.

Once a non-tissue background mask was produced it was used to set all the pixels of the non-tissue background to intensity values of zero.

Table 2. K-means segmentation parameters

Parameter	Value
k	2
Number of clustering repetitions	3
Max Iterations	100
Accuracy Threshold	1.00e-04

Data Augmentation. The images were then converted to patches of size 224×224 for use with the neural network. When converting the images to patches, an overlap of 50% was introduced in both the x and y image directions to conserve spatial information occurring at the borders of the patches [8]. This produces four times more patches than when the overlap is not included so it should be noted that the network is seeing the same data four separate times over one epoch of training on the patches.

A random rotation of the images between 0 and 360°C and a random zoom between 0.8 and 1.5 times was applied to the images after every epoch of training. The zoom here is important as the photograph height above the pathology samples is not standard so there is a variation in the zoom level within the datasets.

2.3   Segmentation Model

For the semantic segmentation task, an ensemble-based deep learning approach was applied. This involved training multiple separate deep-learning models and combining the output segmentations into a single averaged segmentation. This was followed by post-processing of the ensemble model output through background masking and morphological steps to improve the output segmentation. The full workflow of the final model is shown in Fig. 3 and described in the following sections.

Several different network architectures were applied to the problem. Deeplabv3+ [3] was applied with both a ResNet50 [9] and MobileNetv2 [22] backbone as well as UNet [19] with an encoder depth of 4. All of these networks were applied with a binary pixel classification output with the pixel classifications as "tumour" or "background". This means that both healthy lung tissue and the non-biological content (slicing table etc.) of the image were included in the background class. All of the network architectures were trained with both weighted cross entropy and DICE loss functions. All of the models used pretrained weights from ImageNet [5]. As the content of ImageNet is greatly different from the content of the datasets in this study, none of the layers of the models were frozen as is often done for transfer learning.

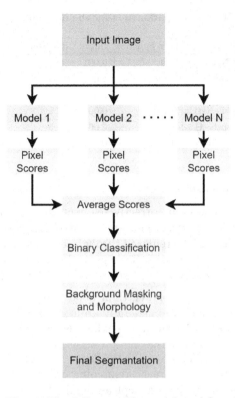

Fig. 3. The full ensemble model workflow.

The dataset was split into training and test sets by patient so that images from one patient were only used for either training or testing. This is important as the different slices from the same patient contain similar features, such as the colour of the tumour and healthy tissue, that would bias the results if they were included in both training and test sets. Using this approach, the data was split into a training set containing 12 patients images and a test set containing 3 patients images. Using this approach causes the number of images in both the training and test sets to change depending on which patients images were used as there were more images available for some patients than others. This method generally created a split of around 96/20 images in the training/test sets. A 5-fold cross-validation approach was applied where all of the models were trained separately on different sets of 12 patients and tested on the three that were not included in the training set with the three test patients changing every fold. This allows for the model to be tested on the full dataset. The training parameters are shown in Table 3. All of the network training was performed on a single NVDIA GeForce GTX 1080 Ti graphics card.

In addition to applying the models individually, an ensemble-based approach was taken. To achieve this the individual pixel prediction probability outputs of each individual network in Table 3 were simply averaged.

Table 3. Network training parameters for the individual models trained.

Model	Network Architecture	Loss Function	Initial LR	Drop Rate	Drop Factor	Epochs
1	DeepLabV3+ ResNet50	BCE	0.001	5	0.2	20
2	DeepLabV3+ ResNet50	DICE	0.001	5	0.2	20
3	DeepLabV3+ MobileNetv2	BCE	0.001	5	0.2	20
4	DeepLabV3+ MobileNetv2	DICE	0.001	5	0.2	20
5	Unet (Encoder Depth: 4)	BCE	0.001	5	0.2	20
6	Unet (Encoder Depth: 4)	DICE	0.001	5	0.2	20

2.4   Image Post-processing

The deep learning models often correctly segment the region of tumour in the input image but also labels some separate erroneous regions as tumour. These incorrect regions can usually be removed through some morphological operations that can be applied based on what is known about the task to improve the segmentation results. The morphology steps are detailed in the list below:

1. Small objects with a size of fewer than 5000 pixels are removed from the image (for reference, pixels are generally around 0.1×0.1 mm).
2. A morphological closing operation is applied using a circular structuring element with a radius of 20 pixels.
3. Any holes in the remaining objects are filled.
4. The total number of pixels in each remaining object is calculated. Only the object made of the most pixels is kept as the final tumour segmentation.

Step 1 is applied to remove small isolated regions that were classified as tumour as these are almost always incorrect classifications, this step also improves the performance of all of the following steps. Steps 2 and 3 in the list above are required because many NSCLC tumours contain necrotic cores. These regions are pathologically and visually different from non-necrotic areas of tumour which, combined with the fact that there are few different patient examples in the datasets, causes them to be often misclassified as non-tumour. Simply closing and filling the tumour region generally fixes this problem. Step 4 can be applied as we know that the images in our dataset are from patients with one large NSCLC tumour.

3   Results

3.1   Segmentation Metrics

The results of the 5-fold cross-validation are shown in Tables 4 and 5 for datasets A and B respectively. Generally for both datasets the ensemble model outperforms the individual models. Including the morphology steps improves the ensemble results across all of the metrics showing that for this particular application it is worthwhile to include them. The results on dataset-B are considerably lower than those in dataset-A. This is expected due to the lower quality of images in dataset-B.

Table 4. Results from testing on dataset-A.

Model	Global Accuracy (%)	Tumour Accuracy (%)	Tumour IoU	Background IoU
1	93.9	69.3	0.486	0.905
2	89.9	58.7	0.332	0.854
3	91.9	57.4	0.396	0.876
4	88.1	52.3	0.262	0.832
5	90.3	56.6	0.366	0.884
6	87.2	52.3	0.296	0.826
Ensemble	95.7	63.8	0.521	0.923
Ensemble + Morphology	**96.8**	**69.7**	**0.616**	**0.940**

Table 5. Results from testing on dataset-B.

Model	Global Accuracy (%)	Tumour Accuracy (%)	Tumour IoU	Background IoU
1	81.5	**68.7**	0.388	0.731
2	84.2	50.0	0.349	0.757
3	85.9	65.7	0.503	0.788
4	86.7	60.3	0.427	0.790
5	81.7	60.4	0.327	0.725
6	80.9	57.1	0.351	0.743
Ensemble	88.8	67.9	0.493	0.817
Ensemble + Morphology	**89.2**	68.6	**0.504**	**0.822**

3.2   Segmentation Examples

There is a large variety in the quality of the segmentation output of the ensemble model depending on the input images, some examples of this are shown in Fig. 4. Figure 4 (a.i) shows an example of a correct segmentation result on a tumour with good contrast between the healthy tissue and tumour tissue. The tumour boundary in image (a.ii) aligns very closely with the ground truth mask producing an IoU of 0.956 for this image.

Figure 4 (b.i) shows an example of a partially correct segmentation where an area of necrosis has caused errors. This example has an IoU of 0.439 for the tumour class. The segmentation contour in this image outlines the region of lighter tissue which corresponds to the living tumour area. The necrotic area is not included in the segmentation output but is part of the ground truth tumour area as seen in image (b.ii). The model tends to misclassify necrotic regions as non-tumour as there are not many examples of heavily necrotic tumours in the training datasets and the coagulated blood that appears in this region also often appears in areas of healthy tissue. In other necrotic examples, this can be fixed by the post-processing morphology steps but in this case, as the living tumour area does not fully enclose the necrotic region, these steps do not solve this problem.

Figure 4 (c.i) shows an example of a failed segmentation with a tumour IoU of only 0.035. Upon analysing this image within the context of the dataset it is seen that image is from one of only two patients that had an adenocarcinoma tumour in the datasets. Adenocarcinoma has a lepidic pattern of growth causing it to be less contrasted against healthy tissue in gross images than other types of NSCLC. All other images from this patient and the other patient with adenocarcinoma have a similarly failed segmentation. It is clear from this that the dataset would need to be expanded to include more adenocarcinoma examples.

4   Discussion

The classification of the entire pathology of the lung into the two categories of tumour and non-tumour is an oversimplification that presents some problems for the segmentation model. This is most notable with adenocarcinoma tumours that are generally not recognised as tumours. Additionally, necrotic regions within the tumour are often misclassified as non-tumour regions. This can generally be fixed through the use of morphological image processing steps but it still highlights a problem with the ground truth data. This would be improved by increasing the dataset size to include more patients as the small dataset used in this study, with only 15 separate patients, included only a few examples of different pathological features such as adenocarcinomas and necrotic regions. A dataset containing a similar number of images that were all from unique patients would likely increase the performance of the trained models as this would allow the model to learn a more comprehensive array of pathological features. In the skin lesion photograph segmentation domain, large datasets such as the HAM10000 dataset [28], which contains 10000 images and ground truth segmentations of skin cancer lesions, allowing for highly accurate models to be produced. In addition to increasing the dataset size, it may be beneficial to increase the number of classes used for the segmentation to include different types of tissue though this would require a time-intensive process of manual segmentation to produce the ground truth labels.

The final results for dataset-A produced better scoring metrics than those produced from dataset-B. This is unsurprising as, for the reasons described in Sect. 2.1, the images in dataset-B are of poorer quality than those in dataset-A. This reduced image quality will increase the difficulty of segmentation first due to the image features being obscured and secondly due to there being fewer of these images of low quality in the overall training datasets. For further development and application of a system for the automatic segmentation of gross pathology photographs, care should be taken to ensure a high image quality by following the photography steps outlined in [17], though the inclusion of lower quality images in the training set may be beneficial to increase the robustness of the model.

For applications in clinical use, it may be beneficial to include some user input to produce a semi-automatic segmentation to decrease the chance of errors and improve overall accuracy. This could involve simply selecting the correct region from the output of the model to remove some of the morphology steps or marking some tumour or background pixels to be input to the network. The

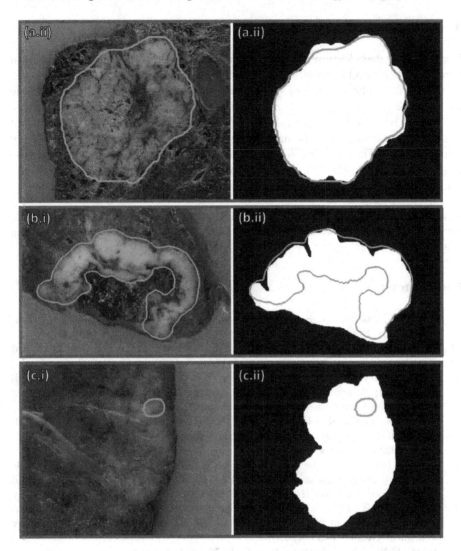

Fig. 4. Three segmentation examples from separate patients are displayed. **(a)** shows a good segmentation example, **(b)** shows a partially failed example due to a necrotic region and **(c)** shows a fully failed segmentation due to the tumour being an adenocarcinoma. Images denoted with **(i)** are the original test images zoomed in on the tumour area and images denoted with **(ii)** are the ground truth tumour segmentations. All images have the automatic segmentation contour overlayed (green line). (Color figure online)

decision to choose a fully or semi-automatic approach would depend on the specific application and pathology workflow that the model is to be included in.

5   Conclusion

In conclusion, deep learning-based methods for semantic segmentation have been applied to the novel application of automatic segmentation of tumour areas in gross pathology photographs of specimens from patients with NSCLC. A pipeline for image preprocessing, model training and post-processing of the segmentation output has been detailed and validated. This work has proven the possibility of achieving this goal as well as highlighting some challenges for producing a fully robust system. The main barrier to improving the performance is a lack of data where the model produced in this work produced good results on more common tumour examples, tumours with less common features were poorly segmented.

Acknowledgements. We would like to thank the Beatson Cancer Charity and UKRI EPSRC for funding this work as well as the CDT in Applied Photonics for facilitating this work.

References

1. Arthur, D., Vassilvitskii, S.: k-Means++: The Advantages of Careful Seeding. Technical Report, Stanford (2006)
2. Baltussen, E.J., et al.: Hyperspectral imaging for tissue classification, a way toward smart laparoscopic colorectal surgery. J. Biomed. Opt. **24**(1), 016002 (2019)
3. Chen, L.C., Zhu, Y., Papandreou, G., Schroff, F., Adam, H.: Encoder-decoder with atrous separable convolution for semantic image segmentation. In: Proceedings of the European Conference on Computer Vision (ECCV), pp. 801–818 (2018)
4. Codella, N., et al.: Skin lesion analysis toward melanoma detection 2018: a challenge hosted by the international skin imaging collaboration (ISIC). arXiv preprint arXiv:1902.03368 (2019)
5. Deng, J., Dong, W., Socher, R., Li, L.J., Li, K., Fei-Fei, L.: Imagenet: a large-scale hierarchical image database. In: 2009 IEEE Conference on Computer Vision and Pattern Recognition, pp. 248–255. IEEE (2009)
6. Ha, Q., Liu, B., Liu, F.: Identifying melanoma images using efficientnet ensemble: winning solution to the SIIM-ISIC melanoma classification challenge. arXiv preprint arXiv:2010.05351 (2020)
7. Hasan, M.K., Ahamad, M.A., Yap, C.H., Yang, G.: A survey, review, and future trends of skin lesion segmentation and classification. Comput. Biol. Med. **155**, 106624 (2023)
8. Hashemi, S.R., Salehi, S.S.M., Erdogmus, D., Prabhu, S.P., Warfield, S.K., Gholipour, A.: Asymmetric loss functions and deep densely-connected networks for highly-imbalanced medical image segmentation: application to multiple sclerosis lesion detection. IEEE Access **7**, 1721–1735 (2018)
9. He, K., Zhang, X., Ren, S., Sun, J.: Deep residual learning for image recognition. In: Proceedings of the IEEE Conference on Computer Vision and Pattern Recognition, pp. 770–778 (2016)
10. Hipp, J.D., Cheng, J.Y., Toner, M., Tompkins, R.G., Balis, U.J.: Spatially invariant vector quantization: a pattern matching algorithm for multiple classes of image subject matter including pathology. J. Pathol. Inform. **2**(1), 13 (2011)

11. Hou, X., Zhang, L.: Saliency detection: a spectral residual approach. In: 2007 IEEE Conference on Computer Vision and Pattern Recognition, pp. 1–8. IEEE (2007)
12. Meyer, C., Ma, B., Kunju, L.P., Davenport, M., Piert, M.: Challenges in accurate registration of 3-d medical imaging and histopathology in primary prostate cancer. Eur. J. Nucl. Med. Mol. Imaging **40**, 72–78 (2013)
13. Milletari, F., Navab, N., Ahmadi, S.A.: V-net: fully convolutional neural networks for volumetric medical image segmentation. In: 2016 Fourth International Conference on 3D Vision (3DV), pp. 565–571. IEEE (2016)
14. Okada, M., et al.: Effect of tumor size on prognosis in patients with non-small cell lung cancer: the role of segmentectomy as a type of lesser resection. J. Thorac. Cardiovasc. Surg. **129**(1), 87–93 (2005)
15. Okasha, O., Kazmirczak, F., Chen, K.H.A., Farzaneh-Far, A., Shenoy, C.: Myocardial involvement in patients with histologically diagnosed cardiac sarcoidosis: a systematic review and meta-analysis of gross pathological images from autopsy or cardiac transplantation cases. J. Am. Heart Assoc. **8**(10), e011253 (2019)
16. Park, H., et al.: Registration methodology for histological sections and in vivo imaging of human prostate. Acad. Radiol. **15**(8), 1027–1039 (2008)
17. Rampy, B.A., Glassy, E.F.: Pathology gross photography: the beginning of digital pathology. Surg. Pathol. Clin. **8**(2), 195–211 (2015)
18. Reines March, G.: Registration of Pre-operative Lung Cancer PET/CT Scans with Post-operative Histopathology Images. Ph.D. thesis, University of Strathclyde (2020)
19. Ronneberger, O., Fischer, P., Brox, T.: U-Net: convolutional networks for biomedical image segmentation. In: Navab, N., Hornegger, J., Wells, W.M., Frangi, A.F. (eds.) MICCAI 2015. LNCS, vol. 9351, pp. 234–241. Springer, Cham (2015). https://doi.org/10.1007/978-3-319-24574-4_28
20. Rubinstein, R.: The cross-entropy method for combinatorial and continuous optimization. Methodol. Comput. Appl. Probab. **1**, 127–190 (1999)
21. Samani, A., Zubovits, J., Plewes, D.: Elastic moduli of normal and pathological human breast tissues: an inversion-technique-based investigation of 169 samples. Phys. Med. Biol. **52**(6), 1565 (2007)
22. Sandler, M., Howard, A., Zhu, M., Zhmoginov, A., Chen, L.C.: Mobilenetv 2: inverted residuals and linear bottlenecks. In: 2018 IEEE/CVF Conference on Computer Vision and Pattern Recognition, pp. 4510–4520 (2018). https://doi.org/10.1109/CVPR.2018.00474
23. Schauerte, B., Stiefelhagen, R.: Quaternion-based spectral saliency detection for eye fixation prediction. In: Fitzgibbon, A., Lazebnik, S., Perona, P., Sato, Y., Schmid, C. (eds.) ECCV 2012. LNCS, vol. 7573, pp. 116–129. Springer, Heidelberg (2012). https://doi.org/10.1007/978-3-642-33709-3_9
24. Schneider, C.A., Rasband, W.S., Eliceiri, K.W.: NIH image to Imagej: 25 years of image analysis. Nat. Methods **9**(7), 671–675 (2012)
25. Sudre, C.H., Li, W., Vercauteren, T., Ourselin, S., Jorge Cardoso, M.: Generalised dice overlap as a deep learning loss function for highly unbalanced segmentations. In: Cardoso, M.J., et al. (eds.) DLMIA/ML-CDS -2017. LNCS, vol. 10553, pp. 240–248. Springer, Cham (2017). https://doi.org/10.1007/978-3-319-67558-9_28
26. Travis, W.D., et al.: The IASLC lung cancer staging project: proposals for coding T categories for subsolid nodules and assessment of tumor size in part-solid tumors in the forthcoming eighth edition of the TNM classification of lung cancer. J. Thorac. Oncol. **11**(8), 1204–1223 (2016)

27. Travis, W.D., et al.: IASLC multidisciplinary recommendations for pathologic assessment of lung cancer resection specimens after neoadjuvant therapy. J. Thorac. Oncol. **15**(5), 709–740 (2020)
28. Tschandl, P., Rosendahl, C., Kittler, H.: The ham10000 dataset, a large collection of multi-source dermatoscopic images of common pigmented skin lesions. Scientific data **5**(1), 1–9 (2018)
29. Wu, P.H., Bedoya, M., White, J., Brace, C.L.: Feature-based automated segmentation of ablation zones by fuzzy c-mean clustering during low-dose computed tomography. Med. Phys. **48**(2), 703–714 (2021)
30. Xie, S., Tu, Z.: Holistically-nested edge detection. In: Proceedings of the IEEE International Conference on Computer Vision, pp. 1395–1403 (2015)
31. Yuan, Y., Chao, M., Lo, Y.C.: Automatic skin lesion segmentation using deep fully convolutional networks with IACCARD distance. IEEE Trans. Med. Imaging **36**(9), 1876–1886 (2017)

Iterative Refinement Algorithm for Liver Segmentation Ground-Truth Generation Using Fine-Tuning Weak Labels for CT and Structural MRI

Peter E. Salah[1], Merna Bibars[1]([✉]), Ayman Eldeib[2],
Ahmed M. Ghanem[3], Ahmed M. Gharib[3], Khaled Z. Abd-Elmoniem[3],
Mustafa A. Elattar[4], and Inas A. Yassine[1]

[1] Systems and Biomedical Engineering, Cairo University, Giza, Egypt
{peter.salah,merna.bibars}@eng.cu.edu.eg
[2] Southern New Hampshire University, Manchester, NH, USA
[3] National Institute of Diabetes and Digestive and Kidney Diseases,
NIH, Bethesda, MD, USA
[4] Center for Informatics Science, Nile University, Cairo, Egypt

Abstract. Medical image segmentation is indicated in a number of treatments and procedures, such as detecting pathological changes and organ resection. However, it is a time-consuming process when done manually. Automatic segmentation algorithms like deep learning methods overcome this hurdle, but they are data-hungry and require expert ground-truth annotations, which is a limitation, particularly in medical datasets. On the other hand, unannotated medical datasets are easier to come by and can be used in several methods to learn ground-truth masks. In this paper, we aim to utilize across-modalities transfer learning to leverage the knowledge learned on a large publicly available and expertly annotated computed tomography (CT) dataset to a small unannotated dataset in a different modality magnetic resonance (MR). Moreover, we prove that quickly generated weak annotations can be improved iteratively using a pre-trained U-Net model and will approach the ground truth masks through iterations. This methodology was proven qualitatively using an in-house MR dataset where professionals were asked to choose between model output and weak annotations. They chose model output 93% ~ 94% of the time. Moreover, we prove it quantitatively using the publicly available annotated Combined (CT-MR) Healthy Abdominal Organ Segmentation (CHAOS) dataset. The weak annotation showed improvements across three iterations from 87.5% to 92.2% Dice score when compared to the ground truth annotations.

Keywords: Deep learning · U-Net · Transfer learning · Iterative Refinement · Semantic Segmentation · Across modalities · Weak labels

G. Waiter et al. (Eds.): MIUA 2023, LNCS 14122, pp. 33–47, 2024.
https://doi.org/10.1007/978-3-031-48593-0_3

1   Introduction

1.1   Background

Liver fibrosis is the accumulation of excess extracellular matrix material. It is a wound-healing response to many liver diseases or infections, including alcoholic liver diseases, nonalcoholic fatty liver, chronic viral infection, and hemochromatosis [1]. Progression of fibrosis over a long period of time leads to cirrhosis, characterized by (1) portal hypertension which can cause internal bleeding, (2) liver disability to perform its usual functions, (3) nodules generation surrounded by fibrotic septa, and (4) complete liver and kidney failure [2].

Chronic liver diseases (CLD) are of particular interest because the liver can restore its initial structure and function if the cause of fibrosis is diagnosed and treated in its early stages [3]. Additionally, CLD is estimated to affect 1.5 billion people worldwide [4]. Time is a key factor in diagnosis. If a liver injury is ignored without effective treatment, fibrosis will take place and could progress into cirrhosis or even hepatocellular carcinoma [5].

Liver segmentation is an essential preprocessing step in grading liver fibrosis. However, manual segmentation of the liver is usually a tedious and time-consuming task, especially in the case of 3D volumes, where segmentation is typically performed in a slice-by-slice manner, and possibly in 3D. The process is sensitive to multiple factors, including imaging parameters, modality variations, and the operator's knowledge. Therefore, it is often prone to intra and interobserver variability. It was found that it takes up to 90 min for an experienced radiologist to manually segment the liver [6].

Deep learning has been lately utilized for liver segmentation [7]. However, deep learning methods are data-hungry and require large amounts of datasets with accurate annotations to be able to learn and generalize well, which is also time-consuming [8,9]. On the contrary, generating weak labels can be accomplished quickly by non-experts or automatic software. These weak labels can then be used by a neural network to produce more accurate labels. We propose an iterative deep learning-based approach for the incremental refinement of quickly created weak labels for the MRI and CT volumes. This approach will radically minimize the efforts and preserve the expert's time in creating ground truth labels.

1.2   Literature Review

Deep neural networks (DNNs) are able to inherently learn hierarchical representations and extract useful features during training. In contrast to machine learning and conventional computer vision approaches, DNNs don't require feature extraction, design, or extensive domain expertise [10]. However, they require large professionally annotated datasets whose availability is limited in the medical field when compared to the non-medical field [8,9,11,12].

In order to deal with the scarcity of expert annotations in medical segmentation, numerous techniques have been developed [13]. When looking at scarce annotations, only some of the data is annotated whether within the same modality or across modalities. Some of the techniques developed to deal with this problem within the same modality cover data augmentation. For example, adding noise, use of unsharp masking, and Gaussian filtering [14], or rotation, scaling, and deformation [15].

Transfer learning and generative modelling are the methods used when there are expert annotations available but within another modality [13]. For example, Wang et al. [16] used transfer learning to fine-tune a model on scarce CT data. They collected 300 MR and 10 CT examinations of the liver and expertly annotated them. They pre-trained a 2D U-Net [15] on the large MR data and fine-tuned the model to the scarce CT data. The Dice score on the CT data reached 94%. On the other hand, Zhou et al. [17] proposed a generative approach to segment the liver from MR images without the presence of expert annotations. They used LiTS [18] and CHAOS [19] for the expertly annotated CT volumes, as well as 16 in-house pre-operative MR volumes. They use a cyclic adversarial network with a 2D U-Net to adapt between the two modalities. They achieved a Dice score of 90.3% on CT and 91.8% on MR.

Unannotated medical datasets can be leveraged with semi-supervised learning techniques where the model learns from annotated data and its own predictions on unannotated data in an iterative manner [13]. Zhang et al. [20] used unannotated chest CT examinations for cyst segmentation. The initially annotated data was generated in an unsupervised manner using K-means clustering and graph cuts with K set to be three (cyst, lung tissue, and others). A U-Net was trained on the initial labels, then used to predict pseudo labels on the entire data. The pseudo labels were then used as ground truth for the model to iteratively improve itself. In three iterations, the model improved the Dice score by 1.2% from the initial K-means output.

Another technique to deal with limited medical annotations is the use of weak labels. Weak labels are characterized by being over or under-segmented with inaccurate boundaries, they may be generated by a model or due to inexpert annotators. Zhang et al. [21] generated weak annotations of a stroke lesion MR dataset from junior trainees. They developed a tri-network architecture to train the model on the weak annotations only. The model consisted of three U-Nets trained simultaneously and each pair chose the confident pixels from the predictions by consensus. The chosen pixels were then used to train the third network. They achieved a Dice score of 68.12%.

In this study, we aim to develop a U-Net based approach that combines the use of transfer learning from large and annotated data to scarce data with the semi-supervised method of iteratively improving upon weak annotations. U-Net architecture was chosen since it is considered the baseline architecture that is widely used in the literature [22,23]. Moreover, it is simpler and easier to be compared with other competing approaches that are close to a vanilla U-Net.

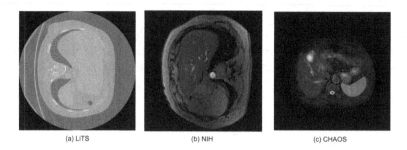

(a) LiTS   (b) NIH   (c) CHAOS

Fig. 1. Examples of the slices used from different datasets: (a) LiTS, (b) NIH, (c) CHAOS

2 Methods

2.1 Datasets Description

Three Datasets were employed in this study: (1) CT LiTS dataset [18], (2) CHAOS dataset [19], and (3) an MR dataset provided by our collaborators at the National Institutes of Health (NIH). In this section, the description of the datasets as well as the data preparation, will be discussed.

Liver Tumor Segmentation (LiTS) Dataset. [18] is formed of 131 abdominal CT scans along with the corresponding liver segmentation annotations. The data were randomly divided by patient into 91 scans for training, 26 scans for validation, and 14 scans for testing. This resulted in 40,812 training slices, 11,543 validation slices, and 6,283 testing slices.

Combined (CT-MR) Abdominal Organ Segmentation (CHAOS) Dataset. [19] includes CT and MR volumes along with the corresponding annotated segmentation for multiple organs, including the liver. In this study, only the T2-weighted MR dataset, formed of twenty volumes, was employed. The number of slices in each volume ranges between 26 and 36 slices. The MR dataset was divided by patient randomly into 14, 3, and 3, for training, validation, and testing datasets, where the total number of slices for each partition was: 458, 82, and 116, respectively. The scan matrix varied between 288×288 and 320×320 pixels per slice.

National Institutes of Health (NIH) Dataset. Twenty-seven patients were scanned in the supine position using 3T MRI (Achieva, Philips Medical Systems, Best, the Netherlands) using a 32-channel phased array cardiac-abdominal coil. The experimental parameters for the ECG-Gated spiral GRE Sequence to acquire T_2^* volumes can be summarized as follows: TR/TE = 20/0.77 msec; FOV = $350 \times 350 \ mm^2$; scan matrix =512×512 voxels; Imaging slice thickness= 8 mm, and flip angles = $30°$. The initial weak annotations for the volumes were generated using MeVisLab software [24]. The details of the weak label annotation are described in the data preparation section.

2.2 Dataset Preprocessing and Weak Label Mask Generation Preparation

Preprocessing of Learning Dataset. Normalization is an essential step to simplify the cross-modality domain adaptation between CT and MR. Min-max normalization was applied to all slices to enhance the contrast and map the gray level range to fall between 0 and 1 across the datasets. The slices, along with the corresponding ground truth masks, have been down-sampled to 128×128 using bilinear interpolation, to match the architecture of the U-Net employed in this study.

Both LiTS and NIH data were batched to 64 slices in each batch before training. However, CHAOS was batched to 16 slices in each batch because of its relatively small size, and it was also enough to ensure well representation of the whole dataset within each batch. Figure 1 shows sample slices for LITS, NIH, and CHAOS datasets. Data augmentation was used to increase the number of training inputs for LiTS dataset. In particular, random contrast and brightness augmentations were applied to each input slice for training, as such: $I_a = C \times I_o + B$, where I_a is the augmented image, I_o is the original image, C and B are random numbers picked from a uniform distribution $C \sim U(1.01, 2)$ & $B \sim U(20, 70)$. The total number of images used for training reached 81,624 slices.

Weak Label Generation Using the Annotated Masks. Artificially weak masks are generated for the CHAOS dataset by deforming the ground-truth masks using random affine and elastic deformations [25]. The employed affine transformation can be described as: $I_b = M \cdot I_a$, where I_a is the original image, I_b is the affine transformed image, and M is the affine transformation matrix. The affine transformation matrix is defined to transform the original coordinates of the image to new random coordinates generated from a uniform random distribution. Whereas the elastic deformation is defined on I_b, where new random coordinates generated from a uniform distribution are filtered by a Gaussian filter with a random α and σ, followed by bilinear transformation to transform the image I_b to the new coordinates resulting in the final deformed image I_c. A sample from the produced weak masks with their true annotations is shown in Fig. 2.

Weak Label Generation Using MeVisLab. MeVisLab is an open-source modular software framework for image processing research with a special focus on medical imaging. It is based on a graphical user interface to form complex image-processing networks. In this study, we have designed a MeVisLab network to create weak label annotation for the Liver. The annotation network consists of two major parallel paths: (1) Level set-based liver segmentation path (2) Scrapping tool path, as shown in Fig. 3.

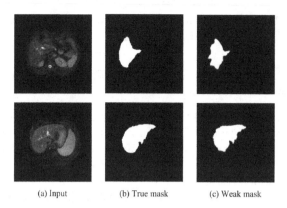

(a) Input   (b) True mask   (c) Weak mask

Fig. 2. Sample weak masks produced for the CHAOS dataset with their ground truth annotations: (a) Input MRI slice, (b) True mask, (c) Generated weak mask

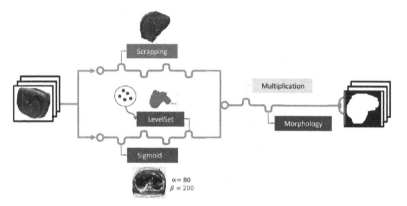

Fig. 3. Weak Label Generation using MeVisLab. A block diagram showing the process used to generate weak labels using MeVisLab for the unannotated NIH dataset

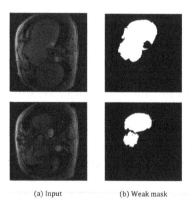

(a) Input   (b) Weak mask

Fig. 4. NIH dataset Sample slices : (a) sample T_2^* MRI slice (b) the generated weak label masks using MeVisLab.

The scrapping tool path is used by non-experts to manually segment the liver across the 3D volume. Whereas, the level set-based segmentation path is initialized by applying a sigmoid filter, Eq. 1, in order to enhance the contrast of the whole volume, where the intensity value of each pixel in the output image is computed according to Eq. 1.

$$f(x) = \frac{max - min}{1 + e^{-\frac{x-\beta}{\alpha}}} + min \tag{1}$$

where x is the input image, max and min are maximum and minimum intensities within the image, and the parameters of the sigmoid function $\alpha = 80$, $\beta = 200$.

The seed points were manually selected, by the observer, and placed throughout the liver slices, where a preliminary liver segmentation was generated after applying the level-set segmentation [26,27]. The output image was then transformed into a binary image, forming the segmented liver from this path. A pixelwise arithmetic multiplication was applied between the outputs of the two paths to form the output fused mask. Morphological operations: dilation and erosion, were applied on the fused mask, to fill any holes and fragmentations inside the segmented mask. The average time for weak label generation per non-expert annotator is 2.9 min while the longest time taken is 3.9 min. A sample of input slices along with the generated weak masks are shown in Fig. 4. It is worth noting that the extracted liver masks, using MeVisLab network, are annotated by non-specialists. Therefore, the resulting masks can be considered as weak labels. However, these are still considered valid masks that could be used to train the model over MR images.

2.3 Model Architecture

U-Net is a convolutional neural network architecture, originally developed for medical image segmentation. It is characterized by its fast and precise segmentation of images in different applications [23]. The basic architecture takes the shape of the letter U. The downward branch is a contractive path that extracts the inherent features of the image, while the upward branch is an expansive path for spatial localization. In this study, we employ a U-Net with four blocks in each path. Each block contains two convolutional modules with a dropout layer with a rate of 20% in between. Each module is made up of a 2D convolutional filter with a 3×3 kernel followed by a rectified linear unit (ReLU) activation, then a batch normalization layer. In the contractive path, the transition downwards from one block to the next is done with a 2×2 max-pooling layer for downsampling. However, in the expansive path, the transition upwards is done using a transpose convolution for up-sampling. A skip connection is added between the two paths such that feature maps from the contractive path are concatenated to the corresponding maps in the expansive path. Finally, a 1×1 convolution with a sigmoid activation is added to the last layer for binary classification. The used architecture is shown in Fig. 5.

The training of the network is done using the Jaccard distance as a loss function to overcome the class imbalance between the foreground and background pixels [28].

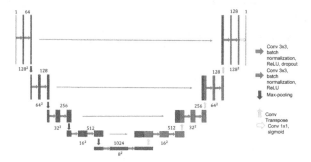

Fig. 5. U-Net architecture

$$Loss(x, \hat{x}) = 1 - \frac{|x \cap \hat{x}|}{|x \cup \hat{x}|} \tag{2}$$

To be able to compare our results with the literature, the Dice metric is used for evaluation as it's widely used in the literature [16, 20, 21].

$$Dice(x, \hat{x}) = 2 \times \frac{|x \cap \hat{x}|}{|x| + |\hat{x}|} \tag{3}$$

2.4 Iterative Refinement Approach

The proposed refinement approach, shown in Fig 6 employs transfer learning in an iterative manner to improve the masks of weakly labeled data. This is done by using a pre-trained model, which is fine-tuned for a small number of epochs using the weakly annotated data. The fine-tuned model is used later to predict new masks for the same weakly annotated data. This process will be further repeated to improve the predicted masks. The hypothesis of this study is that the predicted masks will be more accurate than the input weak annotations for the same iteration. For iteration i, the learned model is called iteration_i model, whereas the predicted mask from the same iteration is named iteration_i masks.

Further improvement can be achieved by retraining, also called fine-tuning, the initial pre-trained model using the weakly annotated data using iteration_1 masks as labels. The recently trained model iteration_2 model can be used to estimate iteration_2 masks, which are hypothetically better than iteration_1 masks and the original weak masks. Repeating the process should improve the extracted later in each iteration.

2.5 Experimentation and Performance Evaluation

The learning process can be described as follows: - First, the U-Net model was pre-trained on the large augmented LiTS dataset for 100 epochs. Adam optimizer was used with a learning rate of 0.001, and Dice metric was used for evaluation. The learned model of this process will be named as "LiTS model" during other

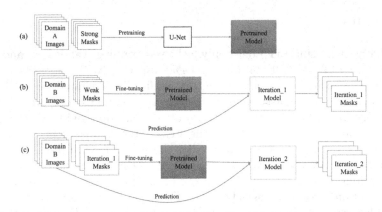

Fig. 6. Iterative Refinement Approach. (a) block diagram describing the learning process needed to generate the pre-trained model that will be used in the following steps. The learned model is training a U-Net from scratch using strongly annotated data CT images. (b) block diagram describing Iteration_1 , employed for fine-tuning and domain adaptation of pre-trained model using weakly annotated MR images. (c) block diagram describing Iteration_2 of pre-trained model fine-tuning to create better stronger segmentation masks for the MR images. The figure shows two iterations only, however, it could be repeated as needed.

experiments. - Second, transfer learning was used to adapt the model from CT domain to MR domain. The LiTS model was fine-tuned using the weak masks of the entire small CHAOS Dataset as well as the NIH data using the same optimizer and learning rate for 15 epochs only.

In order to quantitatively evaluate our hypothesis, the artificially generated weak masks for CHAOS labels are compared with the available ground-truth annotations. Then the iterative refinement process was applied to them to assess the degree of improvement after each iteration. For iteration 1, the LiTS model was fine-tuned using transfer learning on the CHAOS data but with the artificially generated weak masks as the true labels. Then, the iteration_1 model was used to generate predictions for the data, which will be called iteration_1 masks. For iteration_2, the LiTS model used iteration_1 masks as the ground truth for CHAOS data. The output masks for this iteration were iteration_2 masks. The same process was then repeated for iteration_3 to generate iteration_3 masks. The evaluation was done at each iteration to compare its output with the ground truth masks. In order to evaluate the performance of the learned model using NIH test dataset, the input weak masks and the iteration_i mask for iteration i, were randomly ordered and presented to two experts along with the corresponding MRI slice. Each expert must select the best segmentation mask for each slice by selecting (Right, Left, and Tie). The choice of the experts was then collected and evaluated. A sample is shown in Fig. 7.

3   Results

The LiTS model, using LiTS data only, achieved training, validation, and testing Dice score of 96.85%, 90.14%, and 93.4%, respectively. Quantitative evaluation for our hypothesis was done using the artificially generated weak CHAOS masks. First, the weak masks were compared with CHAOS strong ground-truth masks using Dice metric on test partition of the data and giving 87.5%. Then, after fine-tuning the LiTS model to the weak CHAOS masks in *iteration_1*. The *iteration_1* model was used to predict the masks for the test data resulting in *iteration_1masks*. Finally, we compared the *iteration_1* masks with CHAOS strong masks, where a Dice score of 89.3% was achieved, showing an improvement on deformed masks by 1.8%. The same process and evaluation was repeated for two other iterations. In the second iteration, the Dice score between *iteration_2masks* and the strong masks was 91.3% showing an improvement on the original deformed masks by a 3.8%. In the third iteration, the Dice score between iteration_3 masks and the strong masks was 92.2% showing an improvement on the original deformed masks by a 5.2% as seen in Table 1.

Table 1. Quantitative evaluation on CHAOS dataset using LiTS model showing test Dice percentage. For each iteration in the refinement process, the input and output data and the Dice score are specified

Iteration	Input	Output	Model vs. True Masks	% Improvement
iteration_1	Weak masks	iteration_1 masks	89.3	1.8
iteration_2	iteration_1 masks	iteration_2 masks	91.3	3.8
iteration_3	iteration_2 masks	iteration_3 masks	**92.2**	**5.2**

A sample of the weak CHAOS masks is shown in Fig. 8. The weak masks were then used to fine-tune the LiTS model producing iteration_1 weak masks. The process was repeated for three iterations, in each iteration, the weak masks from the previous iteration were used to fine-tune the LiTS model. A sample of the predicted weak mask for each iteration is shown in Fig. 8.

Since there were no strong true masks provided for the NIH dataset, the training Dice score for this step was 94.74%. Figure 7 shows samples of model output and the corresponding weak masks used for transfer learning.

The qualitative evaluation of the NIH dataset, based on the experts' selection, showed that the output segmentation masks - iteration_1 masks - were generally better than the input weak annotations, in 93% ∼ 94% of the time, as seen in Table 2. While in 7.4% ∼ 9.88% the model output and the weak labels were equivalent. It is worth noting that none of the experts reported that the input weak masks are a better segmentation of the liver than the output segmentation masks.

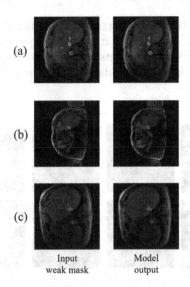

Fig. 7. Three samples (a),(b),(c) given to professionals for qualitative evaluation: in (a), both professionals chose model output, in (b) one of them chose this sample as a tie, in (c) both professionals chose the input weak mask. The output masks from the LiTS model that was fine-tuned using the NIH dataset weak masks are shown on the right, and their corresponding input weak masks are shown left for each sample.

4   Discussion

The quantitative evaluation results, evaluated for the CHAOS dataset, illustrated in Table 1, show that the Dice metric is improving from one iteration to the next, which proves that the iterative approach is heading towards the true mask. The same observation can be drawn from Fig. 8, where masks generated after the first iteration, shown in Fig. 8(d), are closer to the true masks, shown in Fig. 8(b), compared to the weak label masks, shown in Fig. 8(c). The same observation can be drawn when comparing the output mask of the second iteration, shown in Fig. 8(e) compared to the input mask, shown in Fig. 8(d). Moreover, in Fig. 7 although the liver masks in Fig. 7(b) can be equivalent in one of the professional evaluator's opinion, the model output in 7(a) is much better than the weak annotation, the upper boundary is more precise, and the lower boundary is closer to the true boundary. On the other hand, one can argue for 7(c) that although both evaluators chose the weak mask, the model output is still good. This may have happened because the weak mask is not necessarily that weak in all cases, which results in an overall good performance of the proposed method.

Such observations arguably indicate that the lack of ground truth is not a big problem anymore because the ground truth can be reached in a few iterations using weak labels. However, this required the existence of weak annotations which bounded the region in which the model could detect its target and steered the fine-tuning into better generalization. Additionally, the LiTS model's good performance on LiTS data was a necessary step that formed a good start for successful fine-tuning. The pre-trained model was exposed to expert annotations in another modality (CT). The weights learned can represent this knowledge, allowing it to adapt to the other modality (MR) and learn the ground truth from the weak annotations.

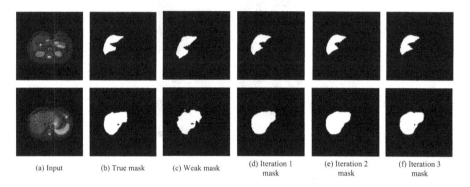

| (a) Input | (b) True mask | (c) Weak mask | (d) Iteration 1 mask | (e) Iteration 2 mask | (f) Iteration 3 mask |

Fig. 8. A sample of the predicted weak masks on CHAOS data throughout the different iterations compared to the generated weak masks and the true strong masks

The qualitative evaluation shown in Table 2, proved the concept that deep learning-based models can be used in an iterative manner to improve their own results rather than a single training on domain-specific data. Iterative learning was able to provide the model the chance to incrementally improve the estimation of the segmentation mask, which guaranteed better long-term results rather than immediate good results, estimated using domain-specific models. Another advantage of the proposed iterative approach is its ability to resolve many problems like the limited amount of data and the lack of expert annotations done by professionals. When compared to the literature, our approach combines two of the approaches used to handle data limitations: transfer learning across modalities, and iterative refinement (semi-supervised learning) from weak or noisy labels. The use of transfer learning from a large professionally annotated data to a small weakly annotated data gives the model more confident weights with which to start learning. This gives better results when compared to approaches using weak labels learning only, such as the one used in [21]. On the other hand, the weak labels done quickly remove the need for expert annotations to fine-tune the LiTS model using transfer learning. This is unlike [16], where the data had to be expertly annotated before applying transfer learning across modalities. It is worth noting that expert liver annotation can take up to 90 min [6] while the longest non-expert annotation using our MeVisLab network took only 3.9 min. This is a tremendous reduction in the time needed for annotation.

Table 2. Qualitative evaluation of experts' choices between model output and weak mask on NIH dataset

ID	Tie	Model	Total	% Model Chosen
A	32	275	324	94
B	24	280	324	**93**

However, our approach is limited to the use of a base architecture vanilla U-Net to prove our hypothesis. More modern architectures using transformers and attention can be studied to see their effect. Moreover, our study is done on liver segmentation only, it would be worthwhile to test our hypothesis on other segmentation tasks.

5   Conclusion

The shortage of expertly annotated medical datasets can be compensated for using weak annotations of data. Moreover, this is possible across modalities. By generating weak annotations of MR data, this paper proved that a U-Net pre-trained on CT domain can be used to refine and improve the weak annotations on MR domain. This was further proven quantitatively by creating deformed annotations of the publicly available and expertly annotated MR dataset CHAOS and comparing the refined masks to the expert annotations. The annotations improved across the iterations. Future work can be done to study this effect on other datasets and to examine the degree of deformation or threshold beyond which the network can't improve the weak annotations.

Acknowledgment. This project was supported by the Science and Technology Fund Institute (STDF), Project ID 45891- EG-US Cycle 20.

References

1. Baues, M.: Fibrosis imaging: Current concepts and future directions. Adv. Drug Deliv. Rev.**121**, 9–26, 2017. ISSN 0169–409X. https://doi.org/10.1016/j.addr.2017. 10.013. Fibroblasts and extracellular matrix: Targeting and therapeutic tools in fibrosis and cancer

2. Bataller, R., Brenner, D.: Liver fibrosis. J. Clin. Invest. **115**, 209–18 (2005). https://doi.org/10.1172/JCI24282

3. Tanaka, M., Miyajima, A.: Liver regeneration and fibrosis after inflammation. Inflamm. Regeneration **36**, 12 (2016). https://doi.org/10.1186/s41232-016-0025-2

4. Moon, A., Singal, A., Tapper, E.: Contemporary epidemiology of chronic liver disease and cirrhosis. Clin. Gastroenterol. Hepatol. **18**, 08 (2019). https://doi.org/10.1016/j.cgh.2019.07.060

5. Acharya, P., Chouhan, K., Weiskirchen, S., Weiskirchen, R.: Cellular mechanisms of liver fibrosis. Front. Pharmacol. **12** (2021). ISSN 1663–9812. https://doi.org/10. 3389/fphar.2021.671640

6. Gotra, A., et al.: Liver segmentation: indications, techniques and future directions. Insights Imaging **8**, 06 (2017). https://doi.org/10.1007/s13244-017-0558-1

7. Gul, S., Khan, M.S., Bibi, A., Khandakar, A., Ayari, M., Chowdhury, M.: Deep learning techniques for liver and liver tumor segmentation: a review. Comput. Bio. Med. **147**, 105620 (2022). https://doi.org/10.1016/j.compbiomed.2022.105620

8. LeCun, Y. Bengio, Y., Geoffrey Hinton, G.: Deep learning. Nature **521**, 436–44 (2015). https://doi.org/10.1038/nature14539

9. Aggarwal, C.C.: Neural networks and deep learning (2018)

10. Janiesch, C., Zschech, P., Heinrich, K.: Machine learning and deep learning. Electron. Mark. **31**(3), 685–695 (2021)

11. Litjens, G., et al.: A survey on deep learning in medical image analysis. Med. Image Anal. **42**, 60–88 (2017). ISSN 1361–8415. https://doi.org/10.1016/j.media.2017.07.005.

12. Willemink, M.J., et al.: Preparing medical imaging data for machine learning. Radiology **295**(1), 4–15 (2020). PMID: 32068507

13. Tajbakhsh, N., Jeyaseelan, L., Li, Q., Chiang, J., Wu, Z., Ding, X.: Embracing imperfect datasets: a review of deep learning solutions for medical image segmentation. Med. Image Anal. **63**, 101693 (2020). https://doi.org/10.1016/j.media.2020.101693

14. Zhang, L., et al.: When unseen domain generalization is unnecessary? rethinking data augmentation (2019)

15. Ronneberger, O., Fischer, P., Brox, T.: U-Net: convolutional networks for biomedical image segmentation. In: Navab, N., Hornegger, J., Wells, W.M., Frangi, A.F. (eds.) MICCAI 2015. LNCS, vol. 9351, pp. 234–241. Springer, Cham (2015). https://doi.org/10.1007/978-3-319-24574-4_28

16. Wang, K., et al.: Automated CT and MRI liver segmentation and biometry using a generalized convolutional neural network. Radiology: Artif. Intell. **1**, 180022 (2019). https://doi.org/10.1148/ryai.2019180022

17. Zhou, B., Augenfeld, Z., Chapiro, S.J., Zhou, K., Liu, C., Duncan, J.S.: Anatomy-guided multimodal registration by learning segmentation without ground truth: Application to intraprocedural CBCT/MR liver segmentation and registration. Med. Image Anal. **71**, 102041 (2021). ISSN 1361–8415. https://doi.org/10.1016/j.media.2021.102041

18. Bilic, P., et al.: The liver tumor segmentation benchmark (lits). arxiv:1901.04056 (2019)

19. Kavur, A.E., Selver, M.A., Dicle, O., Barış, M., Gezer, N.S.: CHAOS - combined (CT-MR) healthy abdominal organ segmentation challenge data (2019). https://doi.org/10.5281/zenodo.3431873

20. Zhang, L., Gopalakrishnan, V., Lu, L., Summers, R.M., Moss, J., Yao, J.: Self-learning to detect and segment cysts in lung CT images without manual annotation. In: 2018 IEEE 15th International Symposium on Biomedical Imaging (ISBI 2018), pp. 1100–1103 (2018). https://doi.org/10.1109/ISBI.2018.8363763

21. Zhang, T., Yu, L., Hu, N., Lv, S., Gu, S.: Robust Medical Image Segmentation from Non-expert Annotations with Tri-network. In: Martel, A.L., et al. (eds.) MICCAI 2020. LNCS, vol. 12264, pp. 249–258. Springer, Cham (2020). https://doi.org/10.1007/978-3-030-59719-1_25

22. Getao, D., Cao, X., Liang, J., Chen, X., Zhan, Y.: Medical image segmentation based on U-Net: a review. J. Imaging Sci. Technol. **64**, 03 (2020). https://doi.org/10.2352/J.ImagingSci.Technol.2020.64.2.020508

23. Siddique, N., Paheding, S., Elkin, C.P., Devabhaktuni, V.: U-net and its variants for medical image segmentation: a review of theory and applications. IEEE Access **9**, 82031–82057 (2021). https://doi.org/10.1109/ACCESS.2021.3086020

24. MeVis Medical Solutions AG. Mevislab logo (2008). https://www.mevislab.de/

25. Simard, P.Y., Steinkraus, D., Platt, J.C.: Best practices for convolutional neural networks applied to visual document analysis. In: Seventh International Conference on Document Analysis and Recognition, 2003. Proceedings, pp. 958–963 (2003). https://doi.org/10.1109/ICDAR.2003.1227801

26. Kass, M., Witkin, A.P., Terzopoulos, D.: Snakes: active contour models. Int. J. Comput. Vision **1**, 321–331 (2004)

27. Osher, S., Sethian, J.A.: Fronts propagating with curvature-dependent speed: algorithms based on hamilton-jacobi formulations. J. Comput. Phys. **79**(1): 12–49 (1988). ISSN 0021-9991. https://doi.org/10.1016/0021-9991(88)90002-2

28. Bertels, J., et al.: Optimizing the dice score and jaccard index for medical image segmentation: theory and practice. In: Shen, D., et al. (eds.) MICCAI 2019. LNCS, vol. 11765, pp. 92–100. Springer, Cham (2019). https://doi.org/10.1007/978-3-030-32245-8_11

M-VAAL: Multimodal Variational Adversarial Active Learning for Downstream Medical Image Analysis Tasks

Bidur Khanal[1]([✉]), Binod Bhattarai[4], Bishesh Khanal[3], Danail Stoyanov[5], and Cristian A. Linte[1,2]

[1] Center for Imaging Science, RIT, Rochester, NY, USA
bk9618@rit.edu
[2] Biomedical Engineering, RIT, Rochester, NY, USA
[3] NepAl Applied Mathematics and Informatics Institute for Research (NAAMII), Lalitpur, Nepal
[4] University of Aberdeen, Aberdeen, UK
[5] University College London, London, UK

Abstract. Acquiring properly annotated data is expensive in the medical field as it requires experts, time-consuming protocols, and rigorous validation. Active learning attempts to minimize the need for large annotated samples by actively sampling the most informative examples for annotation. These examples contribute significantly to improving the performance of supervised machine learning models, and thus, active learning can play an essential role in selecting the most appropriate information in deep learning-based diagnosis, clinical assessments, and treatment planning. Although some existing works have proposed methods for sampling the best examples for annotation in medical image analysis, they are not task-agnostic and do not use multimodal auxiliary information in the sampler, which has the potential to increase robustness. Therefore, in this work, we propose a Multimodal Variational Adversarial Active Learning (M-VAAL) method that uses auxiliary information from additional modalities to enhance the active sampling. We applied our method to two datasets: i) brain tumor segmentation and multi-label classification using the BraTS2018 dataset, and ii) chest X-ray image classification using the COVID-QU-Ex dataset. Our results show a promising direction toward data-efficient learning under limited annotations.

Keywords: multimodal active learning · annotation budget · brain tumor segmentation and classification · chest X-ray classification

1 Introduction

Automated medical image analysis tasks, such as feature segmentation and disease classification play an important role in assisting with clinical diagno-

G. Waiter et al. (Eds.): MIUA 2023, LNCS 14122, pp. 48–63, 2024.
https://doi.org/10.1007/978-3-031-48593-0_4

sis, as well as appropriate planning of non-invasive therapies, including surgical interventions [2,8]. In recent years, supervised deep learning-based methods have shown promising results in clinical settings. However, clinical translation of supervised deep learning-based models is still limited for most applications, due to the lack of access to a large pool of annotated data and generalization capability.

As medical data is expensive to annotate, some methods have explored the generation of synthetic images with ground truth annotation [18]. However, generative models for synthesizing high-quality medical data have not yet achieved a state where their distribution perfectly matches the distribution of real medical data, especially those containing rare cases [1,20]. This limitation can often result in biased models and poor performance. Another approach with a growing interest is semi-supervised learning where very few labeled data along with a large number of unlabeled data is used to train deep learning models [6,13,23]. Nevertheless, in a real-world scenario, semi-supervised learning still requires the selection of a certain number of image samples to be annotated by experts.

Active learning attempts to sample the best subset of examples for annotation from a pool of unlabeled examples to maximize the downstream task performance. Methods to sample examples that provide the best improvement with a limited budget have a long history [12], and several approaches have been proposed in recent years [25]. Active learning (AL) has experienced increased traction in the medical imaging domain, as it enables a human-in-the-loop approach to improve deployed AI models [5]. Recent works have proposed various sampling methods in AL specific to tasks such as classification, segmentation, and depth estimation in medical imaging. Shao et al. [16] used active sampling to minimize annotation for nucleus classification in pathological images. Yang et al. [24] proposed a framework to improve lymph node segmentation from ultrasound images using only 50% training data. Laradji et al. [11] adopted an entropy-based method for a region-based active sampler for COVID-19 lesion segmentation from CT images. Thapa et al. [22] proposed a task-aware AL for depth estimation and segmentation from endoscopic images. Related to our work, Sharma et al. [17] used uncertainty and representativeness to actively sample training examples for 2D brain tumor segmentation. Lastly, Kim et al. [10] studied various AL approaches for 3D brain tumor segmentation task.

Task Agnostic: The methods described earlier require training of downstream task-specific models precluding the possibility of training a task-agnostic model from a given pool of unannotated input images when the intended downstream task is not known a priori. On the other hand, task-agnostic methods could enable the use of the same model for new tasks that may arise in real-world continuous deployment settings. VAAL [19] is a promising task-agnostic method that trains a variational auto-encoder (VAE) with adversarial learning. VAAL produces a low-dimensional latent space that aims to capture both uncertainty and representativeness from the input data, enabling effective sampling for AL.

Sampling from multimodal images: Clinicians rarely reach a clinical decision by looking into a single medical image of a patient. They rely on other information, such as clinical symptoms, patient reports, multimodal images, and auxiliary device information. We believe that sampling methods in AL could benefit from using multimodal information. Although some recent works have explored multimodal medical images in the context of AL [5], none of the existing methods directly use the multimodal image as auxiliary information to learn the sampler.

In this paper, we propose a task-agnostic method that exploits multimodal imaging information to improve the AL sampling process. We modify the existing task-agnostic VAAL framework to enable exploiting multimodal images and evaluate its performance on two widely used publicly available datasets: the multimodal BraTS dataset [14] and COVID-QU-Ex dataset [21] containing lung segmentation maps as additional information.

The contributions of this work are as follows: **1)** We propose a novel multimodal variational adversarial AL method (M-VAAL) that uses additional information from auxiliary modalities to select informative and diverse samples for annotation; **2)** Using the BraTS2018 and COVID-QU-Ex datasets, we show the effectiveness of our method in actively selecting samples for segmentation, multi-label classification, and multi-class classification tasks. **3)** We show that the sample selection in AL can potentially benefit from the use of auxiliary information.

2   Methodology

In active learning, a subset of most informative samples X_s is selected from a large pool of unlabelled set X^* to query labels Y_s. The number of samples selected for label annotation is dictated by the set budget b of the sampler. Let us consider $(x, y) \sim (X, Y)$ as a labeled data pair initially present in the dataset. After sampling b examples from an unlabelled pool X^*, they are labeled and added to the labeled pool (X, Y). The sampling process is iterative, such that b examples are queried at each active sampling round and added to the labeled pool. The labeled samples are used to train a task network at each round by minimizing the task objective function. In our study, we have considered three different downstream tasks: a) semantic segmentation of brain tumors, b) multi-label classification of tumor types, and c) multi-class classification of chest conditions. The overall workflow of our proposed method is shown in Fig. 1.

2.1   Task Learner (A)

Our objective is to enhance the performance of downstream tasks by utilizing the smallest number of labeled samples. The task network is defined by the parameter θ_T. For the segmentation task, we train a U-Net model [15] to generate pixel-wise segmentation maps y', given a labeled sample $(x, y) \sim (X, Y)$, where y is a binary image. We chose the U-Net model due to its simplicity, as our focus is on evaluating the effectiveness of AL, not on developing a state-of-the-art segmentation technique.

Downstream Tasks (A)

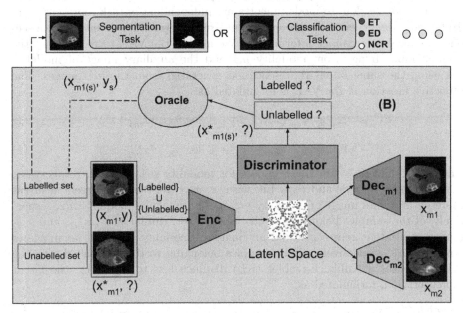

Fig. 1. M-VAAL Pipeline: Our active learning method uses multimodal information ($m1$ and $m2$) to improve VAAL. M-VAAL samples the unlabelled images, and selects samples complementary to the already annotated data, which are passed to Oracle for annotation. Incorporating auxiliary information from the second modality ($m2$) produces a more generalized latent representation for sampling. Our method learns task-agnostic representations, therefore the latent space can be used for both classification and segmentation tasks to sample the best-unlabelled images for annotation. (Refer to 2.2 for the meaning of each notation)

For the classification task, we use a ResNet-18 [9] classifier to predict y', given $(x, y) \sim (X, Y)$ where $y \subseteq \{0, 1, 2, ..\}$. In a multi-label classification setting, y' can contain more than one class, while in a multi-class setting, y' belongs to only one class from the set. Once a task is trained using labeled samples, we move on to the sampler (B) stage to select the best samples for annotation. After annotating the selected samples, we add them to the labeled sample pool and retrain the task learner using the updated sample set. Thus, the task learner is completely retrained for multiple rounds, with an increased training budget each time.

2.2 M-VAAL Sampler (B)

Our proposed M-VAAL method extends the VAAL approach by using an encoder-decoder architecture combined with adversarial learning to generate a lower dimensional latent space that captures sample uncertainty and representativeness. Unlike other AL methods that directly estimate uncertainty or diversity using task representation, VAAL learns a separate task-agnostic representation, providing the freedom to model uncertainty independent of task representation. To further strengthen the representation learning capability, we introduce a second modality as auxiliary information. We modify the VAE to reconstruct the original input image from modality $m1$ and the auxiliary image of modality $m2$ using the same latent representation generated from the $m1$ images. The objective function of the VAE is formulated as:

$$\mathcal{L}_{\text{VAE}}^{m1} = \mathbb{E}\left[\log p_{\theta_{m1}}\left(x_{m1} \mid z\right)\right] + \mathbb{E}\left[\log p_{\theta_{m1}}\left(x_{m1}^* \mid z^*\right)\right] \tag{1}$$

$$\mathcal{L}_{\text{VAE}}^{m2} = \mathbb{E}\left[\log p_{\theta_{m2}}\left(x_{m2} \mid z\right)\right] + \mathbb{E}\left[\log p_{\theta_{m2}}\left(x_{m2}^* \mid z^*\right)\right] \tag{2}$$

where $p_{\theta_{m1}}$ and $p_{\theta_{m2}}$ are the decoders for modality $m1$ and $m2$, respectively, parameterized by θ_{m1} and θ_{m2}. Likewise, z, x_{m1}, and x_{m2} represent a latent representation, an image of modality $m1$, and an image of modality $m2$, respectively, of the samples belonging to the labelled set; similarly, z^*, x_{m1}^*, and x_{m2}^* represent a latent representation, an image of modality $m1$, and an image of modality $m2$, respectively, of the samples belonging to the unlabelled set. The VAE also uses a Kullback-Leibler (KL) distance loss to regularize the latent representation, formulated as:

$$\mathcal{L}_{\text{VAE}}^{KL} = -\,\mathrm{D_{KL}}\left(q_{\phi_{m1}}\left(z \mid x_{m1}\right) \|p(z)\right) - \mathrm{D_{KL}}\left(q_{\phi_{m1}}\left(z^* \mid x_{m1}^*\right) \|p(z)\right) \tag{3}$$

where $q_{\phi_{m1}}$ is the encoder of modality $m1$ parameterized by ϕ_{m1} and $p(z)$ is the prior chosen as a unit Gaussian. It should be noted that only a single encoder is used to generate the latent representation, while there are two decoders.

Finally, an adversarial loss is added to encourage the VAE towards generating the latent representation that can fool the discriminator in distinguishing labeled from unlabelled examples (shown in Eq. 4). We train both the VAE and the discriminator in an adversarial fashion. The discriminator (D) is trained on the latent representation of an image to distinguish if an image comes from a labeled set or an unlabeled set. The loss function for the discriminator is shown in Eq. 5. We used Wasserstein GAN loss with gradient penalty [7].

$$\mathcal{L}_{\text{VAE}}^{adv} = \mathbb{E}\left[D\left(q_{\phi_{m1}}\left(z \mid x_{m1}\right)\right)\right] - \mathbb{E}\left[D\left(q_{\phi_{m1}}\left(z^* \mid x_{m1}^*\right)\right)\right] \tag{4}$$

$$\mathcal{L}_D = -\mathbb{E}\left[D\left(q_{\phi_{m1}}\left(z \mid x_{m1}\right)\right)\right] + \mathbb{E}\left[D\left(q_{\phi_{m1}}\left(z^* \mid x_{m1}^*\right)\right)\right] + \lambda\mathbb{E}\left[\left(\|\nabla_{\hat{x}} D(\hat{x})\| - 1\right)^2\right] \tag{5}$$

where the third term in Eq. 5 is the gradient penalty and $\hat{x} = \epsilon x_{m1} + (1 - \epsilon)x_{m1}^*$ is a randomly weighted average between x_{m1} and x_{m1}^*, such that $0 \leq \epsilon \leq 1$.

During the sampling process, the top samples that the discriminator votes as belonging to the unlabelled set are chosen; our intuition is that these samples contain information most different from the ones carried by the samples already belonging to the labeled set. Figure 2 shows that our method selects the examples that are far from the distribution of labeled examples, thus capturing diverse samples. The overall training and sampling sequence of M-VAAL is shown in Algorithm 1. In the first round, the task network is trained only

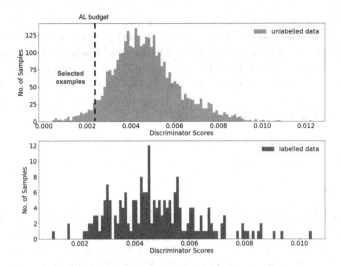

Fig. 2. Histogram of the discriminator scores for unlabeled and labeled data at third AL round. The discriminator is adversarially trained to push unlabeled samples toward lower values and labeled samples toward higher values. Our method involves selecting unlabelled instances that are far from the peak distribution of the labeled data. The number of samples to select is dictated by the AL budget.

Algorithm 1. Multimodal Variational Adversarial Active Learning

---

Given: Hyperparameters: epochs, γ_1, γ_2, γ_3, δ_1, δ_2, δ_3
Input: Labeled data (x_{m1}, y, x_{m2}), Unlabeled data (x^*_{m1}, x^*_{m2})
Initialize: Model parameters θ_T, $\theta_{VAE} = \{\theta_{m1}, \theta_{m2}, \phi\}$, and θ_D
1: **for** $e = 1$ to epochs **do**
2:     sample $(x_{m1}, y, x_{m2}) \sim (X_{m1}, Y, X_{m2})$
3:     sample $(x^*_{m1}, x^*_{m2}) \sim (X^*_{m1}, X^*_{m2})$
4:     $\mathcal{L}_{\text{VAE}} \leftarrow \gamma_1 \mathcal{L}^{adv}_{\text{VAE}} + \gamma_2 \mathcal{L}^{m1}_{\text{VAE}} + \gamma_3 \mathcal{L}^{m2}_{\text{VAE}} + \mathcal{L}^{KL}_{\text{VAE}}$
5:     Update VAE: $\theta'_{VAE} \leftarrow \theta_{VAE} - \delta_1 \nabla \mathcal{L}_{\text{VAE}}$
6:     Update D: $\theta'_D \leftarrow \theta_D - \delta_2 \nabla \mathcal{L}_D$
7:     Train and update T: $\theta'_T \leftarrow \theta_T - \delta_3 \nabla \mathcal{L}_{\text{T}}$

Sampling Phase
Input: b, X_{m1}, X^*_{m1}
Output: X_{m1}, X^*_{m1}
1: Select samples $X^*_{m1(s)}$ with $\min_b\{\theta_D(z^*)\}$
2: $y_s \leftarrow \mathcal{ORACLE}(X^*_{m1(s)})$
3: $(X_{m1}, Y) \leftarrow (X_{m1}, Y) \cup (X^*_{m1(s)}, Y_s)$
4: $X^*_{m1} \leftarrow X^*_{m1} - X^*_{m1(s)}$
5: **return** X_{m1}, X^*_{m1}

---

with the initially labeled samples to minimize the task objective function $\mathcal{L}_T$. Simultaneously, the M-VAAL sampler is also trained independently using all the labeled and unlabelled examples, with the end goal of improving the discriminator at distinguishing between labeled and unlabelled pairs. After the initial round of training, the trained discriminator is used to select the top b unlabelled samples identified as belonging to the unlabelled set. These selected samples are sent to the Oracle for annotation. Finally, the selected samples are removed from the unlabelled set and then added to the pool of labeled sets, along with their respective labels. The next round of AL proceeds in a similar fashion, but with the updated labeled and unlabelled sets.

3   Experiments

3.1   Dataset

BraTS2018: We used the BraTS2018 dataset [14], which includes co-registered 3D MR brain volumes acquired using different acquisition protocols, yielding slices of T1, T2, and T2-Flair images. To sample informative examples from unlabelled brain MR images, we employed the M-VAAL algorithm using contrast-enhanced T1 sequences as the main modality, and T2-Flair as auxiliary information. Contrast-enhanced T1 images are preferred for tumor detection as the tumor border is more visible [4]. In addition, T2-Flair, which captures cerebral edema (fluid-filled region due to swelling), can also be utilized for diagnosis [14]. Our focus was on the provided 210 High-Grade Gliomas (HGG) cases, which included manual segmentation verified by experienced board-certified neuro-radiologists. There are three foreground classes: Enhancing Tumor (ET), Edematous Tissue (ED), and Non-enhancing Tumor Core (NCR). In practice, the given foreground classes can be merged to create different sub-regions such as whole tumor and tumor core for evaluations [14]. The whole tumor entails all foreground classes, while the tumor core only entails ET and NCR.

Before extracting 2D slices from the provided 3D volumes, we randomly split the 210 volumes into training and test cases with an 80:20 ratio. The training set was further split into training and validation cases using the same ratio of 80:20, resulting in 135 training, 33 validation, and 42 test cases. These splits were created before extracting 2D slices to avoid any patient information leakage in the test splits. Each 3D volume had 155 (240 × 240) transverse slices in axial view with a spacing of 1 mm. However, not all transverse slices contained tumor regions, so we extracted only those containing at least one of the foreground classes. Some slices contained only a few pixels of the foreground segmentation classes, so we ensured that each extracted slice had at least 1000 pixels representing the foreground class; any slice not meeting this threshold was discarded. Consequently, the curated dataset comprised 3673 training images, 1009 validation images, and 1164 test images of contrast-enhanced T1, T2-Flair, and the segmentation map.

We evaluated our method on two downstream tasks: whole tumor segmentation and multi-label classification task. In the multi-label classification task, our prediction classes consisted of ET, ED, and NCR, with each image having either one, two, or all three classes. It's worth noting that for the downstream task, only contrast-enhanced T1 images were used, while M-VAAL made use of both contrast-enhanced T1 and T2-Flair images.

COVID-QU-Ex: The COVID-QU-Ex dataset is composed of 256×256 chest X-ray images from different patients that have been categorized into one of three groups: COVID infection, Non-COVID infection, and Normal. The dataset has two subsets: lung segmentation data and COVID-19 infection segmentation data, and the latter was chosen for our experiment. In addition to the X-ray images, the dataset also provides a segmentation mask for each image, which was utilized as an auxiliary modality during training M-VAAL. The dataset has a total of 5,826 images, consisting of 1,456 Normal, 1,457 Non-COVID-19, and 2,913 COVID-19 cases. These images are split into three sets: training, validation, and test, with the training set containing 3,728 images, the validation set containing 932 images, and the test set containing 1,166 images. The downstream task was to classify the input X-ray image into one of three classes.

3.2 Implementation Details

BraTS2018: All the input images have a single channel of size 240×240. To preprocess the images, we removed 1% of the top and bottom intensities and normalized them linearly by dividing each pixel with the maximum intensity, bringing the pixel intensity to a range of 0 to 1. We also normalized the images by subtracting and dividing them by the mean and standard deviation, respectively, of the training data. For VAAL and M-VAAL, the images were center-cropped to 210×210 pixels and resized to 128×128 pixels. For the downstream task, we used the original image size. Furthermore, to stabilize the training of VAAL and M-VAAL under similar hyperparameters as the original VAAL [19], we converted the single-channel input image to a three-channel RGB with the same grayscale value across all channels.

For M-VAAL, we used the same β-VAE and discriminator as in original VAAL [19], but added a batch normalization layer after each linear layer in the discriminator. Additionally, instead of using vanilla GAN with binary-cross entropy loss, we used WGAN with a gradient penalty to stabilize the adversarial training [7], with $\lambda = 1$. The latent dimension of VAE was set to 64, and the initial learning rate for both the VAE and discriminator was $1e^{-4}$. We tested γ_3 in the range $M = 0.2, 0.4, 0.8, 1$ via an ablation study reported in Sec 5, while γ_1 and γ_2 were set to 1. Both VAAL and M-VAAL used a mini-batch size of 16 and were trained for 100 epochs using the same hyperparameters, except for γ_3, which was only present in M-VAAL. For consistency, we initialized the model at each stage with the same random seed and repeated three trials with three different seeds for each experiment, recording the average score.

For the segmentation task, we used a U-Net [15] architecture with four down-sampling and four up-sampling layers, trained with an initial learning rate of $1e^{-5}$ using the RMSprop optimizer for up to 30 epochs in a mini-batch size of 32. The loss function was the sum of the pixel-wise cross-entropy loss and Dice

coefficient loss. The best model was identified by monitoring the best validation Dice score and used on the test dataset. For the multi-label classification,
we used a pre-trained ResNet18 architecture trained on the ImageNet dataset.
Instead of normalizing the input images with our training set's mean and standard deviation, we used the ImageNet mean and standard deviation to make
them compatible with the pre-trained ResNet18 architecture. We used an initial
learning rate of $1e^{-5}$ with the Adam optimizer and trained up to 50 epochs in
a mini-batch size of 32. The best model was identified by monitoring the best
validation mAP score and used on the test set. We used AL to sample the best
examples for annotation for up to six rounds and seven rounds for segmentation
and classification tasks, respectively, starting with 200 samples and adding 100
examples (budget – b) at each round.

COVID-QU-Ex: All the input images were loaded as RGB, with all channels
having the same gray-scale value. To bring the pixel values within the range of 0
to 1, we normalized each image by dividing all the pixels by 255. Additionally, we
normalized the images by subtracting the mean and dividing by the standard
deviation, both with values of 0.5. The same hyperparameters were used for
both VAAL and M-VAAL as those used in BraTS, and we also downsampled
the original 256×256 images to 128×128 for both models. For the downstream
multi-class classification task, we utilized a pre-trained ResNet18 architecture
that was trained on the ImageNet dataset. We trained the model with an initial
learning rate of $1e^{-5}$ using the Adam optimizer and a mini-batch size of 32 for
up to 50 epochs. The best model was determined based on the highest validation
overall accuracy score, and we evaluated this model on the test set. We employed
an AL method that involved sampling up to seven rounds using an AL budget of
100 samples. The initial budget for classification before starting active sampling
was 100 samples.

We have released the GitHub repository for our source code implementation[1].
We implemented our method using the standard Pytorch 12.1.1 framework in
Python 3.8 and trained in a single A100 GPU (40 GB).

3.3   Benchmarks and Evaluation Metrics

For our study, we employed two baseline control methods: random sampling and
VAAL – a state-of-the-art task agnostic learning method [19] that doesn't use
any auxiliary information while training. To quantitatively evaluate the segmentation, multi-label classification, and multi-class classification, we measured the
Dice score, mean Average Precision (mAP), and overall accuracy, respectively.

---

[1] https://github.com/Bidur-Khanal/MVAAL-medical-images.

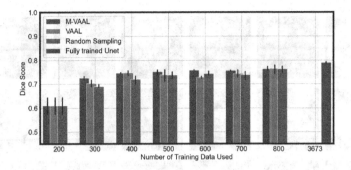

Fig. 3. Whole tumor segmentation performance comparison of proposed (M-VAAL) method against the VAAL and random sampling baselines according to the Dice score.

4   Results

4.1   Segmentation

Figure 3 compares the whole tumor segmentation performance (in terms of Dice score) between our proposed method (M-VAAL), the two baselines (random sampling and VAAL), and a U-Net trained on the entire fully labeled dataset, serving as an upper bound with an average Dice score of 0.789.

As shown, with only 800 labeled samples, the segmentation performance starts to saturate and approaches that of the fully trained U-Net on 100% labeled data. Moreover, M-VAAL performs better than baselines in the early phase and gradually saturates as the number of training samples increase.

Figure 4 illustrates a qualitative, visual assessment of the segmentation masks yielded by U-Net models trained with 400 samples selected by M-VAAL against ground truth, VAAL, and random sampling, at the AL second round. White denotes regions identified by both segmentation methods. Blue denotes regions missed by the test method (i.e., method listed first), but identified by the reference method (i.e., method listed second). Red denotes regions that were identified by the test method (i.e., method listed first), but not identified by the reference method (i.e., method listed second). As such, an optimal segmentation method will maximize the white regions, while minimizing the blue and red regions. Furthermore, Fig. 4 clearly shows that the segmentation masks yielded by M-VAAL are more consistent with the ground truth segmentation masks than those generated by VAAL or Random Sampling.

4.2   Multi-label Classification

In Fig. 6, a comparison is presented of the multi-label classification performance, measured by mean average precision, between the M-VAAL framework and two baseline methods (VAAL and random sampling). The upper bound is represented by a fully fine-tuned ResNet18 network. It should be noted that, on average, M-VAAL performs better than the two baseline methods, particularly when fewer

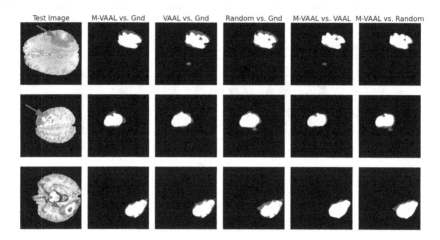

Fig. 4. Qualitative comparison of test segmentation masks generated by U-Nets trained on 400 samples using M-VAAL's selection method at the second round of AL phase. Columns 2–4 compare M-VAAl, VAAL, and random sampling against ground truth. Columns 5–6, compare M-VAAL against VAAL and random sampling. White denotes regions identified by both segmentation methods. Blue denotes regions missed by the test method (i.e., method listed first), but identified by the reference method (i.e., method listed second). Red denotes regions that were identified by the test method (i.e., method listed first), but not identified by the reference method (i.e., method listed second). Additionally, an arrow indicates the features that were segmented.

training data samples are used (i.e. 300 to 600 samples). The comparison models achieve a mean average precision close to that of a fully trained model even with a relatively small number of training samples, suggesting that not all training samples are equally informative. The maximum performance, which represents the upper bound, is achieved when using all samples and is 96.56%.

4.3 Multi-class Classification

The classification performance of ResNet18 models in diagnosing a patient's condition using X-ray images selected by M-VAAL was compared with baselines. Figure 5 illustrates that M-VAAL consistently outperforms the other methods, albeit by a small margin. The maximum performance, which represents the upper bound, is achieved when using all samples and is 95.10%.

5 Ablation Study: M-VAAL

We conducted an ablation study to assess the effect of the multimodal loss component, which involved varying the hyperparameter γ_3 across a range of values ($M = 0.2, 0.4, 0.8, 1.0$). The results suggested that both VAAL and M-VAAL were sensitive to hyperparameters, which depended on the nature of the downstream

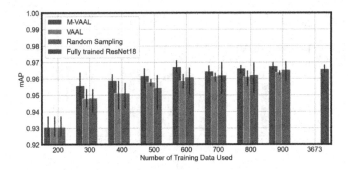

Fig. 5. Comparison of the mean average precision (mAP) for multi-label classification of tumor types between the proposed M-VAAL method and two baseline methods (VAAL and random sampling).

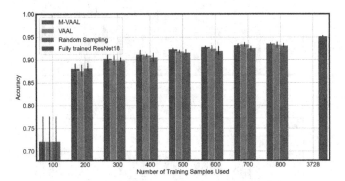

Fig. 6. Comparison of the overall accuracy for multi-class classification of COVID, Non-COVID infections, and normal cases between the proposed M-VAAL method and two baseline methods (VAAL and random sampling).

task (see Fig. 7). For example, M = 0.2 performed optimally for the whole tumor segmentation task, while M = 1 was optimal for tumor-type multi-label classification and Chest X-ray image classification. We speculate that the multimodal loss components (see 1 and 2) in the VAE loss serve as additional regularizers during the learning of latent representations. The adversary loss component, on the other hand, contributes more towards guiding the discriminator in sampling unlabelled samples that are most distinct from the distribution of the labeled dataset.

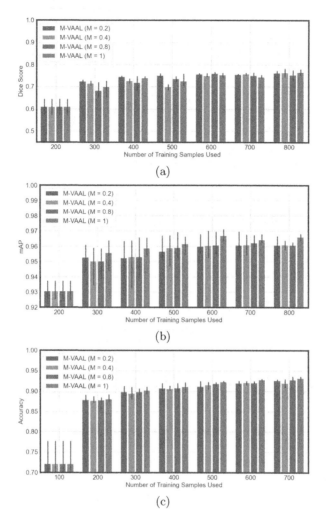

Fig. 7. Comparing the performance of (a) whole tumor segmentation, (b) tumor type multi-label classification, (c) chest X-ray infection multi-class classification with different training budgets, using the samples selected by M-VAAL trained with different values of M.

6   Discussion and Conclusion

M-VAAL, while being sensitive to hyperparameters, samples more informative samples than the baselines. Our method is task-agnostic, but we observed that the optimal value of M differs for different tasks, as shown in Sec. 5. The type of dataset also plays an important role in the effectiveness of AL. While AL is usually most effective with large pools of unlabelled data that exhibit high diversity and uncertainty, the small dataset used in this study, coupled with the nature of the labels, led to high-performance variance across each random initialization of the task network. For instance, in the whole tumor segmentation task, as tumors do not have a specific shape, the prediction was based on texture and contrast information. Similarly, in the Chest X-ray image classification task, the images looked relatively similar, with only subtle minute features distinguishing between classes. In addition, the evaluation test also plays a crucial role in accessing the AL sampler. If the test distribution is biased and does not contain diverse test cases, the effective evaluation of AL will be undermined. We conducted several Student's t-tests to evaluate the statistical significance of the performance results; nevertheless, we observed high variance across runs in downstream tasks when smaller AL budgets are used as shown by the error bars in Fig. 6, 3, and 5, as a result, the scores are not always statistically significant, given the large variance.

In the future, we plan to investigate this further by evaluating the benefit of multimodal AL on a larger pool of unannotated medical data on diverse multimodal datasets that guarantee a diversified distribution. Additionally, we aim to explore the potential of replacing the discriminator with metric learning [3] to contrast the labeled and unlabelled sets in the latent space. There is also a possibility of extending M-VAAL to other modalities. For instance, depth information can serve as auxiliary multimodal information to improve an AL sampler for image segmentation and label classification on surgical scenes with endoscopic images.

In this work, we proposed a task-agnostic sampling method in AL that can leverage multimodal image information. Our results on the BraTS2018 and COVID-QU-Ex datasets show initial promise in the direction of using multimodal information in AL. M-VAAL can consistently improve AL performance, but the hyperparameters need to be properly tuned.

Acknowledgements.. Research reported in this publication was supported by the National Institute of General Medical Sciences Award No. R35GM128877 of the National Institutes of Health, and the Office of Advanced Cyber Infrastructure Award No. 1808530 of the National Science Foundation. BB and DS are supported by the Wellcome/EPSRC Centre for Interventional and Surgical Sciences (WEISS) [203145Z/16/Z]; Engineering and Physical Sciences Research Council (EPSRC) [EP/P027938/1, EP/R004080/1, EP/P012841/1]; The Royal Academy of Engineering Chair in Emerging Technologies scheme; and the EndoMapper project by Horizon 2020 FET (GA 863146).

References

1. Al Khalil, Y., et al.: On the usability of synthetic data for improving the robustness of deep learning-based segmentation of cardiac magnetic resonance images. Med. Image Anal. **84**, 102688 (2023)
2. Ansari, M.Y., et al.: Practical utility of liver segmentation methods in clinical surgeries and interventions. BMC Med. Imaging **22**, 1–17 (2022)

3. Bellet, A., Habrard, A., Sebban, M.: A survey on metric learning for feature vectors and structured data. arXiv preprint arXiv:1306.6709 (2013)

4. Bouget, D., et al.: Meningioma segmentation in T1-weighted MRI leveraging global context and attention mechanisms. Front. Radiol. **1**, 711514 (2021)

5. Budd, S., et al.: A survey on active learning and human-in-the-loop deep learning for medical image analysis. Med. Image Anal. **71**, 102062 (2021)

6. Chen, X., et al.: Semi-supervised semantic segmentation with cross pseudo supervision. In: Proceedings IEEE Computer Vision and Pattern Recognition, pp. 2613–2622 (2021)

7. Gulrajani, I., et al.: Improved training of Wasserstein GANs. In: Advances in Neural Information Processing Systems 30 (2017)

8. Hamamci, A., et al.: Tumor-cut: segmentation of brain tumors on contrast-enhanced MR images for radiosurgery applications. IEEE Trans. Med. Imaging **31**, 790–804 (2011)

9. He, K., Zhang, X., Ren, S., Sun, J.: Identity mappings in deep residual networks. In: Leibe, B., Matas, J., Sebe, N., Welling, M. (eds.) ECCV 2016. LNCS, vol. 9908, pp. 630–645. Springer, Cham (2016). https://doi.org/10.1007/978-3-319-46493-0_38

10. Kim, D.D., et al.: Active learning in brain tumor segmentation with uncertainty sampling, annotation redundancy restriction, and data initialization. arXiv preprint arXiv:2302.10185 (2023)

11. Laradji, I., et al.: A weakly supervised region-based active learning method for COVID-19 segmentation in CT images. arXiv:2007.07012 (2020)

12. Lewis, D.D.: A sequential algorithm for training text classifiers: corrigendum and additional data. In: ACM SIGIR Forum, vol. 29, pp. 13–19. ACM New York, NY, USA (1995)

13. Luo, X., et al.: Semi-supervised medical image segmentation through dual-task consistency. In: AAAI Conference on Artificial Intelligence, pp. 8801–8809 (2021)

14. Menze, B.H., et al.: The multimodal brain tumor image segmentation benchmark (BraTS). IEEE Trans. Med. Imaging **34**, 1993–2024 (2015)

15. Ronneberger, O., Fischer, P., Brox, T.: U-Net: convolutional networks for biomedical image segmentation. In: Navab, N., Hornegger, J., Wells, W.M., Frangi, A.F. (eds.) MICCAI 2015. LNCS, vol. 9351, pp. 234–241. Springer, Cham (2015). https://doi.org/10.1007/978-3-319-24574-4_28

16. Shao, W., et al.: Deep active learning for nucleus classification in pathology images. In: 2018 IEEE 15th International Symposium on Biomedical Imaging (ISBI 2018), pp. 199–202 (2018)

17. Sharma, D., et al.: Active learning technique for multimodal brain tumor segmentation using limited labeled images. In: Domain Adaptation and Representation Transfer and Medical Image Learning with Less Labels and Imperfect Data: MICCAI Workshop 2019, pp. 148–156 (2019)

18. Singh, N.K., Raza, K.: Medical image generation using generative adversarial networks: a review. In: Patgiri, R., Biswas, A., Roy, P. (eds.) Health Informatics: A Computational Perspective in Healthcare. SCI, vol. 932, pp. 77–96. Springer, Singapore (2021). https://doi.org/10.1007/978-981-15-9735-0_5

19. Sinha, S., et al.: Variational adversarial active learning. In: Proceedings of the IEEE/CVF International Conference on Computer Vision, pp. 5972–5981 (2019)

20. Skandarani, Y., et al.: GANs for medical image synthesis: an empirical study. J. Imaging **9**(3), 69 (2023)

21. Tahir, A.M., et al.: COVID-QU-Ex Dataset (2022), https://www.kaggle.com/dsv/3122958

22. Thapa, S.K., et al.: Task-aware active learning for endoscopic image analysis. arXiv:2204.03440 (2022)
23. Verma, V., et al.: Interpolation consistency training for semi-supervised learning. In: Proceedings of the Twenty-Eighth International Joint Conference on Artificial Intelligence, IJCAI-19, pp. 3635–3641 (7 2019)
24. Yang, L., Zhang, Y., Chen, J., Zhang, S., Chen, D.Z.: Suggestive annotation: a deep active learning framework for biomedical image segmentation. In: Descoteaux, M., Maier-Hein, L., Franz, A., Jannin, P., Collins, D.L., Duchesne, S. (eds.) MICCAI 2017. LNCS, vol. 10435, pp. 399–407. Springer, Cham (2017). https://doi.org/10.1007/978-3-319-66179-7_46
25. Zhan, X., et al.: A comparative survey of deep active learning. arXiv:2203.13450 (2022)

BliMSR: Blind Degradation Modelling for Generating High-Resolution Medical Images

Samiran Dey[1]([✉]), Partha Basuchowdhuri[1], Debasis Mitra[2], Robin Augustine[3], Sanjoy Kumar Saha[4], and Tapabrata Chakraborti[5,6]([✉])

[1] School of Mathematical and Computational Sciences, IACS, Kolkata, India
Smcssd2661@iacs.res.in
[2] Department of Electronics and Telecommunication Engineering,
IIEST, Howrah, India
[3] Department of Electrical Engineering, Uppsala University, Uppsala, Sweden
[4] Computer Science and Engineering Department,
Jadavpur University, Kolkata, India
[5] The Alan Turing Institute and University College London, London, UK
[6] Linacre College, University of Oxford, Oxford, UK
tchakraborty@turing.ac.uk

Abstract. A persisting problem with existing super-resolution (SR) models is that they cannot produce minute details of anatomical structures, pathologies, and textures critical for proper diagnosis. This is mainly because they assume specific degradations like bicubic downsampling or Gaussian noise, whereas, in practice, the degradations can be more complex and hence need to be modelled "blindly". We propose a novel attention-based GAN model for medical image super-resolution that models the degradation in a data-driven agnostic way ("blind") to achieve better fidelity of diagnostic features in medical images. We introduce a new ensemble loss in the generator that boosts performance and a spectral normalisation in the discriminator to enhance stability. Experimental results on lung CT scans demonstrate that our model, BliMSR, produces super-resolved images with enhanced details and textures and outperforms recent competing models, including a diffusion model for generating super-resolution images, thus establishing a state-of-the-art. The code is available at https://github.com/Samiran-Dey/BliMSR.

1 Introduction

Computed Tomography (CT) is a widely employed medical imaging modality to diagnose diseases, monitor their progression, and conduct interventional radiology procedures. However, generating high-resolution (HR) CT scans requires advanced hardware components like high-precision detector elements and a fine focal spot radiation tube, which makes the process both expensive and time-consuming. Additionally, it is widely acknowledged that excessive exposure to high-dose of radiation can lead to genetic mutations and an increased risk of

G. Waiter et al. (Eds.): MIUA 2023, LNCS 14122, pp. 64–78, 2024.
https://doi.org/10.1007/978-3-031-48593-0_5

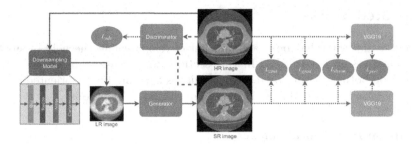

Fig. 1. Overview of the proposed methodology for BliMSR. The LR images are blindly synthesized from their HR counterparts by applying the components of the downsampling model in a random way, such that the degradation applied to each image is different and unknown. The generator is trained to super-resolve the LR images by a scale of 4, using the loss terms shown in ovals. Feature maps extracted from pre-trained VGG19 are used to compute the perceptual loss. The discriminator predicts whether an image is synthesized or real contributing to the adversarial loss

cancer [24]. Therefore, to mitigate the risk, doctors often suggest the use of low-dose computed tomography (LDCT), which generates images of inferior quality compared to that of the high-dose CT images [25] and need to be super-resolved for generating detailed information. We, thus, propose a methodology to perform single image super resolution (SISR) and enhance the details of medical images, as illustrated in Fig. 1.

Most of the existing SR models for medical images are trained on low-resolution (LR) images synthesized from their HR counterparts using bicubic-downsampling [5, 22, 38] and at times the addition of Gaussian noise [7, 19]. But in the real scenario, there are far more complex degradations present in the acquired low-resolution medical images, which need to be considered for accurate training. Our model, BliMSR, thus performs blind super-resolution by attempting to model the unknown and complex degradations while synthesizing the LR images, similar to Real_ESRGAN [18]. In addition, our contributions include -

1. A modified generator architecture with an enhanced attention mechanism to learn the minute details and textures of the HR images.
2. A new loss function for the generator including image gradient loss and dissimilarity loss for better reconstruction of the ground truth images.
3. A stronger discriminator with residual connections and spectral normalization for gaining stability in training.

In Sect. 2, we discuss the related works in the field of super-resolution on both real-world and medical images. In Sect. 3, we describe our methodology including details of the network architecture and the modified loss function. Finally, in Sect. 4, we present an overview of our experiments and results including details regarding the datasets, the adapted downsampling model, the training process, as well as the quantitative and qualitative evaluations on both synthetic and real data.

2   Related Works

In recent years, there has been a significant development in the field of single-image super-resolution (SISR). In this section, we discuss some of the SISR methods, on both real-world and medical images covering PSNR-oriented methods, GAN-based models, diffusion models and Blind SR methods.

2.1   SR on Real-World Images

The first machine learning model to perform SISR was SRCNN [23], proposed by Dong et al. It was followed by models that increased the network depth [28], used dense skip connections [33], deep recursive layers [29], memory blocks [32], and residual dense blocks [40]. EDSR [2] removed batch normalization layers from residual blocks to retain the range flexibility of the features and reduce GPU memory usage. All these methods were PSNR-oriented, trained with MSE or MAE loss and suffered from the drawback of generating over-smoothened images, which were not perceptually convincing. As a solution, GAN models, trained with perceptual loss, were employed for SISR, the pioneering work being SRGAN [12]. The perceptual loss was computed as the L1 distance between SR and HR feature maps obtained from the VGG network pre-trained with the ImageNet dataset. It was followed by ESRGAN [35], where batch normalization layers were removed from the residual blocks of SRGAN to eliminate BN artefacts, thus introducing *Residual* in Residual Dense Blocks (RRDB). ESRGAN also had a relativistic discriminator, which instead of predicting whether an image is real or fake, predicted how much more real or fake is an image in comparison with another, and helped in learning sharper edges and detailed textures. Wang et al. [34] and Sajjadi et al. [15] used spatial feature transformation and texture matching loss respectively in their GAN models to obtain realistic textures. Cheon et al. [14] attempted to build a GAN model based on EUSR, and balance between perceptual performance and distortion, by training with a modified loss function with discrete cosine transform coefficients loss and differential content loss. Some other significant works on GAN models include [16,27]. But as GANs are very unstable to train and susceptible to mode collapse, diffusion models emerged as a replacement for GANs. Recently, Li et al. [8] came up with a diffusion probabilistic model for SISR, which includes an RRDB-based LR encoder for upsampling and a U-Net-based conditional noise predictor for generating image residues.

Blind SR. The above-stated models perform non-blind SR, as they learn the assumed downsampling and noise in the synthesized LRs, which inevitably deviates from the real scenario. Zhang et al. [20] used the classical degradation model comprising blur, downsampling and addition of noise to generate LRs, and proposed a network that takes the concatenated LR image and degradation maps as input to perform blind SR. Gu et al. [10] performed blind SR using iterative kernel correction, whereas, Real_ESRGAN [18] attempted to synthesize the unknown and complex degradations present in real images, by using a second-order degradation model comprising of jpeg compression and sinc filters along with the classical degradation model. We find this model effective for medical images as well and adapt it to synthesize LR images for our experiments.

2.2 SR on Medical Images

None of the works discussed in Sect. 2.1 considered evaluating their models on any modality of medical images, probably because medical images have completely different underlying distributions and textures. Initial works in medical SISR include the use of residual-based transformations from lower-resolution thick-slice images to high-resolution thin-slice images [1], residual deep CNNs to promote high-frequency textures [37], and modified U-Net based architecture for improved deblurring of boundaries of bone structures and air cavities [11]. Georgescu et al. [6] proposed a multimodal multi-head convolutional attention with various kernel sizes to perform SR from medical images of varying contrast. With the advent of GANs, they were being used for medical image super-resolution, to obtain higher perceptual quality. Yang et al. [36] and Wolterink et al. [9] performed denoising of low-dose CTs using GAN frameworks. Gu et al. [7] proposed a generator architecture with Residual Whole Map Attention Blocks (RWMAB) to extract information from different channels, a pairwise discriminator that learns by concatenating feature maps extracted from both LR and SR/HR images and a multi-task loss function to generate accurate textures and patterns of medical images. Zhang et al. [38] proposed a GAN model employing high-resolution representation learning by extracting features from parallel convolutions at different levels and then aggregating them. Bing et al. [22] used improved Squeeze and Excitation (SE) block in their GAN for strengthening important features while weakening non-important ones. Other GAN models for SR on medical images include [3,5]. You et al. [19] proposed a cycleGAN-based model to perform blind SR on medical images using paired and unpaired data. But real paired data like the micro-CT tibia dataset [4] used in GAN-CIRCLE [19] are extremely difficult to obtain and training with synthesized LRs, whether paired or unpaired, is bound to make the model learn the imposed degradations which differ in reality. Hence, we suggest that attempting to model the unknown and complex degradations instead, is a more effective way to blind SR.

3 Methodology

In this section, we first present the proposed network architecture and then discuss our ensemble loss function.

3.1 Network Architecture

The generator architecture of the model is illustrated in Fig. 2. The Residual Whole Map Attention Block (RWMAB) of MedSRGAN [7] is replaced with our

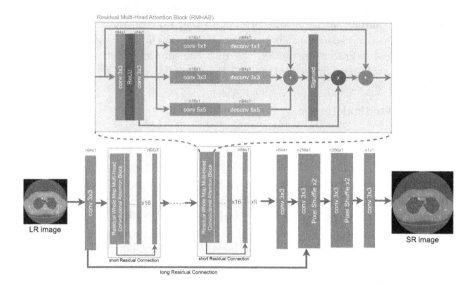

Fig. 2. Architecture of the generator with Residual Multi-Head Attention Block (RMHAB). For convolutions, n denotes the number of channels and s denotes stride. The network produces images super-resolved by a scale of 4

Residual Multi-Head Attention Block (RMHAB). Instead of using a single 1×1 convolution for attention, we use an ensemble of convolution heads with kernels of sizes 1×1, 3×3, and 5×5 as proposed in MMHCA [6]. The convolution and deconvolution layers perform both spatial and channel attention. The reduction in dimensions by convolution is reversed by deconvolution. As the different attention heads allow different levels of information to flow, a sigmoid applied to their summation combines the features from all channels and produces a more accurate attention map than a single 1×1 convolution. The output of RMHABs is convolved and upscaled by a factor of 2 using pixel shuffle which increases the height and width of an image while reducing the number of channels. The long residual connection helps in preserving the information present in the input LR.

Figure 3 shows the discriminator architecture. Unlike previous works [7,12, 19], our discriminator consists of residual blocks similar to ResNet18 to facilitate learning from the residual connections and avoid the degradation problem. As suggested by Lin et al. [13], we use spectral normalization in the discriminator to mitigate the problems of exploding and vanishing gradients, and thus, gain stability in training.

3.2 Loss Function

We propose an ensemble loss function for the generator by including image gradient loss and dissimilarity loss along with adversarial, content and perceptual losses. The adversarial loss helps the generator to produce images closer to the

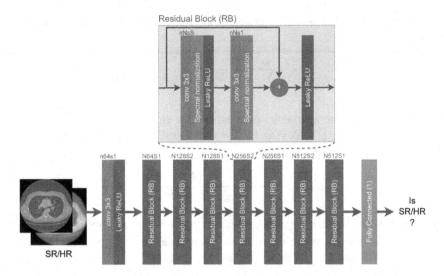

Fig. 3. Architecture of the discriminator network with residual blocks. N and n denote the number of channels for the residual blocks and convolutions respectively. S and s denote the stride of the residual blocks and convolutions respectively

HR ground truth such that it can deceive the discriminator into classifying the synthesized images as real. It is defined as,

$$L_{adv} = E_{HR \sim P_{data}(HR)}[log(D(HR))] + E_{LR \sim P_{data}(LR)}[log(1 - D(G(LR)))] \quad (1)$$

where G denotes the generator and D denotes the discriminator. The content loss is a pixel-wise loss that helps in minimising the distance between the synthesized SRs and the ground truth HRs and is formulated as,

$$L_{cont} = \frac{1}{wh} \sum_{i=1}^{w} \sum_{j=1}^{h} | G(LR)_{ij} - HR_{ij} | \quad (2)$$

The perceptual loss aids in achieving higher perceptual quality by considering the L1 distance between high-dimensional feature maps of SR and HR images, obtained from VGG19 pre-trained on the ImageNet dataset. According to [35], feature maps obtained after activation contain less information and can impact image brightness. Thus, we extract feature maps before activation from the last convolution of the first five VGG blocks, represented as ϕ_k, for the kth block.

$$L_{perc} = \frac{1}{5wh} \sum_{k=1}^{5} \sum_{i=1}^{w} \sum_{j=1}^{h} | \phi_k(G(LR)_{ij}) - \phi_k(HR_{ij}) | \quad (3)$$

The image gradient loss helps in mapping the edges of the synthesized image to the ground truth, thus, preserving the textures and producing sharper images.

It is defined as the L1 difference between the gradients of the SR image and the HR image.

$$L_{grad} = \frac{1}{2wh} \sum_{k=1}^{2} \sum_{i=1}^{w} \sum_{j=1}^{h} | \nabla G(LR)_{ijk} - \nabla HR_{ijk} | \qquad (4)$$

where ∇ represents the gradient operation and k represents the two axes. The dissimilarity loss aids in achieving higher structural similarity with the ground truth HR image and is defined as,

$$L_{dssim} = \frac{1 - ssim(G(LR), HR)}{2} \qquad (5)$$

where SSIM(x,y) [21] $= \frac{(2\mu_x\mu_y+C_1)+(2\sigma_{xy}+C_2)}{(\mu_x^2+\mu_y^2+C_1)(\sigma_x^2+\sigma_y^2+C_2)}$, C_1 and C_2 being two constants to stabilize division by weak denominator and L is the range of pixel values. Finally, the total loss for the generator is given by,

$$L_G = L_{perc} + \lambda_{adv} * L_{adv} + \lambda_{cont} * L_{cont} + \lambda_{grad} * L_{grad} + \lambda_{dssim} * L_{dssim} \qquad (6)$$

where $\lambda_{adv}, \lambda_{cont}, \lambda_{perc}, \lambda_{grad}$ and λ_{dssim} are the coefficients for the respective losses and controls the influence of the loss terms in training the model. The discriminator is trained with the adversarial loss in Eq. 1.

4   Experiments and Results

In this section, we first discuss the dataset used and the degradation model to synthesize the LR images. Subsequently, we describe the training methodology and present the results of qualitative and quantitative evaluations on the LIDC-ICRI dataset. Additionally, we provide results of testing our model on a dataset of compressed lung CT images and verify the effectiveness of the random downsampling model.

4.1   Dataset

Lung CT scans from the Lung Image Database Consortium and Image Database Resource Initiative (LIDC-IDRI) [17] public database has been used for our experiments. 9336 CT slices are used for training and 919 CT slices are used for testing. Following [7], the intensity values of the CT slices are clipped to [-1024, 1024] HU (Hounsfield Unit) and then rescaled to [0,1] for training. All CT slices are of dimension 512×512. They are brought down to 128×128 for a scale of 4 SR. We adapt the second-order random downsampling methodology proposed by Real-ESRGAN [18] for synthesizing the LR images, as illustrated in Fig. 1. For degradation, blur filters are chosen between isotropic and anisotropic with probabilities of 0.7 and 0.3 respectively, the kernel size being randomly selected from {7,9,11,...,21}. Resizing is done using bicubic, bilinear, or area downsampling with probabilities of 0.6, 0.2, and 0.2 respectively. Then, Gaussian noise and

Poisson noise are introduced with a probability of 0.5 each. According to Gravel et al. [25], the noise present in CT scans consists of Gaussian and Poisson noise. Therefore, the use of the random downsampling model ensures accurate simulation of the real-world conditions encountered in medical CT scans. Finally, we apply jpeg compression and sinc filter in our downsampling model with a probability of 0.3 so that the readily available medical images in jpeg or png format can also be super-resolved using our model. To verify it with lung CTs in compressed formats we use 331 images from the Kaggle Chest CT-scan images dataset [26].

4.2   Training Details

The generator is trained using the loss function in Eq. 6 with $\lambda_{adv} = 0.01$, $\lambda_{cont} = 0.1$, $\lambda_{grad} = 1$ and $\lambda_{dssim} = 1$ and the discriminator is trained using the adversarial loss in Eq. 1. Following [7,35], we set the coefficient of content loss larger than adversarial loss and smaller than the perceptual loss to balance the information content and smoothening of images. Both gradient and dissimilarity loss gets equal weightage to the perceptual loss to ensure that we obtain accurate textures and structural information while generating perceptually compelling images. The model is trained for 350K iterations with a learning rate of 1×10^{-5} and Adam optimizer with $\beta_1 = 0.9$ and $\beta_2 = 0.999$ on an NVIDIA a100 GPU.

4.3   Result and Analysis

For evaluations, we compare the performance of our model, BliMSR, with bicubic upsampling, a PSNR-oriented deep CNN, EDSR+MHCA [2,6], two GAN models, MedSRGAN [7], Real_ESRGAN [18] and a diffusion model, SRDiff [8]. All models are trained for 350K iterations with a batch size of 4. The RRDB-based [35] Real_ESRNet from Real_ESRGAN and LR encoder of SRDiff are first pre-trained for 700K iterations with L1 loss, following their training methods. To evaluate how effective our contributions are, we train the discriminator of MedSRGAN paired with our generator and losses, referred to as BliMSR*, and compare its performance with BliMSR.

Qualitative Analysis. Our BliMSR outperforms all other models as observed from Fig. 4. BliMSR can enhance anatomical structures, pathologies and textures more than any other methodology and generate images closest to the HR ground truth. Bicubic upsampling, EDSR+MHCA and MedSRGAN generate blurred and noisy images. Patches of white pixels are generated encircling the CT by EDSR+MHCA and SRDiff. While Real_ESRGAN and SRDiff can achieve some detailing, they distort anatomical structures. The structure of the blood vessel pointed by the red arrow in Fig. 4a has been completely changed by SRDiff and the space between the two blood vessels has been widened by Real_ESRGAN, but BliMSR can generate the blood vessels quite accurately, closest to the ground truth. This proves that the enhanced attention mechanism along with the image

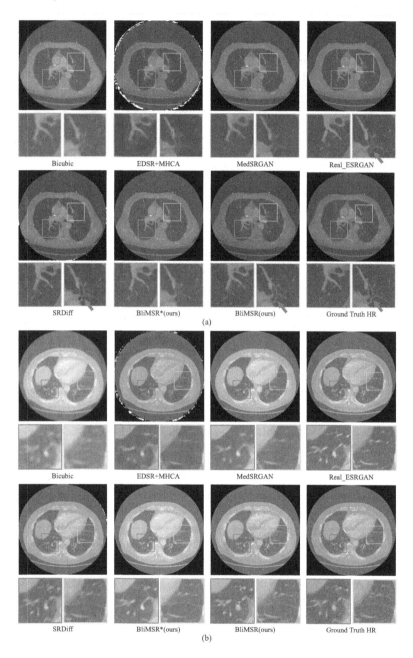

Fig. 4. Qualitative comparison using two representative slices (a) and (b) from the LIDC-IDRI dataset. Different regions of the slices, cropped and zoomed, have been shown in four different coloured boxes. Our BliMSR outperforms other methodologies in enhancing anatomical details and produces images closest to the ground truth. The red arrows in (a) point to a blood vessel that has been distorted by SRDiff and Real_ESRGAN but quite accurately generated by BliMSR (Color figure online)

Table 1. Table of results for the various methods, tested on downsampled LIDC-IDRI lung CTs. ↑ indicates the larger the better and ↓ indicates the smaller the better. The values are expressed in mean and standard deviation. Red denotes the best score and blue denotes the second-best score. To evaluate the significance of the result, p-values are derived for each metric using Welch's T-test by comparing the scores for images generated by BliMSR and the existing leading edge model that performed best in the metric.

Models	PSNR ↑	SSIM ↑	VIF ↑	LPIPS ↓
Bicubic	16.90871 ± 4.50271	0.67431 ± 0.07969	0.23637 ± 0.07000	0.36308 ± 0.06385
EDSR+MHCA	17.59007 ± 2.53675	0.81246 ± 0.04141	0.23906 ± 0.04403	0.26993 ± 0.04845
MedSRGAN	21.96609 ± 5.99931	0.88015 ± 0.05018	0.46330 ± 0.07230	0.22438 ± 0.05001
Real_ESRGAN	25.17862 ± 5.15055	0.88745 ± 0.03723	0.40977 ± 0.05768	0.17462 ± 0.04587
SRDiff	23.37245 ± 4.88581	0.89323 ± 0.03651	0.46370 ± 0.07781	0.17260 ± 0.05933
BliMSR*(ours)	23.62873 ± 6.89547	0.89648 ± 0.05223	0.52129 ± 0.07358	0.18024 ± 0.05544
BliMSR(ours)	26.48864 ± 6.89493	0.90146 ± 0.04583	0.52138 ± 0.07380	0.14351 ± 0.04654
p-value	4.28777e-06	2.18214e-05	7.48292e-56	1.93271e-30

gradient and dissimilarity loss is effective in enhancing details with greater accuracy. Besides, our discriminator helps in generating sharper images by acting as a stronger critic because of the residual connections as evident from Fig. 4. The images generated by BliMSR* have thinner vessels and other minute details blurred than the images generated by BliMSR.

Quantitative Analysis. Given in Table 1 is the quantitative evaluation of the various super-resolution methods. We use PSNR, SSIM [21], VIF [31], and LPIPS [39] as metrics for comparison. PSNR estimates the quality of the image in terms of noise, SSIM measures the structural similarity of the image, VIF estimates how well the image has been reconstructed and LPIPS measures the image's perceptual quality. BliMSR gets the best score in all four metrics. We assume, that on being generated from the same LR, the images produced by all models have the same underlying distribution, and perform Welch's T-test for each metric to further comment on the significance of the result. We compare the scores obtained by the images generated using BliMSR and the corresponding images generated by the model with the second-best score, that is, Real_ESRGAN for PSNR and SRDiff for SSIM, VIF and LPIPS. We do not consider BliMSR* for the test and include the existing models only. We test the null hypothesis that the images generated by the two models have identical average scores and find that the p-values for all metrics are less than $\alpha = 0.05$, as seen in Table 1. Thus, we reject the null hypothesis, concluding that there is a significant improvement in the scores obtained by BliMSR over the second-best model for each metric and hence, any other model having average scores inferior to the second-best model.

Despite diffusion models having the advantages of stable training and no possibility of mode collapse, we find that our GAN model, BliMSR, performs better.

This proves that the spectral normalization used in the discriminator helped in stabilizing the training and our attention-based generator trained with the loss function in Eq. 6 is capable of generating sharper images with minute details, closest to the ground truth. The superior performance of BliMSR to BliMSR* proves that our discriminator forces the generator to reduce noise and obtain higher perceptual quality. Only Real_ESRGAN achieves a higher PSNR than BliMSR* as the model has been pre-trained with L1 loss for 700K iterations. Also, BliMSR* outperforms MedSRGAN in all metrics proving that the modifications made to the attention mechanism of the generator and the ensemble loss function jointly help in obtaining higher fidelity by enhancing greater details and producing perceptually better and sharper images, as can be seen in Fig. 4.

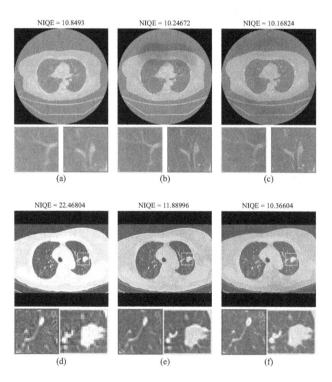

Fig. 5. Result of applying BliMSR on representative real images from (a)LIDC-IDRI dataset [17] and, (d)Kaggle Chest CT-scan images dataset [26], with corresponding NIQE scores. (b) and (e) are super-resolved by BliMSR trained on LR images synthesized using bicubic-downsampling and Gaussian blur only. (c) and (f) are super-resolved by BliMSR trained on LR images synthesized using the adapted second-order random downsampling model. Cropped and zoomed segments of the images are shown in different coloured boxes for better understanding

Analysis on Real Images. We further try to evaluate the effectiveness of BliMSR and the second-order random downsampling model adopted by us, on some lung CTs in png format from the Kaggle Chest CT-scan images dataset and on real CTs from the LIDC-IDRI dataset without imposing any degradation on them. As evident in Fig. 5, BliMSR trained with LR images synthesized using the random downsampling model generates sharper images with less noise and obtains a lower NIQE score [30], indicating higher perceptual quality than BliMSR trained with LR images synthesized using bicubic-downsampling and Gaussian blur only. Additionally, BliMSR reduces the dark patches in certain regions of the SR images synthesized using blur and downsampling only, as can be seen in Fig. 5b,c. It is also evident from Fig. 5f that BliMSR is capable of producing realistic textures and missing details from compressed CTs in png format, which justifies the use of compression in our downsampling model. Having a methodology to super-resolve such LR images would help researchers to use the publicly available compressed datasets as an alternative when raw image files are not available.

5 Conclusion

In this paper, we address a persistent problem of optimal degradation modelling for medical image super-resolution tasks, by taking a data-driven "blind" modelling approach in our proposed GAN-based BliMSR model. Our generator made up of the Residual Multi-Head Attention Blocks (RMHAB), along with the formulated loss function and the ResNet-based discriminator, outperforms other SR models including EDSR+MHCA, MedSRGAN and Real_ESRGAN. Despite diffusion models being state-of-the-art for image generation, our GAN model performs better than SRDiff. The adopted second-order random downsampling methodology for Blind SR also proved to be effective for generating realistic super-resolved medical images. Moreover, the statistical evaluation using Welch's T-test suggests that BliMSR is a significant improvement over the existing methods and hence, can be clinically applied to low-resolution images acquired using a lower dosage of ionizing radiations, like LDCTs, for obtaining high-resolution images with enhanced details, further extending to other modalities of images as well. The code is available at https://github.com/Samiran-Dey/BliMSR.

Acknowledgement. We acknowledge Indo-Swedish DBT-Vinnova project, BT/PR41 025/Swdn/135/9/ 2020, for supporting this research. TC is funded through the Turing-Roche strategic partnership, who are sponsoring the conference registration and attendance costs. TC is also affiliated with Linacre College, University of Oxford through a non-stipendiary EPA Cephalosporin fellowship.

References

1. Chaudhari, A.S., et al.: Super-resolution musculoskeletal MRI using deep learning. Magn. Reson. Med. **80**, 2139–2154 (2018). https://doi.org/10.1002/mrm.27178

2. Lim, B., et al.: Enhanced deep residual networks for single image super-resolution. In: 2017 IEEE Conference on Computer Vision and Pattern Recognition Workshops (CVPRW). IEEE (2017). https://doi.org/10.1109/cvprw.2017.151

3. Chen, Y., et al. Efficient and accurate MRI super-resolution using a generative adversarial network and 3D multi-level densely connected network. arXiv:1803.01417 (2018)

4. Chen, C., et al.: Quantitative imaging of peripheral trabecular bone microarchitecture using MDCT. Med. Phys. **45**, 236–249 (2017). https://doi.org/10.1002/mp.12632

5. Mahapatra, D., et al.: Image super-resolution using progressive generative adversarial networks for medical image analysis. Comput. Med. Imaging Graph. **71**, 30–39 (2019). https://doi.org/10.1016/j.compmedimag.2018.10.005

6. Mariana-Iuliana, G., et al.: Multimodal multi-head convolutional attention with various Kernel sizes for medical image super-resolution. In 2023 IEEE/CVF Winter Conference on Applications of Computer Vision (WACV), pp. 2194–2204. https://doi.org/10.1109/WACV56688.2023.00223

7. Zeng, G.Y., et al.: MedSRGAN: medical images super-resolution using generative adversarial networks. Multimed. Tools Appl. **79**, 21815–21840 (2020). https://doi.org/10.1007/s11042-020-08980-w

8. Li, H., et al.: SRDiff: single image super-resolution with diffusion probabilistic models. Neurocomputing **479**, 47–59 (2022). https://doi.org/10.1016/j.neucom.2022.01.029

9. Wolterink, J.M., et al.: Generative adversarial networks for noise reduction in low-dose CT. IEEE Trans. Med. Imaging **36**, 2536–2545 (2017). https://doi.org/10.1109/tmi.2017.2708987

10. Gu, J., et al.: Blind super-resolution with iterative Kernel correction. In: 2019 IEEE/CVF Conference on Computer Vision and Pattern Recognition (CVPR). IEEE (2019). https://doi.org/10.1109/cvpr.2019.00170

11. Park, J., et al.: Computed tomography super-resolution using deep convolutional neural network. Phys. Med. Biol. **63**(14), 145011 (2018). https://doi.org/10.1088/1361-6560/aacdd4

12. Christian, L., et al.: Photo-realistic single image super-resolution using a generative adversarial network. In: 2017 IEEE Conference on Computer Vision and Pattern Recognition (CVPR), pp. 105–114 (2017). https://doi.org/10.1109/CVPR.2017.19

13. Lin, Z., et al.: Why spectral normalization stabilizes GANs: analysis and improvements (2020)

14. Cheon, M., Kim, J.-H., Choi, J.-H., Lee, J.-S.: Generative adversarial network-based image super-resolution using perceptual content losses. In: Leal-Taixé, L., Roth, S. (eds.) ECCV 2018. LNCS, vol. 11133, pp. 51–62. Springer, Cham (2019). https://doi.org/10.1007/978-3-030-11021-5_4

15. Sajjadi, M.S.M., et al.: EnhanceNet: single image super-resolution through automated texture synthesis. In: 2017 IEEE International Conference on Computer Vision (CCV). IEEE (2017). https://doi.org/10.1109/iccv.2017.481

16. Rad, M.S., et al.: SROBB: targeted perceptual loss for single image super-resolution. In: 2019 IEEE/CVF International Conference on Computer Vision (ICCV), pp. 2710–2719 (2019)

17. Armato, S.G., et al.: The lung image database consortium (LIDC) and image database resource initiative (IDRI): a completed reference database of lung nodules on CT scans. Med. Phys. **38**, 915–931 (2011). https://doi.org/10.1118/1.3528204

18. Wang, X., et al.: Real-ESRGAN: training real-world blind super-resolution with pure synthetic data. In: International Conference on Computer Vision Workshops (ICCVW) (2021)

19. Chenyu, Y., et al.: CT super-resolution GAN constrained by the identical, residual, and cycle learning ensemble (GAN-circle). IEEE Transactions on Medical Imaging (2019). https://doi.org/10.1109/TMI.2019.2922960

20. Zhang, K., et al.: Learning a single convolutional super-resolution network for multiple degradations. In: IEEE Conference on Computer Vision and Pattern Recognition, pp. 3262–3271 (2018)

21. Wang, Z., et al.: Image quality assessment: from error visibility to structural similarity. IEEE Trans. Image Process. **13**(4), 600–612 (2004). https://doi.org/10.1109/TIP.2003.819861

22. Bing, X., Zhang, W., et al.: Medical image super resolution using improved generative adversarial networks. IEEE Access **7**, 145030–145038 (2019). https://doi.org/10.1109/ACCESS.2019.2944862

23. Dong, C., Loy, C.C., et al.: Image super-resolution using deep convolutional networks. IEEE Trans. Pattern Anal. Mach. Intell. **38**, 295–307 (2016). https://doi.org/10.1109/tpami.2015.2439281

24. de González, A.B., Darby, S.: Risk of cancer from diagnostic x-rays: estimates for the UK and 14 other countries. Lancet **363**(9406), 345–351 (2004). https://doi.org/10.1016/s0140-6736(04)15433-0

25. Gravel, P., Beaudoin, G., De Guise, J.A.: A method for modeling noise in medical images. IEEE Trans. Med. Imaging **23**, 1221–1232 (2004). https://doi.org/10.1109/TMI.2004.832656

26. Hany, M.: Chest CT-scan images dataset: CT-scan images with different types of chest cancer. www.kaggle.com/datasets/mohamedhanyyy/chest-ctscan-images (2020)

27. Kim, D., et al.: Progressive face super-resolution via attention to facial landmark (2019)

28. Kim, J., Lee, J.K., Lee, K.M.: Accurate image super-resolution using very deep convolutional networks. In: 2016 IEEE Conference on Computer Vision and Pattern Recognition (CVPR), pp. 1646–1654 (2016). https://doi.org/10.1109/CVPR.2016.182

29. Kim, J., Lee, J.K., Lee, K.M.: Deeply-recursive convolutional network for image super-resolution. In: 2016 IEEE Conference on Computer Vision and Pattern Recognition (CVPR), pp. 1637–1645 (2016). https://doi.org/10.1109/CVPR.2016.181

30. Mittal, A., Soundararajan, R., Bovik, A.C.: Making a "completely blind" image quality analyzer. IEEE Signal Process. Lett. **20**, 209–212 (2013). https://doi.org/10.1109/LSP.2012.2227726

31. Sheikh, H.R., Bovik, A.C.: Image information and visual quality. IEEE Trans. Image Process. **15**(2), 430–444 (2006). https://doi.org/10.1109/TIP.2005.859378

32. Tai, Y., et al.: MemNet: a persistent memory network for image restoration. In: 2017 EEE International Conference on Computer Vision (ICCV). IEEE (2017). https://doi.org/10.1109/iccv.2017.486

33. Tong, T., Li, G., et al.: Image super-resolution using dense skip connections. In: 2017 IEEE International Conference on Computer Vision (ICCV), pp. 4809–4817 (2017). https://doi.org/10.1109/ICCV.2017.514

34. Wang, X., Yu, K., et al.: Recovering realistic texture in image super-resolution by deep spatial feature transform. In: 2018 IEEE/CVF Conference on Computer Vision and Pattern Recognition. IEEE (2018). https://doi.org/10.1109/cvpr.2018.00070

35. Wang, X., Yu, K., et al.: ESRGAN: enhanced super-resolution generative adversarial networks (2018)

36. Yang, Q., Yan, P., et al.: Low-dose CT image denoising using a generative adversarial network with Wasserstein distance and perceptual loss. IEEE Trans. Med. Imaging **37**, 1348–1357 (2018). https://doi.org/10.1109/TMI.2018.2827462

37. Yu, H., Liu, D., Shi, H., et al.: Computed tomography super-resolution using convolutional neural networks. In: 2017 IEEE International Conference on Image Processing (ICIP), pp. 3944–3948 (2017). https://doi.org/10.1109/ICIP.2017.8297022

38. Zhang, L., Dai, H., Sang, Yu.: Med-SRNet: GAN-based medical image super-resolution via high-resolution representation learning. Comput. Intell. Neurosci. **2022**, 1–9 (2022). https://doi.org/10.1155/2022/1744969

39. Zhang, R., Isola, P., et al.: The unreasonable effectiveness of deep features as a perceptual metric. In: CVPR (2018)

40. Zhang, Y., Tian, Y., et al.: Residual dense network for image super-resolution. In: 2018 IEEE/CVF Conference on Computer Vision and Pattern Recognition. IEEE (2018). https://doi.org/10.1109/cvpr.2018.00262

Radiomics, Predictive Models and Quantitative Imaging

Efficient Semantic Segmentation of Nuclei in Histopathology Images Using Segformer

Marwan Khaled[1]([✉]), Mostafa A. Hammouda[1], Hesham Ali[1,2], Mustafa Elattar[1,2], and Sahar Selim[1,2]

[1] School of Information Technology and Computer Science, Nile University, Giza 12677, Egypt
mar.khaled@nu.edu.eg

[2] Medical Imaging and Image Processing Research Group, Center for Informatics Science, Nile University, Giza 12677, Egypt

Abstract. Segmentation of nuclei in histopathology images with high accuracy is crucial for the diagnosis and prognosis of cancer and other diseases. Using Artificial Intelligence (AI) in the segmentation process enables pathologists to identify and study the unique properties of individual cells, which can reveal important information about the disease, its stage, and the best treatment approach. By using AI-powered automatic segmentation, this process can be significantly improved in terms of efficiency and accuracy, resulting in faster and more precise diagnoses. Ultimately, this can potentially lead to better patient outcomes, making it a vital tool for healthcare professionals. In this paper, a novel method is proposed for semantic segmentation of nuclei using Segformer-b0 and Segformer-b4 on the PanNuke dataset. To achieve the most efficient and accurate segmentation, Segformer architecture is used as it combines the advantages of transformers and convolutional neural networks. To evaluate the performance of the models, dice evaluation metric is used. The proposed method achieved state-of-the-art results on the PanNuke dataset, with Segformer-b4 achieving a mean dice score of 0.845, and Segformer-b0 achieving a mean dice score of 0.82. The findings demonstrate the high efficiency of the Segformer architecture for semantic segmentation in histopathology images, as it also highlights the importance of choosing the appropriate model architecture for the task at hand. The proposed method provides a promising approach for accurate and efficient segmentation of histopathology images. The proposed approach outperformed state of the art methods by achieving better results, as it achieved 85.45% average Dice score which is more accurate compared to the RCSAU-Net which achieved 84.82% average Dice Score.

Keywords: Segformer · attention-based model · semantic segmentation · PanNuke dataset

1 Introduction

Medical imaging has become an indispensable tool for diagnosis and treatment of various diseases, including cancer [1]. Among different imaging modalities, histopathology imaging stands out as a powerful technique that provides detailed information about

© The Author(s), under exclusive license to Springer Nature Switzerland AG 2024
G. Waiter et al. (Eds.): MIUA 2023, LNCS 14122, pp. 81–95, 2024.
https://doi.org/10.1007/978-3-031-48593-0_6

tissue morphology, structure, and cellular activity [2]. As the gold standard for cancer diagnosis, histopathology images are routinely used by pathologists to identify and classify different types of cancer cells and tissues. However, the manual analysis of these images is a time-consuming and error-prone process that can lead to inter- and intra-observer variability [3]. Therefore, the development of accurate and efficient automated methods for histopathology image segmentation is critical for improving cancer diagnosis and treatment. Histopathology images are high-resolution images of tissue samples that are captured using a microscope. These images are commonly used in the field of pathology to study and diagnose diseases such as cancer [4]. In histopathology, tissue samples are typically taken from a patient's body, processed, and then stained with dyes to enhance the visibility of certain features. The stained tissue is then placed on a glass slide and examined under a microscope to generate high-resolution images [5]. These images are used by pathologists to visually inspect the tissue and identify any abnormalities or patterns that may be indicative of a disease. With the advent of digital imaging technology, histopathology images can now be digitized, stored, and analyzed using machine learning and deep learning algorithms, allowing for faster and more accurate diagnoses. Histopathological images are commonly used in research and clinical practice. They are also widely used in machine learning and computer vision research to develop algorithms for automated diagnosis and image analysis [6]. Accurate segmentation of nuclei in histopathology images is crucial for the diagnosis and prognosis of cancer and other diseases.

Semantic segmentation in histopathology images is a computer vision technique that divides an image into regions or areas of interest based on semantic content [7]. This technique is used in histopathology to identify and segment different types of tissues, cells, or other features in tissue images, such as tumors, glands, blood vessels, or nuclei. This procedure entails dividing the image into smaller parts or pixels and then classifying each pixel based on its content [8]. Deep learning- based methods have shown promising results for semantic segmentation in histopathology images, with several architectures proposed for this task. Pathologists use semantic segmentation to identify and quantify specific features of interest in tissue images, which can aid in the diagnosis, prognosis, and treatment planning of various diseases, including cancer [7]. Semantic segmentation is a challenging task in histopathology images due to the complex morphology of nuclei and the presence of artifacts such as debris, staining variations, and background noise [9].

In this paper, a novel method is proposed for semantic segmentation of nuclei using Segformer-b0 and Segformer-b4 [10] on the PanNuke dataset [11]. The Segformer is a powerful and efficient semantic segmentation framework which unifies Transformers with lightweight MLP decoders. Segformer has a novel hierarchically structured Transformer encoder which outputs multiscale features, as it also avoids complex decoders. In this paper we aim to prove the efficiency of Segformer in Medical Imaging, since this is one of the first research papers to discuss Segformer with Histopathology images.

We opted to use Segformer for semantic segmentation as it offers improved accuracy compared to CNNs by capturing global context. Unlike CNNs that require increasing kernel size to capture such context, Segformer's hierarchical structure facilitates the integration of global features, making it computationally more efficient. The proposed method involves training Segformer-b0 and Segformer-b4 models [10] on the PanNuke dataset and evaluating their performance using Dice evaluation metric. The Segformer-b0 and Segformer-b4 were compared, to conclude that the larger model is more robust and can generalize better to different tissue types.

2 Related Work

Semantic segmentation of nuclei in histopathological images has been an active area of research in recent years. Deep learning-based methods have shown promising results for this task, with several architectures proposed for semantic segmentation in histopathology images. Recently, transformer- based architectures have shown promising results for semantic segmentation. Deformable DETR [12] is a transformer-based architecture used on the COCO 2017 which is a semantic segmentation benchmark dataset. It has been shown to achieve state-of-the-art results using AP score for evaluation. It uses deformable attention mechanisms to handle the complex morphology of nuclei and capture long-range dependencies.

Segformer and attention-based models didn't get that much attention in the field of digital pathology; therefore, there were only a few related works using a similar approach. One of the previous research papers using similar approaches presented an overview of an end-to-end deep learning strategy that was suggested for segmenting nuclei [13]. They proposed a method known as RCSAU-Net, where patches are entered into the network for feature encoding and decoding after data augmentation for the obtained patch and its label, and the output is a probability map of nuclear contours. In the meantime, patches are fed into a second RCSAU-Net and GAN training network for fine-tuning based on TCGA dataset pre-training, in which RCSAU-Net is employed as the Generator.

To decrease the training process overhead, the model's capacity to distinguish between the nuclei of various organs is improved through adversarial training, and using RCSAU-Net and GAN, they were able to acquire the probability map of the segmented nuclei instances. The two probability maps are then combined using a straightforward threshold fusion technique without any parameters to produce the patch level segmentation results. Finally, the patch level results for the corresponding positions are combined using the voting integration method.

Another research paper proposed the PointNu-Net architecture [14] which is based on Conv block and Upsample Block. Before being delivered to each branch, the retrieved multi-scale features are up sampled to the same scale and concatenated. The results of instance segmentation are then obtained by combining the outputs from the kernel branch and feature branch. PointNu-Net fundamental concept of PointNu-Net is to recast the nuclei instance segmentation as two concurrent prediction problems. Finally, for evaluation they used Mpq evaluation metric. Zhang et al. used the PyTorch toolkit to build the Multi-compound Transformer [15]. They used traditional CNN backbone networks to extract multi-scale feature representations, such as ResNet-34 and VGG-Style encoder. To optimize the network, they used a cross-entropy loss with a weight of 0.1 for auxiliary supervision and dice loss to penalize segmentation training errors.

Moreover, another paper proposed a new scale and region-enhanced network for nuclei classification. The network employs a classical U-shape architecture with a redesigned decoding branch and three techniques for improved performance. The first technique is a regional enhancement module that proposes a nuclei bounding box detection strategy to enhance the representation capability of the feature map for nuclei regions. The second technique is a scale-aware feature fusion module that captures multi-scale contextual information. The third technique is a scale attention module that makes the model automatically adaptive to the most suitable scale of nuclei. In the following subsections, the paper provides details about each module and the loss function [16].

The final paper [17] uses attention-based model. The paper presents a deep learning neural network called TSHVNet, which performs simultaneous nuclear instance segmentation and classification. The network comprises a trunk and three branches, where the trunk is responsible for feature extraction and encoding, and the branches are responsible for feature decoding. The authors have introduced a Transformer attention module to the trunk of the HoVer-Net to explore the potential global dependencies between the network input and outputs, along with several SimAM attention modules added to both the trunk and branches of the HoVer-Net to discover the important information in the channel and space on feature maps. The three branches decode the features in different ways and predict the nuclear pixel, HoVer, and nuclear classification.

In addition to deep learning-based methods, several classical methods have also been proposed for semantic segmentation of histopathology images. These methods include thresholding [18], watershed segmentation [19], and graph-based methods [20]. However, these methods often suffer from low accuracy and require extensive parameter tuning. In summary, deep learning-based methods have shown promising results for semantic segmentation of histopathology images, with transformer-based architectures such as the methods mentioned above showing state-of-the-art performance.

While the related work has achieved promising results in semantic segmentation of histopathology images, there are still some limitations that need to be addressed. The DETR model suffers from slow convergence and limited feature spatial resolution, which may impact its performance in real-world applications. Additionally, the Transformer attention modules used in the DETR model have limitations in processing image feature maps, which may affect its ability to model complex visual features. On the other hand, CNN-based segmentation models like FCN and UNet are limited by inefficient non-local context modeling among arbitrary positions, which can also limit their accuracy in complex views. Therefore, future research should focus on developing more efficient and accurate segmentation models that can better handle complex visual features and improve the performance of semantic segmentation on histopathology images.

In this paper, we propose a novel method for semantic segmentation of nuclei using Segformer-b0 and Segformer-b4 on the PanNuke dataset that has the potential to address the limitations mentioned above and improve the performance of semantic segmentation on histopathology images. To prove the efficiency and accuracy of our proposed methods we compared our results with the state-of-the-art methods.

Our research makes several contributions to the field of multi-class segmentation in medical imaging. First, we propose the use of Segformer, state-of-the-art transformer-based architecture, as a powerful tool for this task. Segformer captures the global context of the image by processing it, rather than relying solely on local features as traditional

CNNs do. This allows for more accurate and consistent segmentation results, particularly in complex images with multiple classes. We used both Segformer-b0 and Segformer-b4 and found that increasing the model parameters led to better results. Second, we explore the use of multiple loss functions to optimize the performance of Segformer. Specifically, we employ focal loss, dice loss, cross entropy, Tversky loss, and weighted loss, all of which have been shown to be effective in various segmentation tasks. By combining these loss functions, we aim to improve the robustness and adaptability of our model to different datasets and imaging modalities. Finally, we compared Segformer to other attention-based models that have been used in the literature on the Panuke dataset. Our results demonstrate that Segformer outperforms these models, achieving higher Dice scores. Taken together, our contributions demonstrate the effectiveness of Segformer, the importance of utilizing a diverse set of loss functions, and the benefits of increasing model parameters for accurate and reliable multi-class segmentation in medical imaging.

3   Methods

The proposed method aims to use Segformer which is a recently proposed deep learning architecture for semantic segmentation, which combines the Transformer architecture with the idea of self- attention in computer vision. Segformer is an image segmenta-tion model that consists of a hierarchical Transformer encoder and a lightweight All-MLP decoder. The encoder generates multi- level multi-scale features given an input image through a process called patch merging, which produces features with both high-resolution coarse features and low-resolution fine-grained features. The self-attention mechanism of the encoder is made more efficient by using the sequence reduction pro-cess introduced in PVT and reducing the length of the sequence input with a reduction ratio. Overlapping patch merging is also used to preserve local continuity around image patches. To avoid the drop of accuracy due to resolution mismatch, Segformer intro-duces a positional-encoding-free design called Mix-FFN, which directly uses a 3x3 Conv in the feed-forward network (FFN) and considers the effect of zero padding to the leak location information. Compared to other transformer- based architectures, Seg-former generates multi-level multi-scale features and introduces several novel features, including overlapped patch merging and positional-encoding-free design.

Once the input image is passed through data augmentation, it is fed into the Segformer model. The Segformer encoder is a hierarchical transformer encoder which consists of several Transformer blocks that are repeated multiple times to form the full architecture. Each block consists of several layers, including the following:

1. Efficient self-attention layer: The hierarchical feature representation in Segformer suffers from a quadratic self-attention complexity due to long sequence inputs from higher resolution features. To address this issue, Segformer adopts a sequence reduc-tion process that reduces the length of the sequence by a reduction ratio R. This reduces the complexity of the self- attention mechanism from $O(N2)$ to $O(N2/R)$
2. The Mix-feedforward network layer applies a nonlinear transformation to the input features, allowing the model to capture more complex patterns and relationships between pixels. Where a 3×3 convolution and an MLP are mixed into each FFN. This approach considers the effect of zero padding to the leak location information by

directly using a 3 × 3 Conv in the feed- forward network (FFN). In experiments, it has been shown that a 3 × 3 convolution is sufficient to provide positional information for Transformers, and depth-wise convolutions can be used to reduce the number of parameters and improve efficiency [10].

3. Overlapping Patch Merging: to combine neighboring patches and obtain hierarchical feature maps.

After passing through these layers, the output is passed through a Lightweight All-MLP Decoder, The All-MLP decoder of Segformer consists of four main steps. Firstly, multi-level features obtained from the MiT encoder are passed through an MLP layer to standardize the channel dimension. In the second step, the features are Up-sampled their original size and concatenated.

Thirdly, an MLP layer is used to merge the concatenated features. Finally, the fused feature is passed through another MLP layer to predict the segmentation mask.

Overall, the Segformer model combines self-attention and feedforward networks to capture both local and global features of the input image, making it a powerful tool for semantic segmentation tasks. To further explain our methods, the dataset that was used, the preprocessing of the input, the model architecture, training, evaluation and the implementation details will be discussed in this section (Fig. 1).

3.1 Dataset

The PanNuke Dataset is used for segmentation and classification with exhaustive nuclei labels across 19 different tissue types including datasets for breast, colon, prostate, ovarian and oral tissue [11]. It consists of 481 visual fields, of which 312 are randomly sampled from more than 20K whole slide images at different magnifications, from multiple data sources. In total the dataset contains 205,343 labeled nuclei, each with an instance segmentation mask. Models trained on this dataset can aid in whole slide image tissue type segmentation and generalize to new tissues. PanNuke dataset demonstrates one of the first successfully semi-automatically generated datasets.

3.2 Preprocessing

Prior to training the models, we preprocessed the Input by reading the folds and defining the masks. In order to increase the diversity of the training data and prevent overfitting, we shuffled the training data and applied data augmentation techniques, including random horizontal flip, shift with limit equals 0.0625, rotation with 45 limit, and scale limit equals to 0.5.

3.3 Model Architecture

We used two different architectures of Segformer for our experiments, namely Segformer-b0 and Segformer-b4 [10]. Segformer-b0 is basically formed of attention heads, transformer encoder layers, and decoder blocks. As for the Segformer-b4 architecture it has the same architecture but with more attention heads, transformer encoder

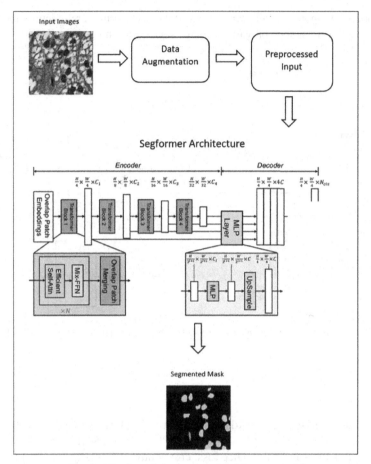

Fig. 1. Pipeline of the proposed Segformer model

layers, and more decoder blocks [10]. Both models use the transformer-based architecture, which has been shown to be efficient and effective in image classification and segmentation tasks. Figure 2 [10] shows the architecture of the Segformer.

In this research paper, two Mix Transformer encoders (MiT) are introduced, MiT-b0 and MiT-b4, as part of their proposed architecture for image segmentation tasks. The Mix Transformer encoders combine the benefits of both traditional convolutional neural networks (CNNs) and transformer-based networks to better capture the spatial dependencies in image data. Specifically, the encoders consist of a stack of multi-head self-attention layers, followed by feedforward layers and convolutional layers. The convolutional layers help to extract local features, while the self-attention layers capture the global context and long-range dependencies. The MiT-b0 and MiT-b4 models differ in the number of layers, width of the hidden layers, and other architectural parameters, allowing for a trade-off between accuracy and computational complexity. It was found that using the

MiT encoders, along with other components of their Segformer architecture, resulted in state-of-the-art performance on various image segmentation benchmarks.

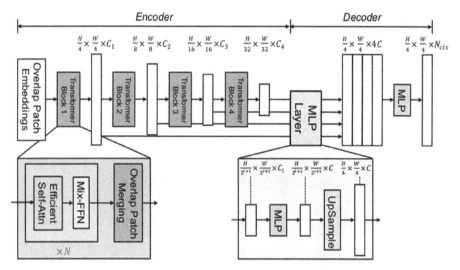

Fig. 2. Architecture of the Segformer model with the Mix Transformer encoders [10]

3.4 Experiments Setup

We trained our models using the Adam optimizer with a learning rate of 0.0001 for 100 epochs. We used a batch size of 10 for Segformer-b0 and 10 for Segformer-b4. We utilized various loss functions including Dice loss, cross entropy, Tversky loss, and focal loss. Specifically, we used a combination of these loss functions during training to optimize model performance. In addition, we experimented with the weighted cross-entropy loss function which assigns higher weights to underrepresented classes in the dataset. We implemented our models using PyTorch deep learning framework and trained them on Colab pro GPU. The code will be made publicly available to encourage reproducibility and further research in this area. We also performed per tissue cross validation on 3 folds. Each fold was set to a number of 50 epochs. Then we compared the results of the Segformer-b0 and Segformer-b4 results to prove the efficiency of Segformer-b4 in semantic segmentation of histopathology images. In addition, we compared our results with the previous state-of-the-art results.

3.5 Performance Metrics

We evaluated the performance of our models using Dice evaluation metric.

The Dice coefficient is calculated as follows:

$$Dice = (2 * |A \cap B|) \big/ (|A| + |B|) * 100$$

A refers to the predicted segmentation mask, B represents the ground truth segmentation mask, |A| denotes the total number of pixels in A, and |B| represents the total number of pixels in B. The term |A ∩ B| quantifies the number of pixels that coincide between the predicted segmentation mask and the ground truth segmentation mask.

4 Results

As mentioned earlier in the method, we used Segformer-b0 and Segformer-b4 for segmentation and compared between them. In addition, we compared our results with the state-of-the-art results to prove that our proposed methods outperformed the state-of-the-art methods. In this section, we will go through the comparison results of the Segformer-b0 and Segformer-b4, as well as the comparison of our proposed method with the results of the state of art methods. Further on, we will show the results of the Segformer-b4 cross validation in detail where the results of each fold will be presented. Finally, we will visualize a sample of the images before and after segmentation.

4.1 Segformer-b0 and Segformer-b4 results

We evaluated the performance of Segformer-b0 and Segformer-b4 on the PanNuke dataset using dice evaluation metric. In addition, we conducted a comparative analysis of the segmentation results produced by the two models on a set of randomly selected images from the PanNuke dataset. The results of the analysis showed that Segformer-b4 outperformed Segformer-b0 in terms of both segmentation accuracy. The results of the cross-validation are presented in Table 1.

Table 1. Performance of Segformer-b0 and Segformer-b4 on different tissue types using the PanNuke dataset based on dice scores.

Tissue	Segformer-b0 Crossvalidation	Segformer-b4 Crossvalidation
Adrenal gland	0.8391	0.8506
Bile-Duct	0.8488	0.8609
Bladder	0.8642	0.8817
Breast	0.8228	0.8454
Cervix	0.8568	0.8630
Colon	0.8127	0.8193
Esophagus	0.8491	0.8619
Head-Neck	0.7906	0.8192
Kidney	0.8141	0.8367
Liver	0.8332	0.8695
Lung	0.7618	0.8103

(continued)

Table 1. (*continued*)

Tissue	Segformer-b0 Crossvalidation	Segformer-b4 Crossvalidation
Ovarian	0.8235	0.8571
Pancreatic	0.8383	0.8466
Prostate	0.8238	0.8509
Skin	0.8228	0.7783
Stomach	0.8407	0.8684
Testis	0.8302	0.8566
Thyroid	0.8550	0.8697
Uterus	0.7849	0.8107
Average	**0.8260**	**0.8451**

As shown in Table 1, Segformer-b4 achieved higher dice scores than Segformer-b0. This indicates that Segformer-b4 is more accurate in semantic segmentation than Segformer-b0. Specifically, Segformer-b4 achieved a mean dice score of 0.845, while Segformer-b0 achieved a mean dice score of 0.826.

4.2 Comparison with Previous State-of-the-Art Methods

To further evaluate the performance of Segformer-b0 and Segformer-b4, we compared our proposed method with the RCSAU-Net [13] results. They only used 9 tissue types, therefore we used only these 9 tissue types for comparison. The results are presented in Table 2. As for the rest of the previous state-of-the-artwork mentioned in the related work section, they either used different datasets or different evaluation metrics, that is why we could not compare our results with them.

Table 2. Performance comparison of Segformer-b4 and RCSAU-Net tissues using dice score

Tissue	Our Segformer-b0 Dice score	Our Segformer-b4 Dice score	RCSAU-Net Dice Score [13]
Bladder	0.8642	**0.8817**	0.8591
Breast	0.8228	**0.8454**	0.8232
Colon	0.8127	**0.8193**	0.8012
Esophagus	0.8491	**0.8619**	0.8389
Kidney	0.8141	0.8367	**0.8393**
Liver	0.8332	**0.8695**	0.8344

(*continued*)

Table 2. (*continued*)

Tissue	Our Segformer-b0 Dice score	Our Segformer-b4 Dice score	RCSAU-Net Dice Score [13]
Ovarian	0.8235	0.8571	**0.8607**
Prostate	0.8238	0.8509	**0.8646**
Stomach	0.8407	**0.8684**	0.8512
Average Dice	0.8315	**0.8545**	0.8482

According to the results presented in Table 2, Segformer-b4 outperformed the previous state- of-the-art methods in terms of overall Dice scores, with Segformer-b4 being the most effective. In their research paper, the authors of RCSAU-Net [13] compared their results with two other approaches, namely StarDist [21] and BRP-Net [22] and found that RCSAU-Net outperformed them. Since our approach achieved better results than RCSAU-Net, it follows that our approach also outperformed StarDist and BRP-Net (Table 3).

Table 3. Comparison of Segformer Models and Related Work Models Based on Overall Dice Score

Model	Dice score
Segformer-b0	0.8260
Segformer-b4	**0.8451**
TSHVNet [17]	0.835
Xiao et al.[16]	0.789

As shown in the table above, the Segformer-b0 model achieved a dice score of 0.8260, while the Segformer-b4 model achieved a higher score of 0.8451. The TSHVNet model achieved a dice score of 0.835, while Model 2 achieved a score of 0.789. Based on these results, we can conclude that the Segformer models outperformed the other models [16, 17] in terms of the dice score on this task.

4.3 Visual Results

Finally, we visualized the segmentation results produced by our models on the PanNuke dataset and observed that our models were able to accurately segment various tissue types, including epithelial, stromal, and inflammatory cells. The following figure (Fig. 3) shows image samples from both models compared to each other.

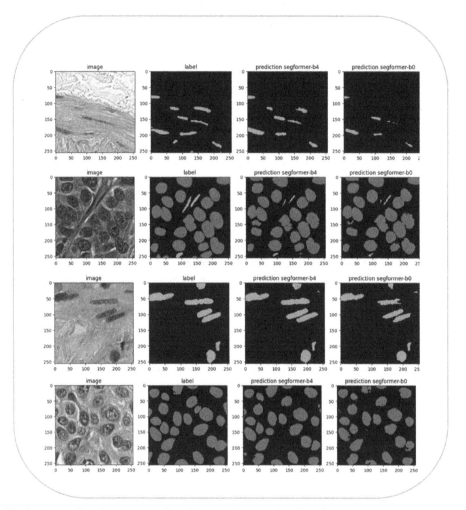

Fig. 3. Comparison between samples of the visualized results of Segformer-b0 and Segformer-b4, showing the image and its corresponding label

5 Discussion

In this study, we proposed a novel methodology for nuclear segmentation using the PanNuke dataset. Our approach employed the Segformer-b0 and Segformer-b4 architectures and evaluated their performance on the dataset. Our results demonstrated that the Segformer-b4 architecture, with a mean Dice score of 0.854 and an average cross-entropy loss of 0.215, outperformed the Segformer-b0 and previous state-of-the-art methods, indicating that deeper and more complex models are better suited for semantic segmentation tasks on the PanNuke dataset.

We compared our methodology with the approaches that utilized the PanNuke dataset for segmentation and evaluated the performance using Dice Score. Our proposed methodology achieved better results than the methods mentioned in the related work section, specifically the RCSAU-Net, StarDist, TSHVNet, Xiao et al. [16], and BRP-Net methods. This indicates the effectiveness of our proposed approach for nuclear segmentation on the PanNuke dataset.

However, we acknowledge that our evaluation was limited to the approaches and did not include all the methods discussed in the related work section. Nevertheless, our comparison with the RCSAU-Net method showed that the Segformer-b4 architecture achieved better results, highlighting the potential of the Segformer approach for semantic segmentation on the PanNuke dataset.

Furthermore, our results indicated that the Segformer-b4 model consistently outperformed the Segformer-b0 across different tissue types. This suggests that the larger and more complex model is more robust and can better generalize to different tissue types, making it a promising approach for nuclear segmentation on the PanNuke dataset.

In summary, our proposed methodology utilizing the Segformer-b0 and Segformer-b4 architectures achieved better results than the state-of-the-art methods for nuclear segmentation on the PanNuke dataset. The Segformer-b4 architecture was identified as a more robust and effective model, and our results demonstrate the potential of the Segformer approach for semantic segmentation tasks on the PanNuke dataset.

In terms of limitations, it is important to note that our proposed method for semantic segmentation on the PanNuke dataset requires many computational resources and memory. This can be a barrier for researchers with limited access to high-performance computing resources. Furthermore, while our method demonstrated state-of-the-art results, there is still potential for further improvement.

Despite these limitations, our experiments highlight the effectiveness of Segformer-b0 and Segformer-b4 for semantic segmentation on the PanNuke dataset. Our proposed method achieved high performance across different tissue types and outperformed existing state-of-the-art methods. These findings hold promise for medical image analysis and can potentially contribute to the development of new diagnostic tools and therapies.

6 Conclusions

In conclusion, this paper has introduced the importance of semantic segmentation of histopathology images using attention-based approaches. The proposed Segformer-b0 and Segformer- b4 models have been implemented to perform semantic segmentation on histopathology images, achieving a dice score of 84.51%, which is better than the previous work, indicating the effectiveness of attention-based models in enhancing accuracy. The proposed approach has shown significant impact compared to the state-of-the-art, demonstrating its potential in advancing the field of medical image analysis.

Although the results achieved are promising, the study has some limitations. The PanNuke dataset requires significant computational resources and memory, which can be a barrier for researchers with limited access to high-performance computing resources.

Additionally, there is still potential for further improvement in the performance of semantic segmentation on the PanNuke dataset. Therefore, future research can explore different model architectures or data augmentation techniques to enhance the segmentation results.

Looking forward, the use of Segformer-b5 could be explored to further enhance the accuracy of semantic segmentation. Furthermore, instance segmentation models can be experimented with to investigate their potential for improving the segmentation of nuclei within the PanNuke dataset. Addressing these areas of future research can help advance the state-of-the-art in medical image analysis and contribute to the development of more accurate and effective diagnostic tools. In conclusion, this study has made significant contributions to the field of medical image analysis, and there is great potential for further advancement through continued research efforts.

References

1. Doi, K.: Computer-aided diagnosis in medical imaging: Historical review, current status and future potential. Comput. Med. Imaging Graph. **31**(4–5), 198–211 (2007). https://doi.org/10.1016/j.compmedimag.2007.02.002

2. Gurcan, M.N., Boucheron, L.E., Can, A., Madabhushi, A., Rajpoot, N.M., Yener, B.: Histopathological image analysis: a review. IEEE Rev. Biomed. Eng. **2**, 147–171 (2009). https://doi.org/10.1109/RBME.2009.2034865

3. Punithavathy, K., Ramya, M.M., Poobal, S.: Analysis of statistical texture features for automatic lung cancer detection in PET/CT images. In: 2015 International Conference on Robotics, Automation, Control and Embedded Systems (RACE) , pp. 1–5. IEEE (2015). https://doi.org/10.1109/RACE.2015.7097244

4. Elazab, N., Soliman, H., El-Sappagh, S., Islam, S.M.R., Elmogy, M.: Objective Diagnosis for histopathological images based on machine learning techniques: classical approaches and new trends. Mathematics **8**(11), 1863 (2020). https://doi.org/10.3390/math8111863

5. Alturkistani, H.A., Tashkandi, F.M., Mohammedsaleh, Z.M.: Histological stains: a literature review and case study. Glob. J. Health Sci. **8**(3), 72 (2015). https://doi.org/10.5539/gjhs.v8n3p72

6. Wu, Y., et al.: Recent advances of deep learning for computational histopathology: principles and applications. Cancers (Basel) **14**(5), 1199 (2022). https://doi.org/10.3390/cancers14051199

7. Kim, H., et al.: Deep learning-based histopathological segmentation for whole slide images of colorectal cancer in a compressed domain. Sci. Rep. **11**(1), 22520 (2021). https://doi.org/10.1038/s41598-021-01905-z

8. Han, C., et al.: Multi-layer pseudo-supervision for histopathology tissue semantic segmentation using patch-level classification labels. Med. Image Anal. **80**, 102487 (2022). https://doi.org/10.1016/j.media.2022.102487

9. Pan, X., et al.: Accurate segmentation of nuclei in pathological images via sparse reconstruction and deep convolutional networks. Neurocomputing **229**, 88–99 (2017). https://doi.org/10.1016/j.neucom.2016.08.103

10. Xie, E., Wang, W., Yu, Z., Anandkumar, A., Alvarez, J. M., Luo, P.: Segformer: simple and efficient design for semantic segmentation with transformers (2021). http://arxiv.org/abs/2105.15203

11. Gamper, J., et al.: PanNuke Dataset Extension, Insights and Baselines (2020). http://arxiv.org/abs/2003.10778

12. Zhu, X., Su, W., Lu, L., Li, B., Wang, X., Dai, J.: Deformable DETR: deformable transformers for end-to-end object detection (2020). http://arxiv.org/abs/2010.04159

13. Wang, H., Xu, G., Pan, X., Liu, Z., Lan, R., Luo, X.: Multi-task generative adversarial learning for nuclei segmentation with dual attention and recurrent convolution. Biomed. Signal Process. Control **75**, 103558 (2022). https://doi.org/10.1016/j.bspc.2022.103558

14. Yao, K., Huang, K., Sun, J., Hussain, A., Jude, C.: PointNu-Net: Simultaneous multi-tissue histology nuclei segmentation and classification in the clinical wild (2021). http://arxiv.org/abs/2111.01557

15. Ji, Y., et al.: Multi-compound transformer for accurate biomedical image segmentation (2021). http://arxiv.org/abs/2106.14385

16. Xiao, S., Qu, A., Zhong, H., He, P.: A scale and region-enhanced decoding network for nuclei classification in histology image. Biomed. Signal Process. Control **83**, 104626 (2023). https://doi.org/10.1016/j.bspc.2023.104626

17. Chen, Y., et al.: TSHVNet: simultaneous nuclear instance segmentation and classification in histopathological images based on multiattention mechanisms. Biomed. Res. Int. **2022**, 1–17 (2022). https://doi.org/10.1155/2022/7921922

18. Yan, W., Qian, Y., Wang, C., Yang, M.: Threshold-adaptive unsupervised focal loss for domain adaptation of semantic segmentation (2022). http://arxiv.org/abs/2208.10716

19. Bai, M., Urtasun, R.: Deep watershed transform for instance segmentation (2016). http://arxiv.org/abs/1611.08303

20. Landrieu, L., Simonovsky, M.: Large-scale point cloud semantic segmentation with superpoint graphs (2017). http://arxiv.org/abs/1711.09869

21. Schmidt, U., Weigert, M., Broaddus, C., Myers, G.: Cell detection with star- convex polygons, pp. 265–273 (2018). https://doi.org/10.1007/978-3-030-00934-2_30

22. Chen, S., Ding, C., Tao, D.: Boundary-assisted region proposal networks for nucleus segmentation, pp. 279–288 (2020). https://doi.org/10.1007/978-3-030-59722-1_27

Cross-Modality Deep Transfer Learning: Application to Liver Segmentation in CT and MRI

Merna Bibars[1]([✉]) [iD], Peter E. Salah[1] [iD], Ayman Eldeib[2] [iD],
Mustafa A. Elattar[3] [iD], and Inas A. Yassine[1] [iD]

[1] Systems and Biomedical Engineering, Cairo University, Giza, Egypt
{merna.bibars,peter.salah}@eng.cu.edu.eg
[2] Southern New Hampshire University, Manchester, NH, USA
[3] Center for Informatics Science, Nile University, Cairo, Egypt

Abstract. Liver diseases cause up to two million deaths yearly. Their diagnosis and treatment plans require an accurate assessment of the liver structure and tissue characteristics. Imaging modalities such as computed tomography (CT) and Magnetic resonance (MR) can be used to assess the liver. CT has better spatial resolution compared to MR, which has better tissue contrast. Each modality has its own applications. However, CT is widely used due its ease of access, lower cost and a shorter examination time. Liver segmentation is an important step that helps to accurately identify and isolate the liver from other organs and tissues in medical images. This can be useful for diagnosing and treating liver diseases. Manual liver segmentation is costly and time-intensive. Machine learning and deep learning algorithms can be used to automate this process. However, deep learning methods require a large amount of training data, which is not available for MR. There are many CT datasets available compared to few MR datasets. The use of transfer learning can help to mitigate the problem of having a small amount of training data. We suggest training a U-Net deep learning model on the large publicly available CT dataset liver tumor segmentation challenge (LiTS), then use transfer learning to fine-tune the model to the smaller available MR datasets Duke liver dataset (DLDS) and Combined (CT-MR) Healthy Abdominal Organ Segmentation (CHAOS). This allows the model to leverage the knowledge it has gained from the larger CT dataset to improve its performance on the smaller MR datasets. The model reached a dice of 0.83 on unseen DLDS data after fine-tuning to the small CHAOS dataset. It also improved the testing dice on CHAOS itself when compared to training the model from scratch on CHAOS only from 0.59 to 0.86. A universal model trained on the combined datasets achieved a testing dice of 0.928 on LiTS, 0.865 on DLDS, and 0.815 on CHAOS.

Keywords: U-Net · Transfer Learning. · Across modalities · Liver Segmentation

1   Introduction

1.1   Background

The risk of liver diseases has increased, causing two million deaths yearly due to several factor such as alcohol consumption, an increase in the number of patients suffering from diabetes, reaching 400 million cases, as well as lifestyle habits and obesity progression [1]. Liver segmentation is an essential pre-processing step for liver diseases' diagnosis, procedures, and treatment planning. Moreover, it is vital for the assessment of the liver morphology and tissue mechanical properties, fibrosis grading, cirrhosis detection, liver transplantation, surgical resection for treatment of hepatocellular carcinoma, and liver surgery planning [2,3].

Several modalities can be employed for liver assessment such as MR and CT. Different varieties of imaging parameters, within the same modality, can be used such as structural properties, injection of contrasts, employment of external force employed for elastography imaging, field of view and resolutions [4].

CT imaging provides better spatial resolution images than MR [5]. It is used for liver transplant planning, surgical operations planning, hepatosteatosis assessment [6,7], liver parenchyma structure assessment, assessment of hepatic tissue pathological changes, hepatic vessels tree and biliary system [8,9]. It also has the advantages of accessibility and shorter examination time, but it suffers from patient radiation exposure [10].

On the other hand, MR imaging has better contrast resolution [5], so it is usually used for tissue characterization. as it can discriminate between tissues with similar pathological tissue properties such as focal nodular hyperplasia, hepatocellular adenomas, and hepatocellular carcinoma, and used in the detection of small hepatic lesions ($< 1\,cm$). Moreover, during MRI imaging, there is no risk of exposure to radiation. However, it suffers from longer examination time and contraindication with pacemaker and metal implants [10].

This results in the higher availability of publicly available CT datasets compared to MR. Until 2022, there has been only one public MR dataset to four public CT datasets [11].

It is worth noting that manual liver segmentation requires medical and radiological expertise to manually and accurately delineate the liver boundaries regardless of modality. This process is very expensive and time-consuming as it could take up to 90 min to segment the liver for a single patient [12]. Furthermore, it's prone to human error and inter-observer variability as it relies on the user to perform the segmentation [3,13] [14]. Additionally, the difference in liver pathology has a drastic effect on the liver tissue morphology and characteristics like liver size and boundaries [15]. This makes liver segmentation more challenging.

Therefore, we propose using a universal automated method for Liver segmentation. The proposed method will require a small data size to be fine-tuned while being minimally affected by the changes occurring across modalities and sequences without being affected by differences due to modality specification or liver characteristics.

1.2 Literature Review

A wide range of machine learning and deep learning algorithms have been introduced to automatically segment the liver. Machine learning-based segmentation requires the extraction of features, multiple pre-processing, and post-processing steps. Pre-processing modules may include different components such as spatial resolution improvement, feature extraction [16–18], gray level enhancement and Edge detection. whereas the post-processing usually includes segmentation enhancement operations using morphological operators [19]. Different machine learning algorithms have been employed in the segmentation task such as Support Vector Machine (SVM) [16], k-means clustering [20], and agglomerative clustering [21].

Deep learning is a subset of machine learning that employs artificial neural networks to imitate the human brain's learning process. It inherently extracts the features during the learning process [22]. Fully convolution neural networks (FCNN) are deep learning architectures that consist of stacked convolution layers, learning low-level features in the first layers and combine them to high level-features in the last layers [23]. Several studies employed FCNN-based architectures, with different modifications to segment the liver [11,24,25].

Aghamohammadi et. al [26] introduced cascaded CNN to segment both the liver as well as the inherent hepatic tumor using two pathways. Each path extracts features from a CT input at different scales. An in-house CT dataset [27] acquired from 1000 patients, forming 20,000 slices was used for this study. Data augmentation was implemented to increase the number of slices to 100,000. Several pre-processing techniques to the input image; Z-score has been used to normalize the input image while ignoring the zero pixels. Moreover, local direction of gradient (LDOG) encoding algorithm was also applied to encode the gradient image output from a non-linear Kirsch filter. Both the LDOG and z-score normalized images are fed to CNN along with the original image. The local patches of size $21 \times 21 \times 3$ and the semi-global patches of size $64 \times 64 \times 3$ are extracted around each pixel of the input images. CNN received three different patches at two different scales for each path. The authors achieved 91% median dice, ranging from 85% to 92%.

Tang et. al [28] divided the CT liver segmentation process into two stages. They used Faster Regions with CNN features (Faster R-CNN) [29] to detect the liver and the result is passed to DeepLab network to segment the liver. The first stage combines region proposal network (RPN) and Fast R-CNN where a VGG-16 [30] pre-trained on ImageNet [31] is used to extract feature maps from the liver which are then mapped to candidate regions proposed by the RPN. The result from this stage is a bounding box surrounding the liver, which is the input to DeepLab V2 [32] network. 3Dircadb [33] and the MICCAI-Sliver07 [34] were used to evaluate proposed algorithm. Average volumetric overlap error (VOE) of 5.06% and 8.67% was achieved using Sliver07 and 3DICARDB respectively.

Khan et. al [35] propose a multiscale 2D U-Net to segment both the liver and the tumor using CT images. Residual blocks were added and dilated convolutions replaced the normal convolutions filters to increase the receptive field with a fewer number of parameters. The loss function used was a combination of dice coefficient and absolute volumetric difference. Additionally, the U-Net is trained and tested on 4CT datasets; named 3DIRCADB [33], SLIVER07 [34], CHAOS [36], and LiTS [37]. Data augmentation was employed to achieve accurate training and minimize the over fitting problem. Mean dice scores of 97.31%, 97.38%, 97.39% and 95.49% was achieved for the 3DIRCADB, LiTS, Sliver07, and CHAOS liver test sets respectively. It is worth noting that all previous studies focus on one segmenting the liver using CT images. Zhou et. al [38] proposed a generative cross modal approach to map liver segmentation from CT to MRI images. Labelled LiTS and CHAOS, CT dataset as well as 16 in-house MRI volumes were employed in this study. A U-Net formed of 5-level U-Net with squeeze-excitation blocks were employed to perform the segmentation task. One drawback of the proposed approach requires a registration framework to map between the volumes of the CT and MRI modalities. A dice score of 90.3% and 91.8% was reported for CT and MRI liver segmentation respectively. Wang et. al [39] utilized cross modality transfer learning using 2D U-Net to segment the liver. They collect a cross-modality dataset, formed of CT and MRI volumes with different acquisition sequences. The network is trained on the MRI dataset, formed of 300 volumes. Transfer learning, of the MRI pre-trained U-Net, employed a total of 60 image sets, consisting of both 10 and 30 CT and MRI volumes with different acquisition sequences. CT volumes were employed in the testing before and after transfer learning, where an 82% and 94% dice accuracy was reported respectively.

2 Methods

2.1 Datasets Description

Three Datasets were employed in this study named LiTS [37], CHAOS [36], CT volumes datasets and DLDS [40], MRI volumes. In this section, the description of the datasets as well as the datasets preparation will be discussed.

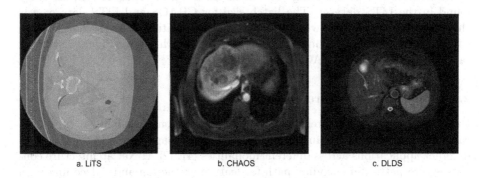

a. LiTS   b. CHAOS   c. DLDS

Fig. 1. A sample image from each dataset

LiTS. Liver Tumor Segmentation dataset [37] includes 131 CT scans for different patients, associated with the corresponding liver segmentation. The data was randomly divided by patient into 91, 26 and 14 patients forming 40,812, 11,543, and 6,283 slices, for training, validation and testing sets respectively to prevent leakage between training and testing.

CHAOS. Combined (CT-MR) Healthy Abdominal Organ Segmentation challenge dataset [36] consists of abdominal CT and MRI volumes for different patients. Only the MR part of the dataset is used throughout this study. The MRI dataset is formed of 40 volumes, where the segmented liver, kidneys and spleen ground truth is found for 20 volumes while the other 20 volumes are not annotated. The annotated volumes were considered in this study. The data was randomly divided into 14, 3, and 3 volumes forming 458, 82, and 116 slices for training, validation, and testing tasks. It is worth noting that the number of slices as well as the spatial resolution of the volumes varies within the dataset. The size of the slices varied between 320×320, 256×256, and 288×288.

DLDS. Duke Liver DataSet [40] contains MRI scans across 95 patients. About 66 patients have manual segmentations of the liver. The data was randomly divided into 54 patients for training, 6 patients for validation and 6 patients for testing; such that the training set contained 3860 slices, the validation set contained 412, while the testing set contained 426 slices. The size of the slices varied in range from 336×320 to 512×512.

2.2 Datasets Preparation

All images were normalized to a consistent intensity range between $(0, 1)$, and down-sampled to 128×128 using bilinear interpolation for anatomical volumes, whereas a 1-nearest neighbor approach was employed for down-sampling of the masks. In the case of CHAOS and DLDS datasets, the slices were rotated $90°C$ around the center of the image to match the supine orientation of the slices found in the LiTS dataset. The batch size for CHAOS and LiTS dataset was chosen to be 16 samples, whereas for DLDS, the batch size was chosen to be 64 to maintain well sampling of the data within each batch. Figure 1 shows a sample slice for LITS, CHAOS and DLDS datasets.

2.3 Model Architecture and Training

U-Net is a widely used architecture for segmentation especially in the medical field. It is also successful in different medical image segmentation tasks in different applications across different modalities [41,42]. U-Net architecture consists of two paths: an encoding path for feature extraction and a decoding path for mask construction using the extracted features. Both the encoder and the decoder consist of four consecutive blocks. The basic building unit of each block is a convolution unit (CONV Unit) that contains: (1) a 2D convolution operation with a 3×3 filter, (2) a rectified linear unit (ReLU) activation, (3) a batch normalization layer, (4) a dropout layer with a rate of 20% to avoid overfitting, (5) a second convolution operation, ReLU activation and batch normalization. Each block of the encoder contains a CONV Unit followed by 2×2 max-pooling layer for down-sampling, whereas in the decoder, the block is composed of a transpose convolution for up-sampling followed by concatenation with the corresponding feature maps from the encoder path and finally a CONV Unit. The final layer is formed of a 1×1 convolution followed by a sigmoid activation for pixel-wise binary classification between the foreground and background. The full architecture is shown in Fig. 2.

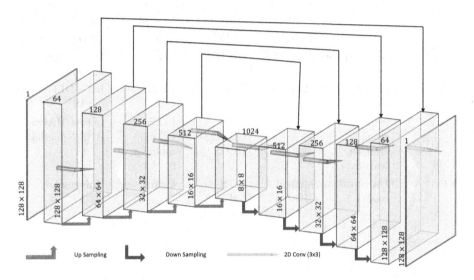

Fig. 2. U-Net architecture on the right with convolution unit (CONV Unit) on the left

For the U-Net training and experimentation throughout the following section, an Adam optimizer with a learning rate of 0.001 was used. The Jaccard distance was used as a loss function

$$loss(x, \hat{x}) = 1 - J(x, \hat{x}) \qquad (1)$$

where J is the Jaccard coefficient, measuring the similarity between the ground truth mask to the segmented region, it is capable of handling the class imbalance between the foreground and background pixels [43].

$$Jaccard(A, B) = \frac{|A \cap B|}{|A \cup B|} \qquad (2)$$

For evaluation, the Dice metric is widely used in the literature [26,35,44].

$$Dice(A, B) = 2 \times \frac{|A \cap B|}{|A| + |B|} \qquad (3)$$

The training was done for 30 epochs using TensorFlow [45] on a device with the following specifications: a Nvidia RTX3090 graphical processing unit with 24GB VRAM, an AMD Ryzen 9 5950X central processing unit with 128GB RAM.

Table 1. A description of the experiments conducted showing the training and testing data, the used model, and the resulting models of each experiment

Experiment number	Experiment title	Training data	Testing data	Used model	Resulting models
#1	Single	LiTS	LiTS	U-Net	LiTS model
	dataset	CHAOS	CHAOS	U-Net	CHAOS model
	training	DLDS	DLDS	U-Net	DLDS model
#2	Single stage	CHAOS	CHAOS	LiTS	LiTS-CHAOS model
	transfer learning	DLDS	DLDS	model	LiTS-DLDS model
#3	Multiple stage transfer learning	DLDS	LiTS CHAOS DLDS	LiTS-CHAOS model	LiTS-CHAOS -DLDS model
#4	Universal model	Shuffled CHAOS DLDS LiTS	LiTS CHAOS DLDS	U-Net	Universal model

2.4 Experimentation

In order to assess the hypothesis of creating an across modalities model for liver segmentation, it was necessary to test every step and for the same dataset as well as across datasets after the transfer learning. Therefore, several experiments have been investigated.

Experiment #1 employs single dataset training and evaluation where U-Net was trained and evaluated on each dataset separately. In this experiment, the U-Net was trained from scratch using each of the aforementioned datasets, resulting in three models: (1) LiTS model, (2) CHAOS model, and (3) DLDS model. Then the best model (LiTS model) was selected to be used in the subsequent experiments. Furthermore, the selected model was evaluated, without being fine-tuned, on the MRI datasets (CHAOS and DLDS) to measure the performance depreciation occurring, due to the change in modalities between the training and evaluation stages.

In experiment #2, the pre-trained model (selected from experiment #1 - LiTS model) was then fine-tuned, via transfer learning, using the datasets from different modalities, where the performance was evaluated and reported.

This experiment is considered to be a single stage transfer learning experiment resulting in two models: (1) LiTS-CHAOS model, and (2) LiTS-DLDS model. Several trials were conducted to decide which block to freeze during fine-tuning. Figure 3 shows the results of each trial. Each trial tested both models on both datasets (CHAOS and DLDS) separately.

It was found that, the results are very close regardless the freezing methodology. However, freezing the encoder only gave a slightly better test Dice score when compared with freezing the last block of the decoder or not applying freezing at all. The training time was not affected that much, only a one minute difference between trials on DLDS dataset, and almost no time difference on CHAOS dataset. This happened because of the relatively small size of those datasets. Therefore, it was decided to use the LiTS model with the encoder frozen, while the decoder layers were set to trainable in the following experiments. This choice can work as a common ground between two extremes, that freezes half the network and train the second half instead of not freezing at all or only training one-eighth of the network. Since the anatomical structure of the liver is the same across modalities, we were interested in updating the last layers only to match the differences between modalities.

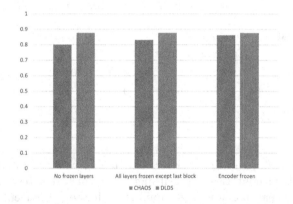

Fig. 3. Experiment # 2 results: Dice scores of test portions of CHAOS and DLDS calculated using different freezing strategies of LiTS pre-trained model after fine-tuning on each dataset separately (LiTS-CHAOS model & LiTS-DLDS model)

Experiment #3 is considered a multiple stage transfer learning experiment, where the pre-trained model using LiTS dataset that was fine-tuned using the CHAOS dataset, then fine-tuned once again using DLDS dataset. This resulted in the LiTS-CHAOS-DLDS model. The freezing strategy used here is encoder freezing according to the results from the previous experiment. CHAOS dataset is much smaller than DLDS dataset, yet sufficient to apply transfer learning across modalities. Afterwards, the model is tested on all three datasets.

In experiment #4, we aimed to train a universal model that can segment the liver across modalities. The training set of this experiment includes shuffled slices from the training portions of CHAOS, LiTS and DLDS. The trained model was then tested using the testing portion for each dataset separately and the performance was reported. Conclusion of all experiments is reported in Table 1.

3   Results

In experiment #1, the model was trained, validated, and tested using each dataset solely to select the best pretrained model to be employed in further experiments. Table 2 lists the achieved Dice scores using the three datasets. LiTS model was able to achieve training, validation, and testing Dice scores of 0.96, 0.92 and 0.94 respectively on LiTS datasets. DLDS model achieved training, validation, and testing Dice scores of 0.89, 0.85, 0.82 respectively on DLDS dataset. Whereas CHAOS model achieved training, validation, and testing Dice scores of 0.76, 0.60, 0.59 respectively on CHAOS dataset. The LiTS pre-trained model was used in the upcoming experiments since it had the best performance among the other models.

Table 2. Experiment #1 results: Training, validation, and testing Dice scores for single dataset training along with the number of training slices within each dataset

Dataset	Dice score			Training Size
	Training	Validation	Testing	
LiTS	0.96	0.92	0.94	40,812
CHAOS	0.76	0.60	0.59	458
DLDS	0.89	0.85	0.82	3,860

In experiment #2, the LiTS pre-trained model was evaluated directly on CHAOS dataset without transfer learning and achieved a testing Dice score of 0.13. Transfer learning was then applied with the chosen encoder-freezing strategy resulting in a 0.96 training Dice score, 0.73 validation Dice score and 0.86 testing Dice score which can be seen in Fig. 3. This model will be called LiTS-CHAOS model.

Before running the multiple stage transfer learning experiment #3, the LiTS-CHAOS pre-trained model was directly tested on DLDS dataset and achieved a Dice score of 0.83 shown in Table 3. Then experiment #3 was run where LiTS-CHAOS model was fine-tuned with encoder-freezing approach using the training portion of DLDS dataset. This model achieved Dice scores of 0.93, 0.90 and 0.88 on training, validation and testing portions of DLDS dataset respectively.

Finally, experiment #4 was run where the U-Net was trained from scratch using a combined shuffled dataset of LiTS, CHAOS, and DLDS. It achieved a Dice score of 0.928, 0.865, and 0.815 for LiTS, DLDS and CHAOS testing portions respectively.

Table 3. Experiments #2 & #3 results: Testing Dice scores for DLDS and CHAOS datasets using the pre-trained U-Net

Pre-trained on	DLDS test	CHAOS test
LiTS	0.16	0.13
LiTS-CHAOS	**0.83**	**0.86**

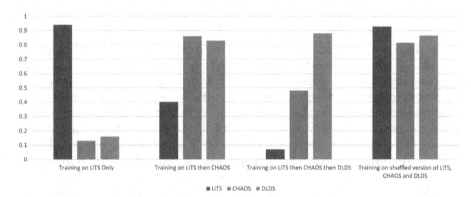

Fig. 4. Results of all experiments: Dice scores for test portions of LiTS, CHAOS, and DLDS datasets calculated using different pre-training data. The dataset training order can tailor the model performance towards a certain outcome

4   Discussion

The performance of the network is dependent on the dataset size. Though, the performance of the learnt model using LiTS dataset outperformed that learnt using CHAOS dataset, as shown in Table 2, since the LiTS dataset is of bigger size compared to the other datasets employed in this study. Moreover, employing a trained model using CT LiTS, was unable to generalize to a dataset of another modality, such as CHOAS to segment the liver with acceptable accuracy. Table 3 shows poor performance on DLDS and CHAOS dataset by the model trained on LiTS dataset only. The performance depreciation was high when using MR datasets to test the model learnt using CT dataset, due to the differences between MR and CT images characteristics and the variety in output scans. However, the performance is highly boosted after employing transfer learning across modality. A drastic jump in test dice score on CHAOS dataset from 0.59 in experiment #1 (trained on CHAOS only) to 0.86 in experiment #2 (LiTS fine-tuned on CHAOS) can be seen. This proves that applying transfer learning from a model trained on large data in one modality to a small data in another modality is possible. Additionally, the relatively high dice score of this fine-tuned model on the DLDS dataset shows that transfer learning can direct the model training toward better generalization. It is worth noting that the fine-tuned models' performance was decreased when testing it using LiTS dataset, as shown in Fig. 4. This can be considered the cost of this generalization or fine-tuning on MR data, the decoder layers of the model learnt MR so well that it now underfits CT data. Moreover, Fig. 4 indicates that the order of the fine-tuning on the datasets affects the performance. It also shows that after training the model sequentially on the CT then the two MR datasets the performance on the latter trained datasets is much higher than the former. While at the cost of fine-tuning to DLDS, the performance on CHAOS decreases. This could be due to training for the same number of epochs in both cases despite the larger size of the DLDS data compared to CHAOS causing the model to perform better on one over the other. Figure 4 also shows that training on all the datasets shuffled gives a good test performance on each dataset on its own. This confirms even a small amount of MRI data is sufficient to achieve good results using transfer learning. A combination of the datasets gives a good generalization performance and prevents the network from overfitting to a single modality. It was also important to study the layers-freezing effect on the overall performance because it helps in speeding up the training significantly, especially with large datasets. It was noted that applying different transfer learning approaches slightly affects the overall performance. However, encoder-freezing approach has generally better results than retraining the whole network or freezing the whole network except for the last block. An interpretation can be that the encoder is the same for all datasets, as it tries to learn a compressed representation and certain features of the liver. On the contrary, the decoder must find the exact location of the liver in spite of the difference of the modality or the image characteristic. Our approach focuses on transfer learning from CT to MR and studying the effect on different datasets unlike [39] that focused on mapping from MR to CT. However, they used 300 MR volumes to train the base architecture when compared to our 131 CT volumes. They also proved that increasing the size of the fine-tuning CT data increases the resulting dice score. On the other hand, we focused on studying the effect of order of datasets used in training and the freezing strategy in transfer learning. Other approaches in the literature like [38] focused on generative modelling to map from CT to MR, but they required a registration framework between CT and MR. Most of the other methods focus on one modality only [26,28,35].

5   Conclusion

Transfer learning for medical image segmentation is possible across modalities. We were able to prove that a small MR data is enough to fine-tune U-Net. Furthermore, the order of data training affects the performance of U-Net, where the dice score of the last trained dataset is always higher at the cost of a decrease in dice on the other datasets. While shuffling multiple datasets together before training will result in a more generalized version of the model that is capable of achieving the target task with reasonable performance on all modalities in spite of any differences between datasets.

Acknowledgment. This project was supported by the Science and Technology Fund Institute (STDF), Project ID 45891- EG-US Cycle 20.

References

1. Asrani, S.K., Devarbhavi, H., Eaton, J., Kamath, P.S.: Burden of liver diseases in the world. J. Hepatol., **70**(1), 151–171 (2019). ISSN 0168–8278. https://doi.org/10.1016/j.jhep.2018.09.014. www.sciencedirect.com/science/article/pii/S0168827818323882
2. Ansari, M., et al.: Practical utility of liver segmentation methods in clinical surgeries and interventions. BMC Med. Imaging, 22 (2022). https://doi.org/10.1186/s12880-022-00825-2
3. Gotra, A., et al.: Liver segmentation: indications, techniques and future directions. Insights Imaging **8**, 06 (2017). https://doi.org/10.1007/s13244-017-0558-1
4. Oliva, M., Saini, S.: Liver cancer imaging: role of CT, MRI, us and pet. Cancer imaging: the official publication of the international cancer imaging society, 4 Spec No A: S42–6 (2004). https://doi.org/10.1102/1470-7330.2004.0011
5. Lin, E., Alessio, A.: What are the basic concepts of temporal, contrast, and spatial resolution in cardiac CT? J. Cardiovasc. Comput. Tomogr. **3**, 403–408 (2009). https://doi.org/10.1016/j.jcct.2009.07.003
6. Lim, M., Tan, C.H., Cai, J., Zheng, J., Kow, A.: CT volumetry of the liver: where does it stand in clinical practice? Clin. Radiol. **69** (2014). https://doi.org/10.1016/j.crad.2013.12.021
7. Duman, D., Celikel, C., Tüney, D., Imeryüz, N., Avsar, E., Tözün, N.: Computed tomography in nonalcoholic fatty liver disease: a useful tool for hepatosteatosis assessment? Digestive Dis. Sci. **51**, 346–51 (2006). https://doi.org/10.1007/s10620-006-3136-9
8. Donato, H., França, M., Candelária, I., Caseiro-Alves, F.: Liver MRI: from basic protocol to advanced techniques. Eur. J. Radiol. **93**, 05 (2017). https://doi.org/10.1016/j.ejrad.2017.05.028
9. Venkatesh, S., Yin, M., Ehman, R.: Magnetic resonance elastography of liver: technique, analysis, and clinical applications. J. Magn. Reson. Imaging: JMRI **37**, spcone (2013). https://doi.org/10.1002/jmri.24092
10. Elbanna, K.Y., Kielar, A.Z.: Computed tomography versus magnetic resonance imaging for hepatic lesion characterization/diagnosis. Clin. Liver Dis. **17**(3), 159–164 (2021). https://doi.org/10.1002/cld.1089
11. Gul, S., Khan, M.S., Bibi, A., Khandakar, A., Ayari, M.A., Chowdhury, M.E.H.: Deep learning techniques for liver and liver tumor segmentation: a review. Comput. Biol. Med. **147**, 105620 (2022). ISSN 0010–4825. https://doi.org/10.1016/j.compbiomed.2022.105620, www.sciencedirect.com/science/article/pii/S0010482522004127
12. Chartrand, G., Cresson, T., Chav, R., Gotra, A., Tang, A., DeGuise, J.: Semi-automated liver CT segmentation using Laplacian meshes. In: 2014 IEEE 11th International Symposium on Biomedical Imaging (ISBI), pp. 641–644. IEEE (2014)
13. Kessler, L.G., et al.: The emerging science of quantitative imaging biomarkers terminology and definitions for scientific studies and regulatory submissions. Statist. Methods Med. Res. **24**(1), 9–26 (2015)

14. Udupa, J.K., et al.: A framework for evaluating image segmentation algorithms. Comput. Med. Imaging Graph. **30**(2), 75–87 (2006)

15. Araújo, J., et al.: Liver segmentation from computed tomography images using cascade deep learning. Comput. Biol. Med. **140**, 105095 (2021). https://doi.org/10.1016/j.compbiomed.2021.105095

16. Luo, S., Jin, J.S., Chalup, S.K., Qian, G.: A liver segmentation algorithm based on wavelets and machine learning. In: 2009 International Conference on Computational Intelligence and Natural Computing, vol. 2, pp. 122–125 (2009). https://doi.org/10.1109/CINC.2009.225

17. Kuo, C.-L., Cheng, S.-C., Lin, C.-L., Hsiao, K.-F., Lee, S.-H.: Texture-based treatment prediction by automatic liver tumor segmentation on computed tomography. In: 2017 International Conference on Computer, Information and Telecommunication Systems (CITS), pp. 128–132 (2017). https://doi.org/10.1109/CITS.2017.8035318

18. Danciu, M., Gordan, M., Florea, C., Vlaicu, A.: 3D DCT supervised segmentation applied on liver volumes. In: 2012 35th International Conference on Telecommunications and Signal Processing (TSP), pp. 779–783 (2012). https://doi.org/10.1109/TSP.2012.6256403

19. Furukawa, D., Shimizu, A., Kobatake, H.: Automatic liver segmentation method based on maximum a posterior probability estimation and level set method (2007)

20. Foruzan, A.H., et al.: Segmentation of liver in low-contrast images using K-means clustering and geodesic active contour algorithms. IEICE Trans. Inf. Syst. E96.D, 798–807 (2013). https://doi.org/10.1587/transinf.E96.D.798

21. Chi, D., Zhao, Y., Li, M.: Automatic liver MR image segmentation with self-organizing map and hierarchical agglomerative clustering method. In: 2010 3rd International Congress on Image and Signal Processing, vol. 3, pp. 1333–1337 (2010). https://doi.org/10.1109/CISP.2010.5648009

22. Janiesch, C., Zschech, P., Heinrich, K.: Machine learning and deep learning. Electron. Mark. **31**(3), 685–695 (2021)

23. Long, J., Shelhamer, E., Darrell, T.: Fully convolutional networks for semantic segmentation. In: Proceedings of the IEEE Conference on Computer Vision and Pattern Recognition (CVPR) (2015)

24. Ben-Cohen, A., Diamant, I., Klang, E., Amitai, M., Greenspan, H.: Fully convolutional network for liver segmentation and lesions detection. In: Carneiro, G., et al. (eds.) LABELS/DLMIA -2016. LNCS, vol. 10008, pp. 77–85. Springer, Cham (2016). https://doi.org/10.1007/978-3-319-46976-8_9

25. Lei, T., Zhou, W., Zhang, Y., Wang, R., Meng, H., Nandi, A.K.: Lightweight V-Net for liver segmentation. In: ICASSP 2020–2020 IEEE International Conference on Acoustics, Speech and Signal Processing (ICASSP), pp. 1379–1383 (2020). https://doi.org/10.1109/ICASSP40776.2020.9053454

26. Aghamohammadi, A., Ranjbarzadeh, R., Naiemi, F., Mogharrebi, M., Dorosti, S., Bendechache, M.: TPCNN: two-path convolutional neural network for tumor and liver segmentation in CT images using a novel encoding approach. Expert Syst. Appl. **183**, 115406 (2021). https://doi.org/10.1016/j.eswa.2021.115406

27. Ranjbarzadeh, R., Saadi, S.: Automated liver and tumor segmentation based on concave and convex points using fuzzyc-means and mean shift clustering. Measurement **150**, 107086 (2019). https://doi.org/10.1016/j.measurement.2019.107086

28. Tang, W., Dongsheng Zou, S., Yang, J.S., Dan, J., Song, G.: A two-stage approach for automatic liver segmentation with faster R-CNN and DeepLab. Neural Comput. Appl. **32**, 06 (2020). https://doi.org/10.1007/s00521-019-04700-0

29. Ren, S., He, K., Girshick, R., Sun, J.: Faster R-CNN: towards real-time object detection with region proposal networks. IEEE Trans. Pattern Anal. Mach. Intell. **39**, 06 (2015). https://doi.org/10.1109/TPAMI.2016.2577031

30. Simonyan, K., Zisserman, A.: Very deep convolutional networks for large-scale image recognition (2014). arxiv.org/abs/1409.1556

31. Deng, J., Dong, W., Socher, R., Li, L.-J., Li, K., Fei-Fei, L.: ImageNet: a large-scale hierarchical image database. In: 2009 IEEE Conference on Computer Vision and Pattern Recognition, pp. 248–255 (2009). https://doi.org/10.1109/CVPR.2009.5206848

32. Chen, L.-C., Papandreou, G., Kokkinos, I., Murphy, K., Yuille, A.L.: Deeplab: semantic image segmentation with deep convolutional Nets, Atrous Convolution, and Fully connected CRFs. IEEE Trans. Pattern Anal. Mach. Intell. **40**(4), 834–848 (2018). https://doi.org/10.1109/TPAMI.2017.2699184

33. Soler, L., et al.: 3D image reconstruction for comparison of algorithm database: a patient specific anatomical and medical image database. IRCAD, Strasbourg, France, Tech. Rep, **1**(1)(2010). www.ircad.fr/research/3dircadb/

34. Bauer, C., et al.: Comparison and evaluation of methods for liver segmentation from CT datasets. IEEE Trans. Med. Imaging **28**(8), 1251–1265 (2009). ISSN 0278–0062. https://doi.org/10.1109/TMI.2009.2013851, https://sliver07.grand-challenge.org/Home/

35. Azam Khan, R., Luo, Y., Wu, F.-X.: RMS-UNet: residual multi-scale UNet for liver and lesion segmentation. Artif. Intell. Med. **124**, 102231 (2022). ISSN 0933–3657. https://doi.org/10.1016/j.artmed.2021.102231, https://www.sciencedirect.com/science/article/pii/S0933365721002244

36. Kavur, A.E., Alper Selver, M., Dicle, O., Barış, M., Sinem Gezer, N.: CHAOS - Combined (CT-MR) Healthy Abdominal Organ Segmentation Challenge Data (2019). https://doi.org/10.5281/zenodo.3431873

37. Bilic, P., et al.: The liver tumor segmentation benchmark (LiTS). CoRR, abs/1901.04056 (2019). arxiv.org/abs/1901.04056

38. Zhou, B., Augenfeld, Z., Chapiro, J., Kevin Zhou, S., Liu, C., Duncan, J.S.: Anatomy-guided multimodal registration by learning segmentation without ground truth: Application to intraprocedural CBCT/MR liver segmentation and registration. Medical Image Anal. **71**, 102041 (2021). ISSN 1361–8415. https://doi.org/10.1016/j.media.2021.102041, www.sciencedirect.com/science/article/pii/S1361841521000876

39. Wang, K., et al.: Automated CT and MRI liver segmentation and biometry using a generalized convolutional neural network. Radiol. Artif. Intell. **1**, 180022 (2019). https://doi.org/10.1148/ryai.2019180022

40. Macdonald, J.A., Zhu, Z., Konkel, B., Mazurowski, M., Wiggins, W., Bashir, M.: Duke liver dataset (MRI), October (2020). https://doi.org/10.5281/zenodo.6328447

41. Getao, D., Cao, X., Liang, J., Chen, X., Zhan, Y.: Medical image segmentation based on u-net: a review. J. Imaging Sci. Technol. **64** (2020). https://doi.org/10.2352/J.ImagingSci.Technol.2020.64.2.020508

42. Siddique, N., Sidike, P., Elkin, C., Devabhaktuni, V.: U-net and its variants for medical image segmentation: a review of theory and applications. IEEE Access **9**, 82031–82057 (2021). https://doi.org/10.1109/ACCESS.2021.3086020

43. Bertels, J., et al.: Optimizing the dice score and Jaccard index for medical image segmentation: theory and practice. In: Shen, D., et al. (eds.) MICCAI 2019. LNCS, vol. 11765, pp. 92–100. Springer, Cham (2019). https://doi.org/10.1007/978-3-030-32245-8_11

44. Meng, L., Zhang, Q., Bu, S.: Two-stage liver and tumor segmentation algorithm based on convolutional neural network. Diagnostics **11**(10) (2021). ISSN 2075–4418. https://doi.org/10.3390/diagnostics11101806. www.mdpi.com/2075-4418/11/10/1806
45. Abadi, M., et al.: TensorFlow: large-scale machine learning on heterogeneous systems (2015). Software available from tensorflow.org

Can SegFormer be a True Competitor to U-Net for Medical Image Segmentation?

Théo Sourget[1], Syed Nouman Hasany[1], Fabrice Mériaudeau[2], and Caroline Petitjean[1(✉)]

[1] Univ Rouen Normandie, Université Le Havre Normandie, INSA Rouen Normandie, Normandie Univ, LITIS UR 4108, 76000 Rouen, France
caroline.petitjean@univ-rouen.fr
[2] ICMUB UMR CNRS 6302, Université Bourgogne, 21000 Dijon, France

Abstract. The U-Net model, introduced in 2015, is established as the state-of-the-art architecture for medical image segmentation, along with its variants UNet++, nnU-Net, V-Net, etc. Vision transformers made a breakthrough in the computer vision world in 2021. Since then, many transformer based architectures or hybrid architectures (combining convolutional blocks and transformer blocks) have been proposed for image segmentation, that are challenging the predominance of U-Net. In this paper, we ask the question whether transformers could overtake U-Net for medical image segmentation. We compare SegFormer, one of the most popular transformer architectures for segmentation, to U-Net using three publicly available medical image datasets that include various modalities and organs: segmentation of cardiac structures in ultrasound images from the CAMUS challenge, segmentation of polyp in endoscopy images and segmentation of instrument in colonoscopy images from the MedAI challenge. We compare them in the light of various metrics (segmentation performance, training time) and show that SegFormer can be a true competitor to U-Net and should be carefully considered for future tasks in medical image segmentation.

Keywords: Medical image segmentation · UNet · transformers

1 Introduction

Since 2015, the U-Net [11] has been established as the state of the art model for medical image segmentation, along with its variants UNet++, nnU-Net [7], V-Net, etc. Its predominance has been challenged by the arrival of transformers in 2021 [2]. Indeed following the excellent results of Transformers on natural language processing problems, transformer-based architectures have been proposed on image processing tasks, starting with the Vision Transformer, for image classification [3]. Transformers process the image as a sequence of patches (typically of size 16 × 16 pixels) and seem to be very efficient thanks to the attention mechanism which allows them to capture long range interaction between the patches - contrary to the reduced receptive field of convolutional kernels.

G. Waiter et al. (Eds.): MIUA 2023, LNCS 14122, pp. 111–118, 2024.
https://doi.org/10.1007/978-3-031-48593-0_8

The transition from the ViT to an architecture with transformers for image segmentation is not obvious. Many architectures have been proposed, such as the Segmenter Transformer SETR [14] or the PVT [12] or SegFormer, whose transformer encoders are based on a pyramid structure, so as to mimic the encoder of a CNN. The recently released "Segment Anything" tool [8], a promptable segmentation system with impressive performance on natural images, is also based on a transformer architecture. Hybrid architectures combining convolutional and transformer blocks such as TransUNet [2], CATS [10] or UNETR [5] have been proposed for medical image segmentation, as underlined by recent reviews of transformers in medical image analysis [1,6]. These models are however often very complex with several tens of millions of parameters and require lot of time to be trained.

The question now is whether a transformer-based architecture can be a true competitor to U-Net, for medical image segmentation. In this study, we decide to focus on SegFormer [13], a lightweight architecture for image segmentation, designed to avoid complex decoders. Its efficiency, accuracy, and robustness have been shown on a variety of datasets such as ADE20K, Cityscapes, and COCO-Stuff; and in particular, it is shown in the paper to outperform the previous best method on ADE20K, the SETR [14] model. In this paper, our goal is to test SegFormer, both pre-trained and trained from scratch, against the U-Net architecture, in terms of segmentation accuracy and compute efficiency and see if SegFormer can be a viable alternative to U-Net for medical image segmentation.

In the following, we first detail the SegFormer architecture. Then we introduce three datasets encompassing different tasks of binary and multilabel segmentation in medical images, and present the experimental protocol to make a fair comparison. We provide both quantitative and qualitative segmentation results of the models under scrutiny.

2   SegFormer

SegFormer has two main modules: a hierarchically structured transformer encoder and an MLP (multi-layer perceptron) decoder (see Fig. 1). One of the key contribution of this architecture is the lightweight All-MLP decoder, resulting in a light architecture in comparison to other transformer architectures for segmentation, e.g. [14]. These aspects of SegFormer lead to multiple benefits. First, as the encoder part produces multiple level feature maps fused in the decoder, the model is able to capture both high and low resolution information. The encoder also relies on a "mix-FFN" (feed-forward network) operation, where a 3×3 convolution with 0-padding and an MLP are mixed into each FFN, to replace the original positional encoding used by other architectures. Moreover, as the model is less complex than other transformer based architectures, it requires less data to be trained and can also be applied to real-time application. Finally, the encoder part of the architecture can be scaled from B0 to B5 by increasing the number of layers or the dimensions of encoder blocks; depending on the need,

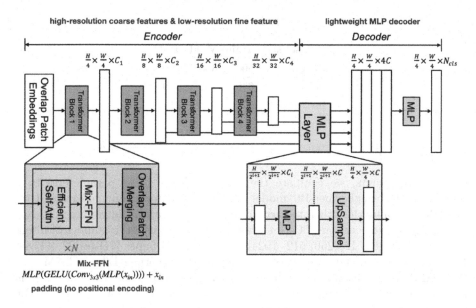

Fig. 1. SegFormer architecture. The two main modules are the hierarchical Transformer encoder, and an all-MLP based decoder merging the features from the encoder. Figure reproduced from [13].

it is possible to favor time efficiency with B0 able to perform real-time applications or performances with B5 obtaining best results on various problems. In this study, we will use the SegFormer-B0 architecture that has 3.1M parameters.

3   Experimentations

3.1   Datasets

We have assessed the two architectures on three different datasets: CAMUS, Polyp, and Instruments. The Cardiac Acquisitions for Multi-structure Ultrasound Segmentation (CAMUS) challenge dataset [9] contains 2D images of cardiac ultrasound images from 500 patients, with the manual segmentation of 4 different classes (Endocardium, Epicardium, Atrium and Background). For each patient, there are 4 different types of images related to different views (2-chambers or 4-chambers) and phase of the heart (diastole and systole). In this experiment, we have used only the diastole images for both training and testing phase. Poly and Instruments are two datasets from the MedAI: Transparency in Medical Image Segmentation challenge. The Polyp Segmentation Task consists in segmenting polyps in endoscopy images from the Kvasir-SEG dataset which contains 1000 images and corresponding binary segmentation masks. The Instrument Segmentation Task is about segmenting instruments present in colonoscopy videos from the Kvasir-Instrument dataset which contains 590 images and corresponding binary segmentation masks.

Table 1. Dataset size

Dataset size	Train	Valid	Test
Camus	500	450	50
Polyp	640	160	200
Instrument	377	95	118

3.2   Protocol

In our experiments, we compare the original U-Net (31M parameters) to 2 versions of the SegFormer-B0 model (3.7M parameters): one pre-trained on ImageNet-1k and the other trained from scratch. We also introduce a U-Net Lite, a lightweight version of U-Net where we have reduced the number of filters per layer from [64, 128,256, 512, 1024] in the original paper [11] to [22, 44, 88, 176, 352] to match the number of parameters of SegFormer-B0, i.e. 3.7M. For the CAMUS dataset the loss function is Cross-Entropy, and the optimizer is Adam with a fixed learning rate, 1e-03 for U-Net, and 1e−04 for SegFormer. For the Polyp and Instrument segmentation tasks, the loss function is an average of Cross-Entropy and Dice, and the optimizer is AdamW with a fixed learning rate of 1e−04 for both the U-Net and the SegFormer.

We use the SegFormer-B0 model from HuggingFace and we applied transfer learning by using encoder's weights trained on Imagenet-1k. For sake of reproducibility, the code used on CAMUS for U-Net can be found here: https://github.com/TheoSourget/UNet_CAMUS and for Seg-Former here: https://github.com/TheoSourget/SegFormer_CAMUS.

Table 1 shows the datasets split into training, validation and test sets. For CAMUS, the images are resized to 256×256 pixels and at every epoch, a random rotation between $-10°$ and $10°$ is applied to perform data augmentation on the dataset. For Polyp and Instrument Segmentation, the images are resized to 224×224 and random flipping (horizontal and vertical) as well as random rotation between $0°$ and $180°$ is applied to perform data augmentation on the dataset.

4   Results and Discussion

4.1   Segmentation Accuracy

The average Dice Scores for each dataset are in Table 2 and Fig. 4. First of all, it is interesting to note that even if the number of parameters of U-Net is drastically reduced, the decrease in accuracy on the CAMUS dataset is not as significant than for Polyp or Instrument. This corroborates the observations made in [4,9] that simpler models can obtain similar results than more complex ones. Not surprisingly, the pre-trained SegFormer is better, sometimes by a large margin, than the SegFormer trained from scratch. This is also visible by the behaviour during training: the pre-trained model seems to converge faster than the one

Table 2. Results: Average Dice scores of U-Net and SegFormer of 3 datasets: CAMUS, Polyp and Instrument. * indicates that the score is significantly different from that of UNet (p<0.05). For graphical representation see Fig. 4

		U–Net	U–Net Lite	SegFormer	SegFormer pre-trained
Pre-trained?		No	No	No	Yes
# param		31M	3.7M	3.7M	3.7M
CAMUS	Endo	0.90	0.90	0.89	**0.91***
	Epi	0.80	0.79	0.81	**0.83***
	Atrium	0.83	0.84	0.81	0.85
Polyp		0.74	0.67	0.60	**0.83***
Instrum		0.79	0.75	0.82	**0.92***

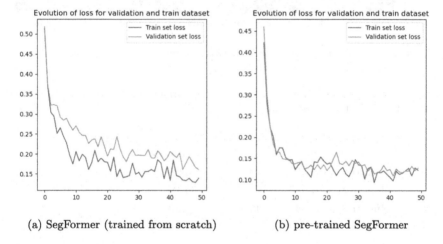

(a) SegFormer (trained from scratch)   (b) pre-trained SegFormer

Fig. 2. Evolution of loss during training for SegFormer with and without pre-training on ImageNet-1K

trained from scratch, as shown by the evolution of loss for the CAMUS dataset in Fig. 2.

Finally, the pre-trained SegFormer performed significantly better than U-Net for almost all segmented regions, i.e. the endocardium and epicardium of CAMUS, Polyp and Instrument, except for the atrium of CAMUS. Statistical significance was achieved using a two-sided Wilcoxon test (p<0.05). We can also see a difference in the training behaviour: while the transformer is able to predict every class at the end of the first epoch, U-Net only predicts the background class during the first two epochs and only afterwards is able to predict relevant classes. This is illustrated in Fig. 3 where the difference of segmentation between U-Net, SegFormer and the pre-trained SegFormer after 1st and 5th epoch on a

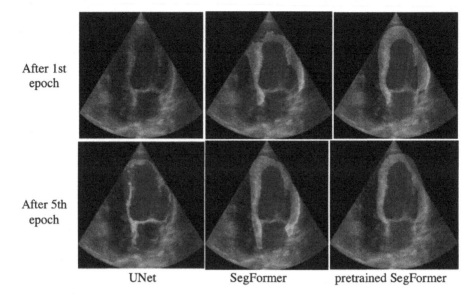

After 1st epoch

After 5th epoch

UNet   SegFormer   pretrained SegFormer

Fig. 3. Comparing segmentation results between U-Net, SegFormer and pre-trained SegFormer after the 1st and 5th epoch

Table 3. Drop in training time of U-Net Lite, SegFormer and pre-trained SegFormer with respect to U-Net's training time.

Model	U-Net Lite	SegFormer	SegFormer pre-trained	Epochs
CAMUS	−57.5%	−49.7%	−53.0%	50
Polyp	−51.2%	−62.3%	−65.1%	80
Instrum	−40.4%	−46.4%	−46.9%	80

validation example from CAMUS dataset can be seen. This Figure also highlights the strong impact of using pre-trained weights for SegFormer.

4.2   Training Time

In Table 3, we display the decrease in training time over the indicated number of epochs of the considered models, with respect to that of U-Net. We can gather from this table that SegFormer is always faster to train than the original U-Net and often faster than the Lite U-Net version, even if it is not pre-trained.

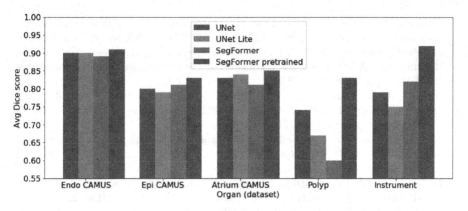

Fig. 4. Average Dice scores of U-Net and SegFormer on test sets of 3 datasets: CAMUS, Polyp and Instrument. Corresponds to results in Table 2.

5   Conclusion

In this study, our goal was to compare the U-Net model to a newly proposed, transformer based model called SegFormer. The underlying aim was to start assessing whether the breakthrough of vision transformer can really offer competitive segmentation performance both in accuracy and training time, on some medical image segmentation tasks, namely here the segmentation of cardiac structured in ultrasound images, of polyp in endoscopy, and of instruments in colonoscopy. On every task, pre-trained SegFormer-B0 obtained on par or better results than U-Net, and in less training time than the original U-Net thanks to its light architecture. Hence, we have shown that even if transformers usually need more data to be trained, they can still be applied on medical imaging tasks and obtained better results with limited dataset, taking advantage of transfer learning. However, this is only a preliminary study, and we are working towards large scale experiments, to include other datasets, to assess whether this findings holds on more segmentation tasks.

Acknowledgments. The authors acknowledge the support of the French Agence Nationale de la Recherche (ANR), under grant Project-ANR-21-CE23-0013 (project MediSEG).

References

1. Azad, R., et al.: Advances in medical image analysis with vision transformers: a comprehensive review (2023). https://doi.org/10.48550/arXiv.2301.03505, arXiv:2301.03505
2. Chen, J., et al.: TransUNet: transformers make strong encoders for medical image segmentation. arXiv preprint arXiv:2102.04306 (2021)
3. Dosovitskiy, A., et al.: An image is worth 16x16 words: transformers for image recognition at scale. CoRR abs/2010.11929 (2020). https://arxiv.org/abs/2010.11929

4. Galdran, A., Anjos, A., Dolz, J., et al.: State-of-the-art retinal vessel segmentation with minimalistic models. Sci. Rep. **12**, 6174 (2022). https://doi.org/10.1038/s41598-022-09675-y

5. Hatamizadeh, A., et al.: UNETR: transformers for 3d medical image segmentation. In: WACV, pp. 1748–1758 (2022)

6. He, K., et al.: Transformers in medical image analysis. Intell. Med. **3**(1), 59–78 (2023). https://doi.org/10.1016/j.imed.2022.07.002, https://www.sciencedirect.com/science/article/pii/S2667102622000717

7. Isensee, F., et al.: nnU-Net: Self-adapting framework for u-net-based medical image segmentation. CoRR abs/1809.10486 (2018). http://arxiv.org/abs/1809.10486

8. Kirillov, A., et al.: Segment anything (2023)

9. Leclerc, S., et al.: Deep learning for segmentation using an open large-scale dataset in 2d echocardiography. IEEE Trans. Med. Imaging **38**(9), 2198–2210 (2019). https://doi.org/10.1109/tmi.2019.2900516

10. Li, H., Hu, D., Liu, H., Wang, J., Oguz, I.: Cats: complementary CNN and transformer encoders for segmentation. In: 2022 IEEE 19th International Symposium on Biomedical Imaging (ISBI), pp. 1–5 (2022). https://doi.org/10.1109/ISBI52829.2022.9761596

11. Ronneberger, O., Fischer, P., Brox, T.: U-Net: convolutional networks for biomedical image segmentation. In: Navab, N., Hornegger, J., Wells, W.M., Frangi, A.F. (eds.) MICCAI 2015. LNCS, vol. 9351, pp. 234–241. Springer, Cham (2015). https://doi.org/10.1007/978-3-319-24574-4_28

12. Wang, W., et al.: Pyramid vision transformer: a versatile backbone for dense prediction without convolutions. CoRR abs/2102.12122 (2021). https://arxiv.org/abs/2102.12122

13. Xie, E., Wang, W., Yu, Z., Anandkumar, A., Alvarez, J.M., Luo, P.: SegFormer: simple and efficient design for semantic segmentation with transformers. CoRR abs/2105.15203 (2021). https://arxiv.org/abs/2105.15203

14. Zheng, S., et al.: Rethinking semantic segmentation from a sequence-to-sequence perspective with transformers. CoRR abs/2012.15840 (2020). https://arxiv.org/abs/2012.15840

Harnessing the Potential of Deep Learning for Total Shoulder Implant Classification: A Comparative Study

Aakriti Mishra[1], A. Ramanathan[5], Vineet Batta[2(✉)], C. Malathy[5], Soumya Snigdha Kundu[3], M. Gayathri[1], D. Vathana[1], and Srinath Kamineni[4]

[1] Department of Computing Technologies, School of Computing, SRM Institute of Science and Technology, Kattankulathur, India
[2] Department of Orthopedics and Trauma, Luton and Dunstable University Hospital NHS Trust, Luton, UK
battavineet@doctors.org.uk
[3] School of Electronic Engineering and Computer Science, Queen Mary University of London, London, UK
[4] Department of Orthopaedic Surgery and Sports Medicine, Elbow Shoulder Research Center, University of Kentucky, UK HealthCare, Lexington, KY, USA
[5] Department of Networking and Communications, School of Computing, SRM Institute of Science and Technology, Kattankulathur, India

Abstract. Orthopedic implant identification is an important and necessary step prior to performing revision surgery of different joints. The inability to identify an implant can lead to significant surgical difficulties with consequent unfavorable outcomes. This paper proposes a novel framework to identify the make and model of seven (7) different total shoulder arthroplasty implants utilizing plain X-ray images and Artificial intelligence. The proposed work classified implants with an accuracy of 91.48% and with an AUC (Area under curve) of 0.9932 showing higher effectiveness in orthopedic implant identification. Further work is required to enhance and progress this work, with a goal of greater accuracy and fewer errors.

Keywords: Shoulder Implant · Orthopedics · Biomedical · Deep Learning · Medical Images

1 Introduction

Shoulder replacements were suggested to patients who were affected by glenohumeral arthropathy with increased pain and reduced range of motion. Surgically the dysfunctional and painful joint is replaced with a prosthetic arthroplasty implant consisting of metallic and non-metallic components [1]. The procedure helps to relieve pain and restore function of the joint [2]. The prostheses were classified in various ways, with the simplest being (a) hemiarthroplasty, (b) anatomic total shoulder arthroplasty (TSA) and (c) reverse total shoulder arthroplasty (RTSA) [1]. Anatomic TSA is performed for patients with glenohumeral arthritis and a functioning rotator cuff. Reverse shoulder arthroplasty (RTSA) is usually performed for patients without a functioning rotator cuff [3]. These

© The Author(s), under exclusive license to Springer Nature Switzerland AG 2024
G. Waiter et al. (Eds.): MIUA 2023, LNCS 14122, pp. 119–132, 2024.
https://doi.org/10.1007/978-3-031-48593-0_9

implants may need to be revised for various reasons, commonly including Infection, glenoid or humeral component loosening, periprosthetic fractures, and instability [2].

The reported prevalence of infection of a joint replacement is between 0–4%. Infection of a joint arthroplasty can present as gross purulence or as non-purulent, indolent component loosening. Periprosthetic fractures contribute to 1.6–2.3% of complications encountered after shoulder replacement while problems due to anterior instability is 0.9–1.8% and posterior instability is 1% [1]. A recent study with 5,379 shoulder joint replacements indicated that the major reasons for revision were infection (16%), rotator cuff failure (17%), instability (19%) and loosening (20%) [4].

Primary and revision shoulder arthroplasty were witnessing a significant increase across most countries including Germany [5], Australia [6], and Italy [7]. The volume of revision shoulder arthroplasty in the United States of America increased by 153% from 2012 to 2018 [8]. Shoulder arthroplasty implants were manufactured by many different companies, all of which have different design features. To revise a prosthesis, due to the failure of one or both components, the surgeon requires information regarding the make and model of the implant [9]. The use of hospital and patients' records were the usual methods for implant identification. However, this is a time-consuming process, which is often unsuccessful due to imperfect or missing records. Inability to identify the implant, and pre-ordering the correct equipment, can lead to increased time of surgery, blood loss, and increased iatrogenic bone loss in these complex surgeries [10].

Thus, identification of implants plays an important role prior to a revision surgery. In this work, we present a framework for the identification of a make and model of total shoulder orthopedic implants from plain x-ray images and using Deep learning. Deep learning is used in many applications of medical imaging including classification and detection of proximal humerus fractures [11] and causes of shoulder pain [12]. We used deep learning based convolutional neural networks (CNN) for identification of implants, which were popularly used algorithms for images analysis [13].

2 Literature Review

Yilmaz used deep learning to classify 597 x-ray images from 4 different manufacturers. Use of multi-channel selection formula produced a classification accuracy of 97.2%. The performance was also compared to machine learning algorithms [9].

Urban et al. used deep learning to classify 16 different shoulder implant models from 4 manufacturers. 10 cross validations produced 80.4% accuracy while other methods achieved 56%. Data augmented training images had a positive effect on the model, with 80.4% using the NASNet neural network [14].

Eric et al. used Deep Learning based DenseNet121 to classify 10 different total and reverse shoulder implants. Accuracy of 93.9 with overall F1 score of 0.94 across 10 different implants. Saliency maps are also generated to show incorrect and correct identification of implants. [15].

Paul et al. used 482 x-ray images of implants with shoulder, reverse shoulder prosthesis and 5 different total shoulder implant models. ResNet supported Deep convolutional neural network (DCNN) based binary classification was used to detect the presence of implants and helped in differentiating 5 total shoulder implant models. The model was also designed to differentiate 5 implant types [16].

3   Dataset Description

We used seven distinct total shoulder replacement models in the current dataset. The database was obtained from an open-source machine learning database maintained by the University of California, Irvine (UCI) and from an independent surgeon. The database includes X-ray images of Anterior-posterior (AP) and Lateral (LAT) views of total shoulder implants. LAT images were discarded due to a low count. For this research, only anterior-posterior (AP) X-ray images were included.

The database was cleared of low-quality images. The data which had been incorrectly labeled when received was labeled accurately with the help of Orthopedic surgeons. The make and model included in this study were listed in the table below, along with the image count. The dataset is fully anonymous. Patient details were not present either directly or indirectly in the dataset. The X-ray images were labeled only for their make and model (Table 1 and Fig. 1).

Table 1. Dataset of Seven Different Make And Models.

Make	Model	Images in AP
Biomet	Biomodular	26
Smith and Nephew	Cofiled II	10
Depuy	Global Advantage	20
Depuy	Global	22
Depuy	Global Fracture	15
Depuy	HRP	37
Zimmer	Bigliani	34

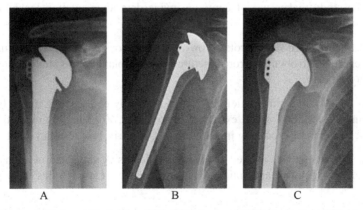

A   B   C

Fig. 1. A-C shows Bimodular, Global Fracture and HRP implants in AP view.

4 Methods and Methodology

4.1 Preprocessing

To adjust the raw data into a format which was acceptable to the network, data pre-processing was the initial required step. The initial preprocessing method was to resize the images into 224 × 224 which is standard Imagenet size. The images were then converted to gray scale to ensure image standardization before entering the images into deep learning networks.

4.2 Training Set

Seventy percent (70%) of the total images taken in random were used for training the models. The model's ability to perform well with hypothetical test images depended on the training process and the level of training quality. Before training began, the total number of images were increased utilizing a multitude of augmentation techniques which were described below.

4.3 Testing Set

The testing set consisted of thirty percent (30%) of the total collection of images that were un-augmented and unseen. The effectiveness of the model's classification performance was evaluated using these images, which were used to evaluate the training quality.

4.4 Augmentation

The process of adding minor changes to existing images in order to produce new images from an existing dataset is known as data augmentation. Data augmentation techniques were only used on the training images in this study; they were not used on the testing images. The study makes use of common image data augmentation techniques like flipping, rotation, and zooming.

Flipping. Images were simply rotated on a horizontal or vertical axis when using the flipping technique. To obtain mirrored pictures of the implant for this study, images were horizontally rotated.

Zooming. The zoom augmentation technique arbitrarily enlarged the image by either zooming in or by adding a few extra pixels around it. Zooming augmentation were carried out by randomly zooming in and out and comparing the results visually.

Rotation. The image was rotated at various angles using the rotation augmentation technique. Both clockwise and anticlockwise rotations were used in this study [17].

4.5 Deep Learning Methods

Convolutional Neural Networks (CNN) based deep learning methods were used for classification. CNN's were similar to Artificial Neural Networks (ANN). The use of CNNs for pattern identification clearly distinguishes them from ANNs. There were four types of layers in CNNs: (1) input layer, (2) convolutional layer, (3) pooling layer, and (4) fully connected layer. The convolutional layer conveys the neuron's output, the pooling layer performs the task of simplification by down sampling, and the fully connected layer receives scores from the activations which were used for classification. The input layer contains the pixel values of the image that is provided as input [18].

Transfer learning (TL) with CNN is recommended because it makes use of prior knowledge from a task that has already been performed to perform better on the new task. Like humans, TL operates by drawing on the prior knowledge it has acquired through training. Huge numbers of real-world photos from the Imagenet collection were used for this training. For the classification of medical pictures using transfer learning, CNN makes use of its understanding of these trained features. [19].

In order to classify shoulder implants with both make and model this research employed a variety of transfer learning techniques using pre-trained models like InceptionV3 [20], VGG16 [21], Xception [22] and InceptionResnetV2 [23].

4.6 Proposed Approach

To help in precise classification of implants as well as to understand the ability of deep learning in classifying across various categories of shoulder implants, we divided the problem of identification into 3 different approaches.

The implant models were separated for (A) Depuy Manufacturer that consists of four (4) total shoulder implant models (B) Non-Depuy Manufacturer that had three (3) different total shoulder implant models and (C) combination of both Depuy and Non Depuy Manufacturers with a total of seven (7) different total shoulder implant models.

Deep Learning algorithms were devised individually to identify Depuy, Non Depuy models. After this, joint classification of both Depuy and Non-Depuy models were also performed.

Depuy Implant Identification. The 4 Depuy implant models (Global, Global Advantage, Global Fracture, HRP) were identified correctly using InceptionResnetV2 deep learning model. The Optimizer used here was 'SGD' [24] with learning rate of 0.0001 as it outperforms Adam [25] and Adagrad Optimizer [26].

Non Depuy Implant Identification. The 3 Non-Depuy implant models (Biomodular, Cofield II and Biglani) were identified correctly using the InceptionResnetV2 deep learning model. Model is tuned for its various hyperparameters with different epochs combination to achieve the best classification that distinguishes 3 Non-Depuy models precisely. The Optimizer used here is 'Adagrad' with a learning rate of 0.0001. Adagrad performs better than SGD and Adam Optimizer.

All 7 Classes Implant Identification. Both Depuy make implants and Non-Depuy make implants together were identified by a single algorithm. The use of InceptionRes-netV2 deep learning model continues to show superiority in classification of all seven classes of total shoulder replacement.

After various fine tuning the best performing optimizer is 'SGD' with learning rate of 0.001 compared to Adam and Adagrad optimizer which were also equally tuned.

Proposed Deep Learning Model. Inception-ResNet-v2 is a convolutional neural network (CNN) architecture for image recognition and classification tasks. It was first introduced in 2016 by researchers from Google, and it is an extension of two popular CNN architectures, the Inception and ResNet networks.

The Inception architecture is known for its use of multiple parallel convolutional layers with different filter sizes, allowing the network to capture features of various scales. The ResNet architecture, on the other hand, introduced the idea of skip connections to improve the training of deep networks. Inception-ResNet-v2 combines these two ideas by using inception modules with residual connections. It also includes additional architectural improvements such as using batch normalization and a modified version of the "bottleneck" structure from the original ResNet.

The network consists of multiple blocks, each containing a combination of convolutional layers, pooling layers, and inception modules with residual connections. The final layer of the network is a fully connected layer with Softmax activation, which produces the classification probabilities for the input image.

The Inception-ResNet-v2 network has multiple layers including stem module, Inception-ResNet-A blocks, Inception-ResNet-B blocks, Inception-ResNet-C blocks, and a final global average pooling layer followed by a fully connected layer for classification. The stem is the initial set of layers in the Inception-ResNet-v2 architecture that process the input image before passing it through the rest of the network. It is designed to perform some basic feature extraction and reduction of the spatial resolution of the input image. The reduction layers are set of blocks which are used to reduce the spatial dimensions of feature maps.

Figure 2 shows the common basic block diagram of IncetionResnetV2.

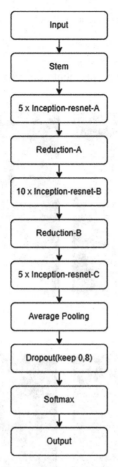

Fig. 2. Common basic block diagram for InceptionResnetV1 and InceptionResnetV2

4.7   Performance Metrics

All the results were evaluated using Confusion Matrix Accuracy, Precision, F1 Score, Recall, and the Area Under Curve (AUC) [27]. These values were usually obtained from a confusion matrix which has samples of images after classification that have been categorized as False Negative (FN), True Negative (TN) and False Positive (FP) and True Positive (TP) respectively [28].

5   Results and Discussion

5.1   Data Augmentation

The Fig. 3 and Fig. 4 shows the data augmentation applied on various implant models.

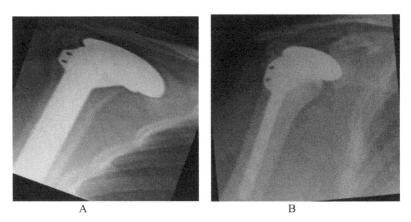

A   B

Fig. 3. A-B shows Shows Biglani and Cofield II rotated at various degrees.

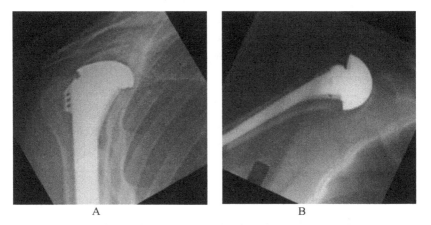

A   B

Fig. 4. A-B shows Shows Global and Global Advantage rotated at various degrees.

5.2   Deep Learning Results

Table 2 shows the best obtained results across various deep learning models in classification of Depuy's implants models.

Table 3 shows best obtained performance metrics such as Precision, recall, F1 Score and AUC across various deep learning models in classification of Depuy's implants models.

Table 2. Best obtained results for Depuy Implant Identification

Model	Epochs	Training Loss	Training Accuracy (%)	Testing Loss	Testing Accuracy (%)
VGG 16	15	0.0149	99.94	1.1223	68.52
Xception	30	0.0306	99.90	0.4757	81.48
InceptionV3	20	0.0236	99.94	0.4543	88.89
Inception ResnetV2	40	0.0283	99.94	0.3318	88.89

Table 3. Best obtained performance metrics for Depuy Implant Identification

Model	Epochs	Precision	Recall	F1 Score	AUC
VGG 16	15	0.7066	0.6851	0.6838	0.8834
Xception	30	0.8196	0.8148	0.8076	0.9644
InceptionV3	20	0.8921	0.8888	0.8869	0.9596
Inception ResnetV2	40	0.9169	0.8888	0.8866	0.9853

It is clear from Table 2 and Table 3 that Deep learning model InceptionResnetV2 performs better than other tested models and achieves an accuracy of 88.89 and also displays better performance metrics in Depuy implants classification. The InceptionResnetv2 model performs better than InceptionV3 in terms of testing loss, precision and AUC.

Table 4 shows the best obtained results across various deep learning models in classification of Non-Depuy's implants models.

Table 4. Best obtained results for Non-Depuy's implants identification

Model	Epochs	Training Loss	Training Accuracy (%)	Testing Loss	Testing Accuracy (%)
VGG 16	20	0.1127	99.17	0.4818	82.50
Xception	30	0.0245	99.95	0.4770	85.00
InceptionV3	30	0.0206	99.88	0.6007	82.49
Inception ResnetV2	30	0.0171	99.95	0.2986	87.50

Table 5 shows best obtained performance metrics such as Precision, Recall, F1 Score and AUC across various deep learning models in classification of non-Depuy's implants models.

Values shown in Table 4 and Table 5 show Deep learning model Inception ResnetV2 efficiency in classification of Non-Depuy implants.

Table 5. Best obtained performance metrics across various deep learning models in classification of non-Depuy's implants models.

Model	Epochs	Precision	Recall	F1 Score	AUC
VGG 16	20	0.8218	0.8250	0.8164	0.9258
Xception	30	0.8777	0.8500	0.8580	0.9329
InceptionV3	30	0.8429	0.8250	0.8057	0.9201
Inception ResnetV2	30	0.8861	0.8750	0.8729	0.9777

Table 6 shows the best obtained results across various deep learning models in classification of all 7 implants models.

Table 6. Best obtained results for all 7 implants identification

Model	Epochs	Training Loss	Training Accuracy (%)	Testing Loss	Testing Accuracy (%)
VGG 16	20	0.0293	99.98	0.7588	70.21
Xception	50	0.0015	99.97	0.4091	90.42
InceptionV3	20	0.0225	99.94	0.6843	72.34
Inception ResnetV2	20	0.0025	99.99	0.2920	91.48

Table 7 shows the best obtained performance metrics such as Precision, Recall, F1 Score and AUC across various deep learning models in classification of all 7 implants models.

Table 7. Best obtained performance metrics for all 7 implants identification

Model	Epochs	Precision	Recall	F1 Score	AUC
VGG 16	20	0.7148	0.7021	0.6954	0.9562
Xception	50	0.9121	0.9042	0.9051	0.9890
InceptionV3	20	0.7444	0.7234	0.7177	0.9652
Inception ResnetV2	20	0.9180	0.9148	0.9131	0.9932

Values shown in Table 6 and Table 7 shows that Deep learning model Inception ResnetV2 higher efficiency in classification of all 7 implants.

Figure 5 shows train and test accuracy obtained for InceptionResnetV2 for all 7 Total shoulder replacements.

Figure 6 shows train and test loss obtained for InceptionResnetV2 for all 7 Total shoulder replacements.

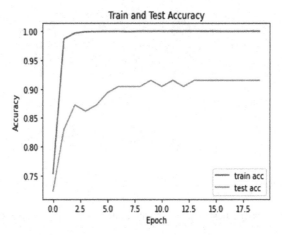

Fig. 5. Train and Test Accuracy of InceptionResnetV2 for all 7 implant classes.

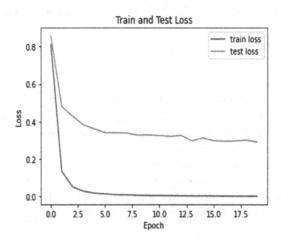

Fig. 6. Train and Test Loss of InceptionResnetV2 for all 7 implant classes.

Figure 7 shows the confusion matrix obtained for InceptionResnetV2 for all 7 Total shoulder replacements.

From the tables it is clear that the split of implants across both Depuy and Non Depuy categories provides a better picture of classification of implants. Joint classification of implants for both make and model was successfully achieved using pretrained InceptionResnetV2. Commonly used deep learning models were selected for classification and their hyperparameters such as Learning rate, Batch size, Regularizers, and Dropouts

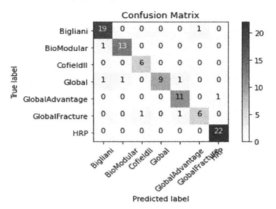

Fig. 7. Confusion matrix for best obtained IncpetionResNetV2 in classification of all 7 total shoulder replacements

were optimized for various epochs [29] to get the best result across each deep learning model.

6 Conclusion

The proposed work uses deep learning in Orthopedic implant identification. Depuy models were classified with an accuracy of 88.89% and Non- Depuy models were classified with an accuracy of 87.50% respectively. The proposed work also demonstrates the ability of deep learning models to observe finer details across all 7 implant classes, thereby classifying them with a dominant accuracy of 91.48%.

The proposed novel study outperforms various other studies [9, 14, 30–32] as this study identifies both make (Company/Manufacturer) and model of the implant in contrast to other existing works which predominantly focus solely on make.

References

1. Mattei, L., Mortera, S., Arrigoni, C., Castoldi, F.: Anatomic shoulder arthroplasty: an update on indications, technique, results and complication rates. Joints **3**(02), 72–77 (2015)
2. Sanchez-Sotelo, J.: Total shoulder arthroplasty. Open Orthopaed. J. **5**, 106 (2011)
3. Jensen, A.R., Tangtiphaiboontana, J., Marigi, E., Mallett, K.E., Sperling, J.W., Sanchez-Sotelo, J.: Anatomic total shoulder arthroplasty for primary glenohumeral osteoarthritis is associated with excellent outcomes and low revision rates in the elderly. J. Shoulder Elbow Surg. **30**(7), S131–S139 (2021)
4. Ravi, V., Murphy, R.J., Moverley, R., Derias, M., Phadnis, J.: Outcome and complications following revision shoulder arthroplasty: a systematic review and meta-analysis. Bone Joint Open **2**(8), 618–630 (2021)
5. Klug, A., Herrmann, E., Fischer, S., Hoffmann, R., Gramlich, Y.: Projections of primary and revision shoulder arthroplasty until 2040: facing a massive rise in fracture-related procedures. J. Clin. Med. **10**(21), 5123 (2021)

6. Gill, D.R., Page, R.S., Graves, S.E., Rainbird, S., Hatton, A.: The rate of 2nd revision for shoulder arthroplasty as analyzed by the Australian Orthopaedic Association National Joint Replacement Registry (AOANJRR). Acta Orthop. **92**(3), 258–263 (2021)
7. Longo, U.G., et al.: Shoulder replacement: an epidemiological nationwide study from 2009 to 2019. BMC Musculoskel. Disord. **23**(1), 1–12 (2022)
8. Best, M.J., Wang, K.Y., Nayar, S.K., Agarwal, A.R., McFarland, E.G., Srikumaran, U.: Epidemiology of revision shoulder arthroplasty in the United States: incidence, demographics, and projected volume from 2018 to 2030. In Seminars in Arthroplasty: JSES, vol. 33, no. 1, pp. 53–58. WB Saunders (2023)
9. Yılmaz, A.: Shoulder implant manufacturer detection by using deep learning: proposed channel selection layer. Coatings **11**(3), 346 (2021)
10. Borjali, A., Chen, A.F., Muratoglu, O.K., Morid, M.A., Varadarajan, K.M.: Detecting total hip replacement prosthesis design on plain radiographs using deep convolutional neural network. J. Orthopaed. Res.® **38**(7), 1465–1471 (2020)
11. Chung, S.W., et al.: Automated detection and classification of the proximal humerus fracture by using deep learning algorithm. Acta Orthop. **89**(4), 468–473 (2018)
12. Grauhan, N.F., et al.: Deep learning for accurately recognizing common causes of shoulder pain on radiographs. Skeletal Radiol. 1–8 (2021)
13. Dhillon, A., Verma, G.K.: Convolutional neural network: a review of models, methodologies and applications to object detection. Prog. Artif. Intell. **9**(2), 85–112 (2020)
14. Urban, G., Porhemmat, S., Stark, M., Feeley, B., Okada, K., Baldi, P.: Classifying shoulder implants in X-ray images using deep learning. Comput. Struct. Biotechnol. J. **18**, 967–972 (2020)
15. Geng, E.A., et al.: Development of a machine learning algorithm to identify total and reverse shoulder arthroplasty implants from X-ray images. J. Orthop. **35**, 74–78 (2023)
16. Yi, P.H., et al.: Automated detection and classification of shoulder arthroplasty models using deep learning. Skeletal Radiol. **49**, 1623–1632 (2020)
17. Khalifa, N.E., Loey, M., Mirjalili, S.: A comprehensive survey of recent trends in deep learning for digital images augmentation. Artif. Intell. Rev. **55**, 1–27 (2022)
18. O'Shea, K., Nash, R.: An introduction to convolutional neural networks. arXiv preprint arXiv: 1511.08458 (2015)
19. Kim, H.E., Cosa-Linan, A., Santhanam, N., Jannesari, M., Maros, M.E., Ganslandt, T.: Transfer learning for medical image classification: a literature review. BMC Med. Imaging **22**(1), 69 (2022)
20. Szegedy, C., Vanhoucke, V., Ioffe, S., Shlens, J., Wojna, Z.: Rethinking the inception architecture for computer vision. In: Proceedings of the IEEE Conference on Computer Vision and Pattern Recognition, pp. 2818–2826 (2016)
21. Simonyan, K., Zisserman, A.: Very deep convolutional networks for large-scale image recognition. arXiv preprint arXiv:1409.1556 (2014)
22. Chollet, F.: Xception: deep learning with depthwise separable convolutions. In: Proceedings of the IEEE Conference on Computer Vision and Pattern Recognition, pp. 1251–1258 (2017)
23. Szegedy, C., Ioffe, S., Vanhoucke, V., Alemi, A.: Inception-v4, inception-resnet and the impact of residual connections on learning. In: Proceedings of the AAAI Conference on Artificial Intelligence, vol. 31, no. 1 (2017)
24. Ruder, S.: An overview of gradient descent optimization algorithms. arXiv preprint arXiv: 1609.04747 (2016)
25. Kingma, D.P., Ba, J.: Adam: a method for stochastic optimization. arXiv preprint arXiv:1412. 6980 (2014)
26. Lydia, A., Francis, S.: Adagrad—an optimizer for stochastic gradient descent. Int. J. Inf. Comput. Sci. **6**(5), 566–568 (2019)

27. Vakili, M., Ghamsari, M., Rezaei, M.: Performance analysis and comparison of machine and deep learning algorithms for IoT data classification. arXiv preprint arXiv:2001.09636 (2020)
28. Ramanathan, A., Christy Bobby, T.: Classification of corpus callosum layer in mid-saggital MRI images using machine learning techniques for autism disorder. In: Saha, S., Nagaraj, N., Tripathi, S. (eds.) MMLA 2019. CCIS, vol. 1290, pp. 78–91. Springer, Singapore (2020). https://doi.org/10.1007/978-981-33-6463-9_7
29. Yu, T., Zhu, H.: Hyper-parameter optimization: a review of algorithms and applications. arXiv preprint arXiv:2003.05689 (2020)
30. Sultan, H., Owais, M., Park, C., Mahmood, T., Haider, A., Park, K.R.: Artificial intelligence-based recognition of different types of shoulder implants in X-ray scans based on dense residual ensemble-network for personalized medicine. J. Pers. Med. 11(6), 482 (2021)
31. Zhou, M., Mo, S.: Shoulder implant x-ray manufacturer classification: exploring with vision transformer. arXiv preprint arXiv:2104.07667 (2021)
32. Karaci, A.: Detection and classification of shoulder implants from X-ray images: YOLO and pretrained convolution neural network based approach. J. Fac. Eng. Archit. Gazi Univ 37, 283–294 (2022)

Deep Facial Phenotyping with Mixup Augmentation

Jonathan Campbell[1,2,3](✉)📷, Mitchell Dawson[1]📷, Andrew Zisserman[1]📷,
Weidi Xie[1,4]📷, and Christoffer Nellåker[2,3]📷

[1] Visual Geometry Group, University of Oxford, Oxford, England
[2] Big Data Institute, University of Oxford, Oxford, England
[3] Nuffield Department of Women's & Reproductive Health, University of Oxford,
Oxford, England
jcampbell@robots.ox.ac.uk
[4] Cooperative Medianet Innovation Center, Shanghai Jiao Tong University,
Shanghai, China

Abstract. The classification of genetic disorders from face images has
the potential to assist with early diagnosis and effective treatment. A
key objective in this field is to enhance the accuracy and robustness of
deep learning models in classifying genetic disorders from face images for
disorders that are not represented in the training data. In this paper, we
propose the use of input mixup augmentation to improve few-shot classification performance on this task. Furthermore, we present a specialised
version of mixup that warps together face images using face keypoints,
and show that this improves performance further. The motivation for
using keypoint guided mixup is to align face structure and produce an
intermediate image between two disorders. We compare the performance
of our proposed method with the baseline model and demonstrate significant improvements in accuracy, with a classification accuracy of 31.3%
on the GMDB-Rare benchmark dataset. Our results show that incorporating input mixup augmentation and face keypoint-based mixup can
enhance the ability of deep learning models to identify genetic disorders
from face images, providing a promising approach for future research in
this area.

Keywords: Genetic disorders · Phenotypes · Deep learning ·
Computer vision · Few-shot · Image classification · Representation
learning

Jonathan Campbell is supported by the EPSRC Center for Doctoral Training in Health
Data Science (EP/S02428X/1). Computation used the Oxford Biomedical Research
Computing (BMRC) facility, a joint development between the Wellcome Centre for
Human Genetics and the Big Data Institute supported by Health Data Research UK
and the NIHR Oxford Biomedical Research Centre. The views expressed are those of
the author(s) and not necessarily those of the NHS, the NIHR or the Department of
Health.

G. Waiter et al. (Eds.): MIUA 2023, LNCS 14122, pp. 133–144, 2024.
https://doi.org/10.1007/978-3-031-48593-0_10

1 Introduction

Approximately 1 in 17 people have a rare disorder, and 80% of rare disorders are genetic [14]. A disease is considered rare in the European Union if it has a prevalence of less than 1:2000, and more than 6000 rare diseases have been recorded in the Orphanet rare disease database [16] as of 2020. However, a significant challenge in managing rare genetic disorders is obtaining an accurate and timely diagnosis. Due to the rarity of these disorders, patients may go undiagnosed for years, leading to delayed treatment and potential harm to their health.

Facial dysmorphology can be a strong indicator of the presence of a genetic disorder, and is present in 30–40% of genetic disorders [9]. For some disorders, clinical geneticists are able to make a confident diagnosis based on facial phenotype alone, or otherwise identify a set of disorders which are candidates for diagnosis [4]. While genetic testing can diagnose many disorders, objective phenotyping, including analysis of facial dysmorphology, remains an important feature for an accurate diagnosis.

If sufficiently developed, computer analysis of face phenotypes could be used in diagnostic pathways prior to testing. Specifically, deep learning for computer vision could identify facial dysmorphology and compare it to known features associated with genetic disorders. This would help clinicians narrow down the list of candidate disorders and make a more accurate diagnosis. Many recent studies use deep learning for computer vision to assist in diagnosis of genetic diseases, using both 2D face images [6–8,12] and 3D head scans [20].

One approach to this problem is to train a model to classify images into specific disorders. Machine learning models struggle to perform well with a large number of classes and a low number of examples for each class, as is the case with genetic disorders. Due to the long tail of the distribution of disorders in the population, it is not feasible to collect enough examples of each disorder, especially for the rarest disorders. Ideally we could build a model that represents the variation of all face phenotypes, which would allow a phenotype of a disorder not seen by the model to be meaningfully represented. This is the goal of building a Clinical Face Phenotype Space (CFPS), which is a vector space that represents the variation in phenotype such that more similar phenotypes are closer together. The process of building this feature space is referred to as representation learning. Examples from a disorder the model has not been trained on can be projected into the CFPS, and classified based on their proximity to labelled examples. However, it is a challenge to build a CFPS that represents the wide range of face phenotypes using the limited data available.

Contributions of this Paper. This paper investigates innovative methods to improve generalisation of the CFPS to outside of the training set. Our hypothesis is that a smooth feature space is essential to represent examples from classes with a small number of examples. To achieve this, we demonstrate the effectiveness of input mixup as an augmentation method. Furthermore, we propose keypoint guided mixup, which incorporates domain-specific knowledge to warp face images together for a realistic depiction of the intermediate between two faces. Our method outperforms the baseline fine-tuned model by 4.8% on the GMDB-Rare subset, indicating that it has significant potential to improve CFPS models. Our findings have practical implications for improving the diagnosis and treatment of a wide range of disorders.

1.1 Digital Facial Phenotyping

Ferry et al. [7] first introduce the concept of a CFPS. They also develop a model to map phenotype information from face images to the CFPS.

DeepGestalt [8] is a convolutional neural network (CNN) model trained on 502 images of 92 disorders. However, it is restricted to only classifying disorders represented in the training set. Hsieh et al. present an extension of this method, called GestaltMatcher [12] that is able to make predictions for disorders not seen by the model during training, by creating a CFPS. In the GestaltMatcher model, feature vectors are generated in the CFPS for both a *gallery* set, containing examples used to support prediction, and a *query* set, containing examples used for evaluation. The query set images are assigned the label of the closest gallery set image in feature space. On a query set of 816 unseen disorders, the method had a top-1 accuracy of 13.66%. This method is less restrictive in supported disorders and therefore more useful in practice, since in a clinical setting it is very likely some patients will have a disorder that the model hasn't been trained to predict. Recently, Hustinx et al. [13] demonstrate that the GestaltMatcher approach can be improved further by including test-time augmentation and model ensembles, and training with the ArcFace loss [5]. They also demonstrate that the performance of the model is strongly influenced by the dataset used for pre-training. The state of art is an ensemble of three models, an iResNet-100 and iResNet50 both fine-tuned on disorder data, along with a face recognition model, trained on the Glint360K dataset [1].

1.2 Few-Shot Image Classification

is the task of classifying images into predefined categories given only a few labeled examples of each category. There are many possible approaches to few shot classification [3]. Meta-learning [18] is a popular approach, which aims to train a model on multiple tasks which are similar to the target task, so that the model is well initialised to learn the new task quickly with a small number of examples. Hospedales et al. [11] presents a survey of recent meta-learning work.

2 Data

We use two different datasets in our experiments, an in-house data that we have collected, and the GestaltMatcher DB (GMDB) dataset. When training the models we combine the two datasets. The two datasets are disjoint in terms of images.

2.1   Our Dataset

Our in-house dataset contains contains 10,495 image with 322 distinct syndromes. Approximately half of the images were sourced from annotated figures in journal articles on PubMed and half were collected through a number of clinical collaborations. Due to the personal nature of this data, which was collected with rigorous ethical approval by local ethics boards, we are unable to make this data publicly available. While there are no annotations for age or ancestry for most of the images, however the majority of the data is from clinics in Europe and therefore a bias towards European ancestry is likely. Many of the images contain artefacts such as blocked out eyes for anonymisation and parts of other figures or captions from the articles they are cropped from. Furthermore some images are of poor quality. It appears there may be an issue of similarity between photos of patients of the same disorder, which is spurious due to having the same source, rather than due to similarity based on face structure.

Data Split. The dataset is split into two groups based on the number of images representing each class. For syndromes with fewer than 25 images, the syndrome is placed in the *novel set* group, and it is not used during training. For syndromes with 25 or more images, they are in the *base set* group and are classes seen by the model during training. The base set classes are further divided into a training and test set. The split is calculated on a per-class basis. For each disorder in the base set, 80% are part of the training set and 20% are part of the held-out test set. For the novel set, 80% of the images for each class are used as the support set, and 20% are used as the query set. The support set are the examples used to guide the classification of the examples in the query set. The novel classes are used for few shot learning.

2.2   GDMB

GMDB [12] is a database of face images of patients with genetic disorders, available to researchers who are approved after an application process, and it is updated frequently. We used GMDB v1.0.3, which contains 7,459 images of 449 disorders. The base set is *GMDB-Frequent* and the novel set is *GMDB-Rare*, with all disorders that have 6 or fewer patients being assigned to GMDB-Rare. GMDB-Frequent contains 6,354 images of 204 disorders, and GMDB-Rare contains 1,105 images of 245 disorders. Images are sourced from publications and from clinics. The GMDB-Rare set is evaluated with 10-fold cross validation, where the support and query sets are split in 10 different permutations, and the results are averaged across the 10 folds. This dataset is the focus of our experiments, however we supplement the training data with the data that we have collected (Table 1).

Table 1. The splits for each data set used in our experiments. The GMDB-Rare set is evaluated using 10-fold cross-validation, so there is not a single held-out test set

Dataset	Training/Support	Test/Query	Combined (10-fold CV)
Combined base set	7296	885	N/A
Our novel set	2023	332	N/A
GMDB-Rare	N/A	N/A	1105

2.3 Pre-training Datasets

We use the VGGFace2 dataset [2] pre-train the models. The dataset contains 3.31 million images of faces of 9,131 individuals, with varying pose, age and ethnicity. The training task is to perform identity recognition among the different individuals.

2.4 Image Pre-processing

All images are preprocessed before the model forward pass. The preprocessing starts with normalised face alignment using facial feature prediction, using the model provided in the dlib library [15]. This is to ensure that the face of each subject is in approximately the same location and at the same scale and rotation in each image. The images are then resized to 256 pixel wide squares.

3 Methods

3.1 Backbone Model

We use the ResNet-50 model architecture [10], and use the VGGFace2 face recognition dataset for pre-training, given the similarity of the data domain to this task. This model can be employed in two ways: (1) by using transfer learning to classify based on the feature vectors while keeping the weights fixed, so the backbone model is not trained, and (2) by fine-tuning the model, which involves training all parameters of the model at once to further enhance its performance. This is sensitive to the learning rate, therefore a low learning rate is used to avoid overfitting.

3.2 Classification

Classification is done differently on the base and novel sets. The base set classifier can be trained either together with the visual encoder or with a fixed visual encoder (transfer learning). We train this with the softmax cross-entropy loss. For novel set classification, we use a k nearest neighbors classifier with $k = 1$. The distance metric used in all experiments is the cosine distance. For a given feature space, a query example is classified based on the closest labelled example from a support set.

3.3 Mixup-Based Methods

We explore three mixup augmentation methods, input mixup, cross-domain mixup and keypoint guided mixup.

Input Mixup. Input mixup [21] is a data augmentation technique whereby input examples are linearly interpolated, and the corresponding labels are also interpolated by the same amount. A factor $\lambda \in [0, 1]$ is selected from a uniform probability distribution for each batch. λ is used to combine the inputs and labels with this formula:

$$\tilde{x} = \lambda x_i + (1 - \lambda)x_j, \quad \tilde{y} = \lambda y_i + (1 - \lambda)y_j \tag{1}$$

where x_i, x_j is the input image pair and y_i, y_j are the labels.

Cross-Domain Mixup. As an alternative to the standard form of input mixup, we also use an alternative form of mixup we call "cross-domain mixup" With this method, each image in the batch is interpolated in image space with an image which is out of domain. We use the ImageNet dataset to sample these out of domain images. The label of the image is unchanged and corresponds to the genetic disorder of the patient. This method is intended to act as regularisation, to introduce variation in the image appearance without attempting to guide the model towards learning intermediate representations, as is the case with input mixup. Previous work has explored forms of cross-domain mixup in many different applications [17,22], however we are using with with no regard for performance in the auxiliary domain.

Keypoint Guided Mixup. We use an extension of Mixup augmentation using warping of images based on face keypoint detection.

First, 68 keypoints are identified on the face using the dlib face keypoint detector, we then compute the average location of all keypoints on our dataset, and use the Delaunay triangulation algorithim to generate a set of non-overlapping triangles for the 68 keypoints. Each set of face keypoints can be mapped to these triangles.

Similar to input mixup, we use λ to weight the average of two sets of face keypoints for each image pair. This weighted average is the target that each image will we warped to. Using the triangles mapped from the average keypoint locations, we warp the triangles using the inverse affine transformation, using cubic interpolation sampling. Finally, the two resulting images are interpolated in image space using the same λ.

The result is an image where the face structure of the image pair is aligned. The motivation to use this form of augmentation is that the resulting images are more realistic than those produced by input mixup, but still allow for visually interpolating between disorders. Similar approaches have been demonstrated to work well on deep learning tasks with face data, for example in emotion recognition [19]. The result is demonstrated in Fig. 1

4   Experiments

We train on the combined training data of our training set and GMDB-Frequent training data. When testing the accuracy on base classes we use a combined test set of both our data and GMDB-Frequent data, described in Sect. 2. For the novel classes we evaluate performance separately on our novel set, shown in table 3, and then on the GMDB-Rare set, shown in table 4. In this area availability of data is a key issue, therefore we report performance on a test set that is available for others to benchmark on, while still making use of as much data as possible for training.

Fig. 1. Image mixup: From left to right: Image A, Image B, Input Mixup, Keypoint Guided Mixup

4.1   Training Details

Models were trained on a single NVIDIA Tesla K80 GPU. We train for a maximum of 100 epochs, with early stopping based on the maximum accuracy of the validation set, with a patience of 20 epochs. We use the AMSGrad optimiser and softmax cross-entropy loss. For the first 10 epochs only the final layer is trained with a learning rate (LR) of 1×10^{-3}, and then the model is trained end-to-end until 50 epochs with a LR of 1×10^{-4}, and then until 100 epochs with a LR of 1×10^{-5}. The batch size is 64.

We use a standard data-augmentation pipeline which includes horizontal flipping with a probability of 0.5, followed by Gaussian blur, Gaussian noise, random brightness and contrast, JPEG compression artefacts, and greyscale, each with independent probability of 0.2. These augmentations were selected since they approximate the variation in images that is expected in the dataset.

4.2   Novel Set Evaluation

In our evaluation, we report top-1 accuracy, top-10 accuracy, and the mean average precision (mAP). The top-k accuracy is the percentage of samples for which a point with the same label as the query point is among the k closest points in the gallery. This metric is useful for distance-based classifiers and image retrieval tasks. In a clinical setting, a clinician may want to see a ranking of possible syndromes that could be a match.

5   Results

In this section we present the results of our experiments with each method. We compare transfer learning and fine-tuning, and the three mixup methods described in Sect. 3. We evaluate first on the base set, then on our novel set, then on the GMDB-Rare novel set.

Table 2. Performance of each method on the combined base set reported as classification accuracy. Aside from transfer learning, all models are trained using the combined training set and evaluated on the combined test set of base classes

Model	Base set accuracy
Transfer learning	39.0%
Fine-tuning	64.3%
Input mixup	64.3%
Cross-domain mixup	**65.1%**
Guided mixup	62.7%

5.1   Results on the Combined Base Set

Table 2 contains the base set accuracy of the 5 methods: transfer learning, fine-tuning, inout mixup, cross-domain mixup, and guided mixup.

Cross-domain mixup achieved the highest base set accuracy of 65.1%, followed by fine-tuning and input mixup with 64.3% accuracy. Transfer learning achieved the lowest base set accuracy of 39.0%. The effectiveness of cross-domain mixup is likely due to the introduction of spurious features from the ImageNet images preventing overfitting. However, input mixup within the same domain does not affect the level of overfitting, perhaps because the model is able to learn to identify the features from each class separately. Guided mixup is likely falling behind on the base set because the warping of the image is slightly destructive in terms of image quality and results in blurring or image artifacts, or fails to produce realistic looking images in some cases.

5.2   Results on Our Novel Set

Table 3 shows results on the novel set generated from our data, reported as top-1, top-10 and mAP. Considering only the top-1 accuracy, fine-tuning the pre-trained model improves accuracy on the novel set by 3.2%. Input mixup improves novel set accuracy by a further 1.7%, with accuracy of 34.3% compared to 32.6% without. Input mixup falls below the fine-tuned model in Top-10 accuracy, but is better in terms of mAP, with an mAP of 0.158 compared to 0.154 with the fine-tuned model. Guided mixup does not improve results, performing worse than the fine-tuned model by 0.6%. Interestingly, the cross-domain mixup results in

Table 3. Performance of each method reported on our novel set only. Aside from transfer learning, all models are trained using the combined training set and evaluated on query examples from our dataset, using support examples from the novel classes

Model	Novel set Top-1	Novel set Top-10	Novel set mAP
Transfer learning	29.4%	53.9%	0.132
Fine-tuning	32.6%	**58.1%**	0.154
Input mixup	**34.3%**	56.6%	**0.158**
Cross-domain mixup	24.3%	50.3%	0.127
Guided mixup	32.0%	54.8%	0.152

the best performance on the base set, with 65.1% accuracy as shown in table 2, but is by far the worst performance on novel classes in table 3, with only 24.3% accuracy. This is evidence that the regularising effect of cross-domain mixup does not generalize to novel classes.

5.3 Results on the GMDB-Rare Set

Table 4. Performance of each method on the GMDB-Rare novel set. Aside from transfer learning, all models are trained using the combined training set and evaluated on GMDB-Rare query examples, using GMDB-Rare support examples, in 10-fold cross valdiation

Model	GMDB-rare Top-1	GMDB-Rare Top-10	GMDB-Rare mAP
Transfer learning	24.7%	55.0%	0.256
Fine-tuning	26.5%	60.8%	0.284
Input mixup	29.5%	60.7%	0.292
Cross-domain mixup	29.5%	56.6%	0.299
Guided mixup	**31.3%**	**62.5%**	**0.311**

Table 4 summarises the results of the different methods on the GMDB-Rare test set, with the same metrics as in table 3. These models are trained with the same base set classes as in table 3 on our test set. When evaluating the models on the novel disorders provided in GMDB-Rare, guided mixup is the best performing method, outperforming the next best performing method,input mixup, by 1.8% in top-1 accuracy. This is evidence for the effectiveness of adding face warping to standard input mixup, since the guided mixup model has a clear advantage in performance. The GMDB-Rare set results are likely more robust than those from our own test set in table 3, due to the 10-fold cross-validation split.

5.4   Comparison to Other Methods

When compared to previous work we are able to provide a significant improvement when applying keypoint guided mixup given the same dataset for pretraining. Fine-tuning with guided mixup on a VGGFace2 pretrained model achieves 31.3% top-1 accuracy on the GMDB-Rare test set, whereas the transfer learning VGGFace2 model from Hustinx et al. has accuracy of 20.3%. Our result is comparable to their 33.3% best Top-1 result when using the significantly larger (>17 million images) Glint360k dataset for pretraining' (Table 5).

Table 5. Our best performing model (R-50), pre-trained on the VGGFace2 dataset and fine-tuned with guided mixup, compared to GestaltMatcher (GM) from Hsieh et al. 2022, pre-trained on the CASIA-WebFace dataset, and the iResnet-50 model (iR-50) pre-trained on VGGFace2 from Hustinx et al. 2023

Method	GMDB-Rare Top-1
Hsieh et al. 2022 (GM + CASIA, fine-tuned)	19.3%
Hustinx et al. 2023 (iR-50 + VGGFace2, transfer learning)	20.3%
Ours (R-50 + VGGFace2 + guided mixup, fine-tuned)	31.3%

6   Discussion

The results show that using mixup augmentation in the fine-tuning of models for prediction of genetic disorders has a positive effect on performance. On our dataset, input mixup gave the highest perfomance when evaluating on unseen disorders, and our cross-domain version of mixup gave the highest performance on disorders in the base set that were previously seen by the model during training. When we apply the models to the GMDB-rare dataset, our keypoint guided mixup method improves performance over standard input mixup by 1.8%, and outperforms the standard fine-tuned model by 4.8%. This shows that the ability of the CFPS to represent unseen disorders has been improved with the addition of our keypoint guided mixup.

Data augmentation can represent the variation in real-world data, but also reduce overfitting by adding noise to input data. Despite advanced methods like Mixup not being representative of real data, they can still improve model performance. Mixup augmentation goes beyond standard data augmentation, as it actively affects the feature space that is created during training. For a model to be able to accurately represent an example of a class it has not been trained on, it needs to be able to embed this in an area of feature space that is not covered by other classes already. It is difficult to directly train a model to have this property. Mixup, and our extension of that in the form of keypoint guided mixup, is a step in this direction, as we are training the model to understand what the areas between the cluster centres of each class might look like.

A limitation of this work is that the pre-training dataset restricts performance. Hustinx et al. report that accuracy of a transfer learning model on the GMDB-Rare set improves from 20.3% when pre-training with VGGFace2 to 33.0% when pre-training with Glint360K. The difference is due to the scale of the training dataset. VGGFace2 contains 3.3M images of over 9K individuals, whereas Glint360K contains 17.1M images of over 360K individuals, giving much more variation. This change seems to have a significant effect on the perfomance of a single model, in the absence of test-time augmentation or model ensembles. Performance improves further when fine-tuning a larger iResnet-100 model with the ArcFace loss, reaching 33.3% accuracy. Therefore future work should investigate the use of Mixup and Guided Mixup in combination with the best-performing pre-trained model.

The ability of deep learning models to generalise to unseen categories is important not just in this area of face phenotyping, but in the wider field of medical image analysis. In most medical image settings, there is not a bounded set of categories that the model will encounter, so it is vital that methods are explored which promote learning of more general and transferable concepts and features. This is also important when transferring models between different data sources or applying them in different centres, as this will be necessary for widespread adoption of deep learning in healthcare.

References

1. An, X., et al.: Partial FC: training 10 million identities on a single machine. In: Proceedings of the IEEE/CVF International Conference on Computer Vision (ICCV) Workshops, pp. 1445–1449 (2021)
2. Cao, Q., Shen, L., Xie, W., Parkhi, O.M., Zisserman, A.: VGGFace2: a dataset for recognising faces across pose and age. In: 2018 13th IEEE International Conference on Automatic Face & Gesture Recognition (FG 2018), pp. 67–74 (2018). https://doi.org/10.1109/FG.2018.00020
3. Chen, W.Y., Liu, Y.C., Kira, Z., Wang, Y.C., Huang, J.B.: A closer look at few-shot classification. In: International Conference on Learning Representations (2019)
4. Cianci, P., Selicorni, A.: "Gestalt diagnosis" for children with suspected genetic syndromes. Ital. J. Pediatr. 41(Suppl 2), A16 (2015). https://doi.org/10.1186/1824-7288-41-S2-A16, https://www.ncbi.nlm.nih.gov/pmc/articles/PMC4707582/
5. Deng, J., Guo, J., Xue, N., Zafeiriou, S.: ArcFace: additive angular margin loss for deep face recognition. In: Proceedings of the IEEE/CVF Conference on Computer Vision and Pattern Recognition (CVPR) (2019)
6. van der Donk, R., et al.: Next-generation phenotyping using computer vision algorithms in rare genomic neurodevelopmental disorders. Genet. Med. 21(8), 1719–1725 (2019). https://doi.org/10.1038/s41436-018-0404-y, https://www.sciencedirect.com/science/article/pii/S1098360021016129
7. Ferry, Q., et al.: Diagnostically relevant facial gestalt information from ordinary photos. eLife 3, e02020 (2014). https://doi.org/10.7554/eLife.02020, publisher: eLife Sciences Publications Ltd
8. Gurovich, Y., et al.: Identifying facial phenotypes of genetic disorders using deep learning. Nat. Med. 25(1), 60–64 (2019)

9. Hart, T., Hart, P.: Genetic studies of craniofacial anomalies: clinical implications and applications. Orthod. Craniofac. Res. **12**(3), 212–220 (2009). https://doi.org/10.1111/j.1601-6343.2009.01455.x

10. He, K., Zhang, X., Ren, S., Sun, J.: Deep residual learning for image recognition. In: Proceedings of the IEEE Conference on Computer Vision and Pattern Recognition, pp. 770–778 (2016)

11. Hospedales, T., Antoniou, A., Micaelli, P., Storkey, A.: Meta-learning in neural networks: a survey. IEEE Trans. Pattern Anal. Mach. Intell. **44**(9), 5149–5169 (2022). https://doi.org/10.1109/TPAMI.2021.3079209, conference Name: IEEE Transactions on Pattern Analysis and Machine Intelligence

12. Hsieh, T.C., et al.: GestaltMatcher facilitates rare disease matching using facial phenotype descriptors. Nature Genet. **54**(3), 349–357 (2022). publisher: Nature Publishing Group US New York

13. Hustinx, A., et al.: Improving deep facial phenotyping for ultra-rare disorder verification using model ensembles. In: 2023 IEEE/CVF Winter Conference on Applications of Computer Vision (WACV), pp. 5007–5017. IEEE, Waikoloa, HI, USA (2023). https://doi.org/10.1109/WACV56688.2023.00499, https://ieeexplore.ieee.org/document/10030218/

14. Jackson, M., Marks, L., May, G.H., Wilson, J.: The genetic basis of disease. Essays Biochem. **62**(5), 643–723 (2018). https://doi.org/10.1042/EBC20170053, https://www.ncbi.nlm.nih.gov/pmc/articles/PMC6279436/

15. King, D.E.: Dlib-ml: a machine learning toolkit. J. Mach. Learn. Res. **10**, 1755–1758 (2009)

16. von der Lippe, C., Diesen, P.S., Feragen, K.B.: Living with a rare disorder: a systematic review of the qualitative literature. Mol. Genet. Genomic Med. **5**(6), 758–773 (2017). https://doi.org/10.1002/mgg3.315

17. Luo, C., Song, C., Zhang, Z.: Generalizing person re-identification by camera-aware invariance learning and cross-domain mixup. In: Vedaldi, A., Bischof, H., Brox, T., Frahm, J.-M. (eds.) ECCV 2020. LNCS, vol. 12360, pp. 224–241. Springer, Cham (2020). https://doi.org/10.1007/978-3-030-58555-6_14

18. Thrun, S., Pratt, L. (eds.): Learning to Learn. Kluwer Academic Publishers, USA (1998)

19. Vonikakis, V., Dexter, N.Y.R., Winkler, S.: MorphSet: augmenting categorical emotion datasets with dimensional affect labels using face morphing. In: 2021 IEEE International Conference on Image Processing (ICIP), pp. 2713–2717 (2021). https://doi.org/10.1109/ICIP42928.2021.9506566, iSSN: 2381-8549

20. White, J.D., et al.: Insights into the genetic architecture of the human face. Nat. Genet. **53**(1), 45–53 (2021). https://doi.org/10.1038/s41588-020-00741-7, http://www.nature.com/articles/s41588-020-00741-7

21. Zhang, H., Cisse, M., Dauphin, Y.N., Lopez-Paz, D.: Mixup: beyond empirical risk minimization. In: International Conference on Learning Representations (2018). https://openreview.net/forum?id=r1Ddp1-Rb

22. Zhuo, L., Fu, Y., Chen, J., Cao, Y., Jiang, Y.G.: TGDM: target guided dynamic mixup for cross-domain few-shot learning. In: Proceedings of the 30th ACM International Conference on Multimedia, pp. 6368–6376 (2022)

Image Classification

Context Matters: Cross-Domain Cell Detection in Histopathology Images via Contextual Regularization

Ziqi Wen, Qingzhong Wang$^{(\boxtimes)}$, Jiang Bian, Xuhong Li, Yi Liu, and Haoyi Xiong

Baidu, Inc., Beijing, China
ziqiwen_cumt@163.com, qingzwang@outlook.com,
{bianjiang03,lixuhong,liuyi22}@baidu.com, haoyi.xiong.fr@ieee.org

Abstract. Deep learning-based cell detectors have shown promise in automating cell detection in histopathology images, which can aid in cancer diagnosis and prognosis. Nevertheless, the color variation in stain appearance across histopathology images obtained from diverse locations might degrade the accuracy of cell detection. The limitations of a basic cell detector on a dissimilar target dataset are attributed to its emphasis on domain-specific features while overlooking domain-invariant features. Thus, a cell detector that is trained on a particular source dataset may not perform optimally on a different target dataset. In this work, we propose a domain generalization method *contextual regularization* (CR) for cell detection in histopathology images, which is derived from the basic object detector and focuses on domain-invariant features to enhance detection performance across varying datasets. Specifically, we involves a reconstruction task that involves masking the high-level semantic features either stochastically or adaptively. Then, a transformer-based reconstruction head is designed to recover the original features based on partial observations. Additionally, CR can be seamlessly integrated with various deep learning-based cell detectors without further modifications, and it does not request additional cost in the inference time. The proposed method was validated on a publicly available dataset that comprises histopathology images acquired at different sites, and the results show that our method can effectively improve the generalization of cell detectors to unseen domains.

Keywords: Cell detection · Domain Generalization · Contextual regularization

1 Introduction

Automated detection of cells using deep neural networks in histopathology images holds potential for assisting pathologists with disease diagnosis, grading, and quantification. This approach has demonstrated high accuracy and

---

Z. Wen and Q. Wang—Equal contribution.

© The Author(s), under exclusive license to Springer Nature Switzerland AG 2024
G. Waiter et al. (Eds.): MIUA 2023, LNCS 14122, pp. 147–156, 2024.
https://doi.org/10.1007/978-3-031-48593-0_11

effectiveness in previous studies [6]. Deep learning-based techniques have been extensively investigated for automated cell detection and medical image analysis, showing promising results [9,11,21]. In particular, tailored deep networks have been employed to locate cell nuclei in some earlier works [4,18]. More recent works have utilized or adapted advanced object detectors like Faster R-CNN [14] to detect cells [15,19]. These detectors provide bounding boxes to precisely locate the cells of interest.

Deep learning-based cell detectors face a challenge when working with histopathology images acquired from various sites. This challenge arises due to differences in staining techniques, resulting in variations in the appearance of these images. Consequently, a detection model trained on one source dataset may not generalize well to a different target dataset [17]. To ensure the successful deployment of deep learning-based cell detectors in real-world scenarios, it is crucial to address this domain generalization issue.

In this work, we focus on modern object detectors [10,14], which the input images are subsequently processed by a feature extractor and an object detection head. Compared to the original input image, the high-level semantic information extracted by the feature extractor contains more domain-invariant features, which can facilitate better generalization across different domains. In other words, these features are less sensitive to domain-specific variations and can be more easily applied to new, unseen domains. To this end, the emphasis on high-level semantic information has the potential to improve the model's ability to generalize across different domains.

To further investigate the problem of domain generalization in cell detection for histopathology images, we propose *contextual regularization* (CR) to enhance the basic object detector by emphasizing the importance of domain-invariant features during the model training process. As in [7], we also adapt a masking strategy to remove certain elements and then reconstruct them. However, unlike [7], we do not directly mask patches of the input image and reconstruct the missing pixels in a self-supervised learning framework. Instead, we propose to mask high-level semantic features during the supervised training of object detection to emphasize the domain-invariant features. Specifically, We propose two masking strategies to create a reconstruction task for context regularization: the stochastic masking approach, which is straightforward and easy to implement, and the adaptive masking approach, which is both trainable and efficient. Depending on how the mask is applied, *contextual regularization* (CR) can be divided into *stochastic contextual regularization* (SCR) and *adaptive contextual regularization* (ACR). Then, a transformer-based reconstruction head is designed that reconstructs the original features based on partial observations. The masking of high-level semantic features creates a challenging reconstruction task that cannot be easily solved by extrapolating from visible neighboring features, thus enable the model to learn more domain-invariant features and improve the domain generalization for automated cell detection in histopathology images. Moreover, our proposed method is model-agnostic which can be applied to different deep learning-based cell detectors without requiring any modifications or retraining. Additionally, our approach is only incorporated in the training process and does not increase the inference time. To evaluate our method, experiments were performed for a mitosis detection tasks [1], where CR was integrated with the widely used detection framework Faster R-CNN [14] and improve the detection accuracy.

2 Methods

2.1 Problem Formulation

The primary objective of this work is to improve the generalizability of cell detection in histopathology images by developing a model trained on a source domain and testing it on an unseen target domain. We begin by collecting histopathology images annotated with cells of interest from a source dataset. Using these images, we train a deep learning-based cell detection model that can be utilized for automated cell detection in unlabeled test histopathology images. However, in practice, it is often necessary to perform cell detection on a set of histopathology images from a different target dataset, acquired at a different site. The histopathology images obtained from different sites may exhibit variations in staining color, resulting in differences in image appearance between the source and target datasets, known as domain shift. Therefore, to improve the generalization of the model to the target domain, we need to address this domain shift issue.

In this work, our focus is on modern object detectors [10,14], which consist of a feature extractor and an object detection head. The feature extractor obtains high-level semantic features $x \in \mathbb{R}^{H \times W \times C}$ from the source training data, where (H, W) denotes the spatial resolution of the features and C is the number of channels. These features are then used for object classification and localization, and the loss function can be described as the sum of two terms:

$$\mathcal{L} = \mathcal{L}_{cls} + \mathcal{L}_{loc}, \tag{1}$$

where $\mathcal{L}_{cls}$ and $\mathcal{L}_{loc}$ are loss terms for object classification and localization, respectively. In order to improve the generalization of the cell detector to unseen target domains, we propose a novel approach that involves masking the high-level semantic features and using a transformer-based reconstruction head to reconstruct the original features from partial observations. An overview of our proposed framework is shown in Fig. 1, and we present the detailed design of our approach below.

2.2 Masking Strategy

Motivated by the success of MAE [7] in self-supervised learning, where patches of the input image are masked to reconstruct the missing pixels, we propose to use a similar approach to mask high-level semantic features during the supervised training of object detection. By doing so, we aim to emphasize the domain-invariant features, and thus enhance the generalization of the cell detector to the

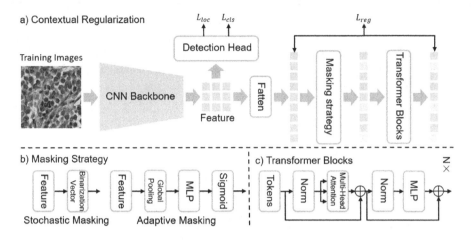

Fig. 1. Method overview: (a) cell detection framework with the proposed *contextual regularization* (CR), (b) two specific masking strategies employed in CR, and (c) the standard transformer blocks used in CR.

unseen target domain. Specifically, we propose two masking strategies to create a reconstruction task for context regularization: the stochastic masking approach, which is straightforward and easy to implement by randomly masking features, and the adaptive masking approach, which is both trainable and efficient by using a learnable gating mechanism to decide which features to mask. To this end, *contextual regularization* (CR) can be divided into two types based on how the mask is applied: *stochastic contextual regularization* (SCR) and *adaptive contextual regularization* (ACR). To process the feature map x in high-dimensional, we reshape the original feature into $x \in \mathbb{R}^{C \times P}$, where P represents the spatial dimensions $H \times W$.

To implement the stochastic masking strategy in SCR, we begin by randomly generating a binarization vector, denoted as k_p, with a specified zero ratio of q. Then, the masked feature $\widetilde{x}_{c,p}^s$ in SCR can be obtained as:

$$\widetilde{x}_{c,p}^s = x_{c,p} k_p. \tag{2}$$

While using random masks for high-level features is generally efficient and easy to implement, it's worth noting that such an approach may be influenced by random factors. To this end, ACR incorporates an adaptive masking strategy to selectively apply masking to the most relevant features, thereby improving its efficacy. To capture the overall spatial context, we utilize global average pooling on features $x \in \mathbb{R}^{H \times W \times C}$ and extract channel-wise statistics. This is accomplished by aggregating x along its spatial dimensions $H \times W$ to create a channel descriptor. Formally, we generate the pooled feature $v \in \mathbb{R}^C$ as follows:

$$v_c = \frac{1}{H \times W} \sum_{i=1}^{H} \sum_{j=1}^{W} x_{c,h,w}. \tag{3}$$

Following the squeeze operation to aggregate information, we utilize a gating mechanism with sigmoid activation to compute the adaptive mask weight. Our approach involves a bottleneck architecture, comprising two fully-connected layers around the non-linearity, to reduce model complexity and enhance generalization. The architecture consists of a dimensionality-reduction layer, which reduces the dimensionality with a reduction ratio of r, and a dimensionality-increasing layer that restores the feature resolution. The adaptive mask weight s_p can be calculated as:

$$s_p = \delta(W_2(W_1 v_c)), \tag{4}$$

where δ refers to the sigmoid function, $W_1 \in \mathbb{R}^{\frac{C}{r} \times C}$ and $W_2 \in \mathbb{R}^{P \times \frac{C}{r}}$. The adapted mask feature $\widetilde{x}^a_{c,p}$ can be obtain by multiply the shaped feature x with the activation s_p:

$$\widetilde{x}^a_{c,p} = x_{c,p} s_p. \tag{5}$$

Our proposed learnable adaptive masking approach allows the network to autonomously discover and learn the optimal positions and intensities for masking, effectively eliminating the ambiguity and hyperparameter selection issues that can arise with random masking. This empowers the network to perform more robust and efficient contextual regularization, thereby enhancing its ability to generalize to unseen target domains.

2.3   Reconstruction Head

To reconstruct the original signal from its partially observed counterpart $\widetilde{x}_{c,p}$, we incorporate a reconstruction head into our architecture. The reconstruction head is composed of a series of transformer blocks that process the masked feature. Each transformer block consists of two parts: an attention part and a *multi-layer perceptron* (MLP) part. The attention part includes LayerNorm and *multi-headed self-attention* (MSA) layers with a head size of m.

$$\widetilde{x}_{c,p} = \mathbf{F}_{\text{MSA}}(\widetilde{x}_{c,p}). \tag{6}$$

Self-attention is based on the idea of using the attention mechanism to model the relationships among different positions within a sequence. In the context of our work, the sequence of masked features $\widetilde{x} \in \mathbb{R}^{C \times P}$ is transformed into query $Q \in \mathbb{R}^{C \times P}$, key $K \in \mathbb{R}^{C \times P}$, and value $V \in \mathbb{R}^{C \times P}$ matrices using a linear transformation. The entire process can be expressed mathematically as follows:

$$\text{Attention}(Q, K, V) = \text{Softmax}\left(\frac{QK^T}{\sqrt{d}}\right)V. \tag{7}$$

where d_k is the dimensionality of the key vectors. In addition to the attention part, each transformer block also contains a *Feed-Forward Network* (FFN), also referred to as an MLP layer:

$$\widetilde{x}_{c,p} = \mathbf{F}_{\text{FFN}}(\widetilde{x}_{c,p}) \tag{8}$$

After being masked, the feature is reconstructed by passing it through a series of N transformer blocks and then reshaped to its original form. During training, the model is optimized using the mean squared error loss function, which measures the difference between the reconstructed feature and the original feature in pixel space:

$$\mathcal{L}_{reg} = \frac{1}{C \times P} \sum_{c=1}^{C} \sum_{p=1}^{P} |x_{c,p} - \widetilde{x}_{c,p}|^2. \tag{9}$$

Then, the complete loss function for incorporating the proposed contextual regularization can be obtained by combining the standard detection loss function with the reconstruction loss of the masked feature. The complete loss function $\mathcal{L}$ can be defined as:

$$\mathcal{L} = \mathcal{L}_{cls} + \mathcal{L}_{loc} + \lambda \mathcal{L}_{reg}, \tag{10}$$

where λ are weights for the loss terms $\mathcal{L}_{reg}$. During the testing phase, the reconstruction blocks are not used, and the enhanced feature extractor is directly employed for cell detection. Therefore, our approach can boost the detection accuracy without increasing the test time, making it an efficient and practical solution for cell detection in histopathology images.

2.4 Implementation Details

We used the Faster R-CNN [14] with FPN [12] and ResNet50 [8] implemented in MMDetection[1] as the backbone network, which is a common option for cell detection [16,20,22]. To facilitate better model initialization, we pre-trained the Faster R-CNN on the COCO dataset [13]. For each detection model, we used the SGD optimizer with an initial learning rate of 10^{-3} and trained it for 32 epochs to ensure training convergence. To optimize computation efficiency, we have integrated the proposed module into FPN architecture, where the channel compression occurs after the feature extraction stage. In SCR, we randomly mask half of the features with a probability of $q = 0.5$. In ACR, we set the reduction rate to $r = 16$. We use $N = 2$ standard transformer blocks for image restoration and set the head size $m = 8$ in the multi-head attention, following the recommendation of [5]. To simplify the training process, we set the loss weights to $\lambda = 1$. All other training settings were kept as default in MMDetection. Since the MIDOG datasets contain large images, we cropped them into 512×512 patches before feeding them into the detection model to optimize computational efficiency.

---

[1] https://github.com/open-mmlab/mmdetection.

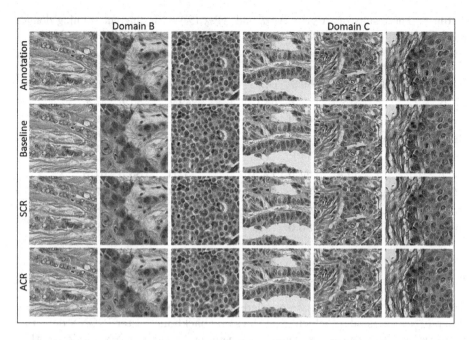

Fig. 2. Examples of detection results achieved with baseline, SCR and ACR on test images (achieved with Domain A for training, Domain B and Domain C for testing) are shown. The annotation is also shown for reference.

3   Results

3.1   Dataset Description

The evaluation dataset used in our experiments is the MIDOG dataset provided by the MIDOG challenge [1]. This dataset is specifically designed for detecting mitosis in breast cancer cells. There are 150 *whole-slide images* (WSIs) publicly available in the MIDOG dataset, with mitosis annotations. These images were acquired using three different scanners, with 50 WSIs associated with each scanner. The first 50 WSIs, referred to as Domain A, have an image size of 7215×5412. The second and third 50 WSIs, referred to as Domain B and Domain C, respectively, have image sizes of 7094×5370 and 6475×4840, respectively. We utilized the MIDOG dataset to evaluate the detection performance when the training and testing images were acquired on different scanners but belonged to the same category. We considered three different scenarios: 1) using Domain A for training and Domain B and Domain C for testing, 2) using Domain B for training and Domain A and Domain C for testing, and 3) using Domain C for training and Domain A and Domain B for testing.

Table 1. The results of $AP_{50:95}$, AP_{50} and AP_{75} (%) with different training domain and test domain on the MIDOG dataset. The best results are highlighted in bold.

Method	Domain	$AP_{50:95}$	AP_{50}	AP_{75}	Domain	$AP_{50:95}$	AP_{50}	AP_{75}
Baseline	A→B	53.9	80.3	66.5	A→C	48.9	78.3	56.6
SCR		54.9	82.3	65.1		49.9	80.4	57.8
ACR		**56.0**	**83.9**	65.9		**51.0**	**81.1**	**59.5**
Baseline	B→A	58.0	84.0	68.9	B→C	50.6	82.6	59.3
SCR		59.1	84.1	71.9		51.8	82.8	60.8
ACR		**59.4**	**84.9**	**72.3**		**52.1**	**83.5**	**61.3**
Baseline	C→A	58.4	85.3	71.8	C→B	55.0	80.4	67.5
SCR		58.7	85.5	73.7		56.0	82.1	68.9
ACR		**59.7**	**86.0**	**74.0**		**57.0**	**83.4**	**69.9**

3.2   Evaluation Results

To ensure fair comparison, we applied SCR and ACR separately to the three different scanners, all of which used the same backbone Faster R-CNN detection network as the proposed method. The vanilla Faster R-CNN model method without SCR and ACR is referred to as the baseline method for convenience. The results on the three scanners are presented next individually.

To evaluate the performance of the detection methods on the MIDOG dataset, we conducted a qualitative analysis of the results. We visually present examples of the detection outputs on test images that were obtained by training the models with Domain A and testing with Domains B and C in Fig. 2. The ground truth annotations are also provided for reference, along with the confidence scores of the predicted bounding boxes. In the examples presented, the use of SCR resulted in more accurate detection of mitosis compared to the baseline model. This is reflected in the avoided false negatives and higher confidence values of true positives. Moreover, the application of ACR further improved the performance of the model, resulting in better detection results. In two of the cases (third and sixth columns), SCR and ACR successfully removed a false positive that was present in the baseline model's results.

We used the *average precision* (AP) as the evaluation metric to quantitatively assess the detection performance, which is a common practice in object detection research [2,3,13]. We computed the AP at two IoU thresholds of 0.5 and 0.75, referred to as AP_{50} and AP_{75}, respectively. Additionally, we computed the average precision over multiple IoU thresholds ranging from 0.5 to 0.95 with an increment of 0.05, denoted as $AP_{50:95}$. Table 1 summarizes the quantitative results. We observed that applying SCR and ACR consistently improved the AP_{50} and $AP_{50:95}$ for all training and testing domain combinations. Most cases show that ACR improves AP_{75}, except for Domain A for training and Domain B for testing where AP_{75} scores with or without SCR and ACR are comparable. The comparison of SCR and ACR reveals that ACR provides further improvements to the AP scores, highlighting the benefits of the proposed adaptive masking approach. These results indicate that the use of SCR and ACR can enhance the domain generalization performance for cell detection in histopathology images.

4   Conclusion

In summary, we presented a domain generalization method called *contextual regularization* (CR), which aims to improve the performance of cell detectors across different datasets by focusing on domain-invariant features. Our proposed approach involves a reconstruction task that masks high-level semantic features either stochastically or adaptively, and we employ a transformer-based reconstruction head that reconstructs the original features based on partial observations. The main advantage of our proposed method is its ability to seamlessly integrate with various deep learning-based cell detectors without requiring any modification, while also not increasing the inference time. The experimental results on a publicly available dataset demonstrate that our approach significantly improves the generalization of cell detectors to unseen domains. These results suggest that our proposed method holds promise for improving cancer diagnosis and prognosis using histopathology images.

References

1. Aubreville, M., et al.: Mitosis domain generalization in histopathology images-the MIDOG challenge. Med. Image Anal. **84**, 102699 (2023)
2. Chang, J.-R., et al.: Stain Mix-Up: unsupervised domain generalization for histopathology images. In: de Bruijne, M., et al. (eds.) MICCAI 2021. LNCS, vol. 12903, pp. 117–126. Springer, Cham (2021). https://doi.org/10.1007/978-3-030-87199-4_11
3. Chen, T., et al.: A task decomposing and cell comparing method for cervical lesion cell detection. IEEE Trans. Med. Imaging **41**(9), 2432–2442 (2022). https://doi.org/10.1109/TMI.2022.3163171
4. Dong, B., Shao, L., Da Costa, M., Bandmann, O., Frangi, A.F.: Deep learning for automatic cell detection in wide-field microscopy zebrafish images. In: 2015 IEEE 12th International Symposium on Biomedical Imaging (ISBI), pp. 772–776. IEEE (2015)
5. Dosovitskiy, A., et al.: An image is worth 16×16 words: transformers for image recognition at scale. arXiv preprint arXiv:2010.11929 (2020)
6. Ducret, A., Quardokus, E.M., Brun, Y.V.: MicrobeJ, a tool for high throughput bacterial cell detection and quantitative analysis. Nat. Microbiol. **1**(7), 1–7 (2016)
7. He, K., Chen, X., Xie, S., Li, Y., Dollár, P., Girshick, R.: Masked autoencoders are scalable vision learners. In: Proceedings of the IEEE/CVF Conference on Computer Vision and Pattern Recognition, pp. 16000–16009 (2022)
8. He, K., Zhang, X., Ren, S., Sun, J.: Deep residual learning for image recognition. In: IEEE Conference on Computer Vision and Pattern Recognition, pp. 770–778 (2016)

9. Huang, Y., Wang, Q., Omachi, S.: Rethinking degradation: radiograph super-resolution via AID-SRGAN. In: Lian, C., Cao, X., Rekik, I., Xu, X., Cui, Z. (eds.) Machine Learning in Medical Imaging, MLMI 2022. Lecture Notes in Computer Science, vol. 13583, pp. 43–52. Springer, Cham (2022). https://doi.org/10.1007/978-3-031-21014-3_5

10. Jiang, P., Ergu, D., Liu, F., Cai, Y., Ma, B.: A review of yolo algorithm developments. Procedia Comput. Sci. **199**, 1066–1073 (2022)

11. Liao, W., et al.: Muscle: multi-task self-supervised continual learning to pre-train deep models for x-ray images of multiple body parts. In: Wang, L., Dou, Q., Fletcher, P.T., Speidel, S., Li, S. (eds.) MICCAI 2022, Part VIII. LNCS, vol. 13438, pp. 151–161. Springer, Cham (2022). https://doi.org/10.1007/978-3-031-16452-1_15

12. Lin, T.Y., Dollár, P., Girshick, R., He, K., Hariharan, B., Belongie, S.: Feature pyramid networks for object detection. In: Proceedings of the IEEE Conference on Computer Vision and Pattern Recognition, pp. 2117–2125 (2017)

13. Lin, T.-Y., et al.: Microsoft COCO: common objects in context. In: Fleet, D., Pajdla, T., Schiele, B., Tuytelaars, T. (eds.) ECCV 2014. LNCS, vol. 8693, pp. 740–755. Springer, Cham (2014). https://doi.org/10.1007/978-3-319-10602-1_48

14. Ren, S., He, K., Girshick, R., Sun, J.: Faster R-CNN: towards real-time object detection with region proposal networks. In: Advances in Neural Information Processing Systems, vol. 28 (2015)

15. Sadafi, A., et al.: Multiclass deep active learning for detecting red blood cell subtypes in brightfield microscopy. In: Shen, D., et al. (eds.) MICCAI 2019. LNCS, vol. 11764, pp. 685–693. Springer, Cham (2019). https://doi.org/10.1007/978-3-030-32239-7_76

16. Sun, Y., Huang, X., Molina, E.G.L., Dong, L., Zhang, Q.: Signet ring cells detection in histology images with similarity learning. In: International Symposium on Biomedical Imaging, pp. 37–48 (2020)

17. Torralba, A., Efros, A.A.: Unbiased look at dataset bias. In: IEEE Conference on Computer Vision and Pattern Recognition, pp. 1521–1528 (2011)

18. Xu, J., et al.: Stacked sparse autoencoder (SSAE) for nuclei detection on breast cancer histopathology images. IEEE Trans. Med. Imaging **35**(1), 119–130 (2016)

19. Yang, S., Fang, B., Tang, W., Wu, X., Qian, J., Yang, W.: Faster R-CNN based microscopic cell detection. In: 2017 International Conference on Security, Pattern Analysis, and Cybernetics (SPAC), pp. 345–350. IEEE (2017)

20. Zhang, J., Hu, H., Chen, S.: Cancer cells detection in phase contrast microscopy images based on Faster R-CNN. In: International Symposium on Computational Intelligence and Design, pp. 363–367 (2016)

21. Zhang, Y., et al.: Video4MRI: an empirical study on brain magnetic resonance image analytics with CNN-based video classification frameworks. arXiv preprint arXiv:2302.12688 (2023)

22. Zhao, Z., Pang, F., Liu, Z., Ye, C.: Positive-unlabeled learning for cell detection in histopathology images with incomplete annotations. In: de Bruijne, M., et al. (eds.) MICCAI 2021. LNCS, vol. 12908, pp. 509–518. Springer, Cham (2021). https://doi.org/10.1007/978-3-030-87237-3_49

TON-ViT: A Neuro-Symbolic AI Based on Task Oriented Network with a Vision Transformer

Yupeng Zhuo[1], Nina Jiang[1], Andrew W. Kirkpatrick[2], Kyle Couperus[3,5], Oanh Tran[3,5], Jonah Beck[3,5], DeAnna DeVane[3,5], Ross Candelore[3,4], Jessica McKee[2], Chad Gorbatkin[3], Eleanor Birch[3], Christopher Colombo[3,5], Bradley Duerstock[1], and Juan Wachs[1(✉)]

[1] School of Industrial Engineering, Purdue University, West Lafayette, IN, USA
jpwachs@purdue.edu
[2] University of Calgary, Calgary, Canada
[3] Madigan Army Medical Center, Joint Base Lewis-McChord, USA
[4] William Beaumont Army Medical Center, Fort Bliss, USA
[5] The Geneva Foundation, Tacoma, USA

Abstract. The objective of this paper is to present a neuro-symbolic AI based technique to represent field-medicine knowledge, referred as to TON-ViT. TON-ViT integrates a Deep Learning Model with an explicit symbolic manipulation, a task graph. This task graph describes the steps of each trauma resuscitation as denoted by a verb and noun pair. Through this representation, symbolic processing and manipulation on task graphs, we can find stereotypical procedures, regardless of style of the performer. Furthermore, we can use this technique to find differences in styles, errors, shortcuts and generate procedures never seen before. When used in combination with a transformer, it can help recognize actions in egocentric vision datasets. Last, through symbolic manipulations on the graph, it is possible to generate medical knowledge which the model has not seen before. We present preliminary results after testing the TON-ViT with the Trauma Thompson Dataset.

Keywords: task graph · knowledge graph · semantic understanding · vision transformer · neuro-symbolic AI · medical procedures

1 Introduction

Medical students and residents are taught how to complete trauma resuscitation using textbooks, guides, videos and demonstrations. Data science has offered a variety of techniques to support the assessment of knowledge of trainees through

---

Disclaimers: The views expressed are those of the author(s) and do not reflect the official policy of the Department of the Army, the Department of Defense, or the U.S. Government. The investigators have adhered to the policies for the protection of human subjects as prescribed in 45 CFR 46.

© The Author(s), under exclusive license to Springer Nature Switzerland AG 2024
G. Waiter et al. (Eds.): MIUA 2023, LNCS 14122, pp. 157–170, 2024.
https://doi.org/10.1007/978-3-031-48593-0_12

autonomous benchmarks that collect spatiotemporal observations of practitioners while at work, and can output a proficiency score based on performance. These methods are based on Deep Learning Models for action recognition, and the score computed is based on a variety of factors. Due to the nature of the Machine Learning (ML) model, a black box model, we cannot tell the reason for the assigned score. Not only mistakes affect performance, but also individual style, shortcuts, and drastic variations. Nevertheless, variations, style and shortcuts have been found to be essential in field medicine when treating casualties in the field. To provide interpretability to such systems, we discuss our neuro-symbolic Artificial Intelligence (AI) based technique, referred as TON-ViT, which complements action recognition models with knowledge representation in the form of task graphs.

1.1   Knowledge Graphs

A knowledge graph is a structured representation of knowledge that captures information about entities, relationships and attributes [1,2]. At its core, a knowledge graph consists of nodes, which represent entities, and edges, which represent the relationships between them. Nodes can be anything from people, places, and things to concepts, ideas, and events [3]. Edges describe the nature of the relationship between the nodes.

A task graph is a knowledge graph that is applied to task representation. In this study, a bipartite graph-based task graph, Task-Oriented-Network (TON), is developed to represent medical procedures. We used a bipartite graph because there are two types of nodes, verb nodes and noun nodes, in our case and each node only connects to the other type of node. There have been several knowledge graphs research works developed for learning and modeling system behaviors [4–8]. In most cases, these graphs show in a graphical form concurrent behavior in their operation. The network consists of two types of nodes and directed edges that run from one type of nodes to the other.

Most recent works model activity in the form of motion and object, object and object [9], as well as motion and state [10]. An example of this is Functional Object-Oriented Network (FOON) which is a graphical representation method that aims to facilitate object affordance-based learning and to comprehend human behavior and learning [10]. It shows the relationship between objects and their associated functional motions and has been specifically designed to facilitate manipulation tasks. In FOON, transitions are equivalent to motion nodes, which represent the movement of objects from one state to another. On the other hand, conditions or places are equivalent to object nodes, which represent the physical objects involved in the manipulation task.

In this paper, we extend the FOON method, so it is suitable for modeling trauma resuscitation tasks that are recurrent, sequential, and structured. In addition, the method presented can be mapped into matrix operations and lists sequences that are interchangeable so to optimize the computing operations. At the basic level, it can perform addition to merge medical procedures, subtraction to detect variations between medical procedures, and average to find common medical practices. Besides generating basic medical knowledge, it can also be applied to find mistakes, different styles, shortcuts, proficiency in medical practices. When it is used in conjunction with machine learning algorithms, it can assist in various classification tasks and even generate novel complex medical knowledge.

1.2 Knowledge Graphs in Neural Network

The use of knowledge graphs in neural networks has been shown to enhance explainability, trustworthiness, and classification performance [11,12]. This is because knowledge graphs provide additional context and information to the black model and can be used to explain the relationship between output classes of the network [13]. First, knowledge graphs can be used to represent information about entities and their relationships in a structured way. Such structure can assist to enhance the representation of input information [14]. Second, they can be used to transfer knowledge from one domain to another, such as transferring medical knowledge between pathologies to improve the performance of the diagnosis of diseases. Third, knowledge graphs can make neural networks more transparent and explainable by representing the reasoning behind their decisions in a structured way, which can help build trust in the network's predictions [15]. Last, knowledge graphs can be used to generate additional training data for neural networks by using the relationships between entities to infer new examples, which is particularly useful in domains with limited training data [16]. Thus, knowledge graphs are a powerful tool for enhancing the performance and interpretability of neural networks by providing additional context and information.

1.3 Neuro-Symbolic Artificial Intelligence (NeSy AI)

Neuro-symbolic artificial intelligence is a subfield of artificial intelligence that combines neural and symbolic approaches [17]. The neural approaches are based on artificial neural networks, particularly deep learning, while symbolic approaches rely on the explicit representation of knowledge using formal languages, including formal logic [18,19]. The study of neuro-symbolic integration is motivated by two reasons: to advance the understanding of the human mind and to overcome the weaknesses of either approach. Recent developments in deep learning have led to a renewed emphasis on neuro-symbolic AI research and a significant increase in research papers and meetings. The major current research directions in neuro-symbolic AI include solving symbolic problems with deep learning, integrating neural and symbolic models, developing explainable AI, and incorporating symbolic knowledge into neural models [20–22]. While deep learning has had a major impact on neuro-symbolic AI, it has also led to a realization that pure deep-learning-based methods may be insufficient for certain types of problems that are now being investigated from a neuro-symbolic perspective. TON-ViT is developed for this purpose by graph-based algorithms.

2 Methods

Our method for knowledge representation and activity recognition is referred as to TON-ViT. The input of TON-ViT is a video sequence, and the output is a knowledge graph with metrics associated to it, such as the accuracy of the activity recognition and the distance between the procedure constructed from the recognized actions and the corresponding average procedure. This distance can evaluate how similar between the recognized actions and the common practices of this procedure are. In this paper, the emphasis is on the metrics and knowledge representation of the system. A pipeline of this approach is shown in Fig. 1.

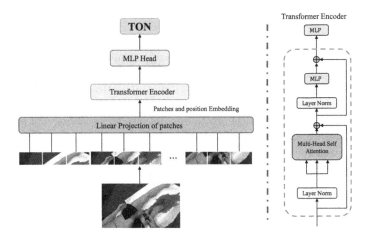

Fig. 1. System diagram of TON-ViT

A vision transformer (ViT) is a neural network designed for image classification and object detection tasks [23]. At the output, there is a TON knowledge graph. It is used to provide additional context and information about objects in the image, transfer knowledge from one domain to another and explain the reasoning behind the network's output. By using a knowledge graph containing information about the relationships between different actions, the network's understanding of complex procedure can be enhanced. Knowledge transfer can be achieved by using a knowledge graph of medical procedures to improve the ViT's performance in recognizing actions. Representing the network's decision-making using a knowledge graph can make it easier to understand and debug the network's behavior [24]. Thus, connecting a knowledge graph to the output of a ViT can improve the network's performance, interpretability, and explainability.

2.1 Activity Recognition of Emergent Procedures

The video stream is processed onboard through a ViT. The algorithm outputs a sequence of recognized actions with TOP 1 and TOP 5 probabilities [25]. When the ViT output is connected to the TON and a pre-trained Support Vector Machine (SVM) model that can classify the medical procedures in the dataset, weight-adjusted TOP 1 and TOP 5 probabilities can be produced to improve the action recognition result. The TON is a powerful tool for enhancing the performance of ViTs in activity recognition tasks by providing additional context and the system diagram in Fig. 1 shows the network of the ViT.

2.2   Task Oriented Network Construction

While the procedure is in operation, an AI system can use the TON to inform a user of what to do next. However, the first step is to create these graphs. The graphs are automatically generated from annotations of the referencing videos or from the result of machine learning algorithms that recognize and predict next steps (see discussed below). Based on the information provided by a graph, the corresponding procedure can be recreated. First, the sequence of actions is parsed, sequentially by a text parser. The first action, made of a verb and a noun (e.g. "make incision"), is converted to a node and an edge. Using a coloring convention, for example, green nodes can be used to denote objects (nouns) used in the procedure and the red nodes for denoting the verbs that are associated with the objects. Each graph starts from a "start" node. Then, a verb and noun pair form a step. The numbers on the edges describe the chronological order of the procedure. By following the numbers, a procedure can be recreated. From the pedagogical stand point, task graphs allow medics have a visual understanding of the procedures and a more immersive instructional experience to improve their performance. From the technical standpoint, this tasks graphs can support real-time action recognition, prediction, mistakes recognition, style characterization and dynamic knowledge generation.

The TON is a multi-edge directed bipartite graph, represented as $G = (U, V, E)$. An action is denoted as a and the set of actions is as A. A verb node is denoted as u and the set is U. A noun node is denoted as v and the set is V. The number of actions in a procedure is denoted as N_A. To solve the cycling issue due to repeating actions in a medical procedure, such as "take swab", a counter j is introduced for each action to distinguish different occurrences and the vocabulary set is denoted as $M(A(j))$. To create a TON, the action sequences are converted to metrical representations, adjacency matrix P and ordinal matrix Q. Then, the two matrices together can generate a corresponding TON graph with ordinal values on the edges to indicate the temporal workflow of the procedure. Algorithm 1 shows the process to generate a TON graph.

2.3   Stereotypical Procedures (Average)

Normally, the process of determining standard operating procedures in the field of trauma management and resuscitation involves collaboration among surgical professionals, medical experts, and regulatory bodies [26]. The process includes reviewing existing literature and research, developing guidelines and recommendations based on factors such as patient safety, efficiency, and cost-effectiveness, and continuous review and revision of these guidelines as new research and technologies become available [17].

The average procedure is to find a common practice of a type of medical procedures in the medical field. Although various practitioners execute a given procedure, dissimilarities exist due to factors such as instrument accessibility,

---

Algorithm 1. TON Generation

$i, j \leftarrow 0, 0$
$A(i), U(i), V(i), M(A(j)) \leftarrow \emptyset, \emptyset, \emptyset$
while $i \leq N_A$ **do**
  if $a \notin A$ **then**
    $M(a) \leftarrow 1$
  else
    $M(a) \leftarrow M(a) + 1$
  end if
  $A(i) \leftarrow$ append a
  $u, v \leftarrow$ parse a
  $U(i) \leftarrow u(j)$
  $V(i) \leftarrow v$
  $E \leftarrow$ append $(V(i-1), U(i))$
  $E \leftarrow$ append $(U(i), V(i))$
end while
$P, Q \leftarrow E$

---

individual inclinations, patient circumstances, safety, and risk control. Standardizing medical procedures is crucial, and as such, securing expert consensus on such matters is crucial. In this regard, an algorithm capable of calculating the average outcome of several instances of the same medical procedure would be highly advantageous in quantitatively establishing expert agreement.

In the dataset used, five life saving procedures are identified to be included in the system. They are (1) interosseous insertion (IO), (2) tube thoracostomy (CT), (3) cricothyroidotomy (Cric), (4) needle thoracostomy (ND), and (5) tourniquet application. Thus, five average procedures were generated and studied. Let K denote the number of instances of a procedure. Let W denote the word set that includes all verbs and nouns and its number of elements is N_W. Let $\bar{P}$ and $\bar{Q}$ denote the averaged adjacency and ordinal matrix. Let P and Q denote the sets of all adjacency matrices and ordinal matrices of a procedure. Let α denote a threshold value that determines the occurrence percentage of a verb or noun in how many instances of the procedure to be averaged. Algorithm 2 shows the process to generate an average procedure.

For the evaluation of the created average procedures, they are evaluated through the questionnaire discussed below. This is the technique to generate implicit consensus procedures without the need of subjective panels or complex review processes.

2.4 Procedure Category, Mistakes and Shortcut Recognition

One important application of the task graphs is to compare procedures. By establishing an average procedure $\bar{M}$, it can be treated as a standard of the procedure. Then, instances M_i of it can be compared to this average procedure for the detection of variations from the standard. This difference can be treated as with-class changes. The variation can be further interpreted to detect mistakes

---

Algorithm 2. Average procedure generation

$P_1, P_2, ..., P_K, Q_1, Q_2, ..., Q_K \leftarrow$ all instances of a procedure
for $i \leftarrow 1$ to N_W **do**
    for $j \leftarrow 1$ to N_W **do**
        $weights \leftarrow \boldsymbol{P}(i, j)$
        $values \leftarrow \boldsymbol{Q}(i, j)$
        if sum of $weights > 0$ **then**
            $\bar{P}(i, j) \leftarrow$ average of nonzero elements of $weights$
        end if
        if $\bar{P}(i, j) < \alpha$ **then**
            $\bar{P}(i, j) \leftarrow 0$
        else
            $\bar{P}(i, j) \leftarrow 1$
        end if
        if sum of $values > 0$ **then**
            $\bar{Q}(i, j) \leftarrow$ average of $values$
        end if
        $\bar{Q}(i, j) \leftarrow \bar{P}(i, j) * \bar{Q}(i, j)$
    end for
end for

---

and innovations while performing this procedure. Furthermore, by computing the difference of two instances belonging to two different procedures, the dissimilarity between the two can be quantified and this difference can be treated as between-class difference. By the within-class and between-class differences, a threshold value can be determined to classify whether two procedure instances are members of the same procedure, or by looking at the N-closest neighbor.

To compare two procedure instances, the ordinal matrices Q_1 and Q_2 are flattened to produce two ordinal vectors q_1 and q_2. Then, the two vectors are divided by their respective lengths as a normalization method, to remove the effect of procedure complexity, and yields two unit vector u_{q_1} and u_{q_2}. We use their Euclidean distance ($d = ||u_{q_1} - u_{q_2}||$) to quantify their difference.

To classify a procedure, following the concept of clustering, we convert all observations into vectors. To reduce the dimensionality of the vectors, principle component analysis (PCA) is used to represent all vectors in a lower dimensional space. By the found mean vectors (stereotypical graphs), we determine whether a new observation x belongs to a procedure C_i by its Mahalanobis distance. Given all instances of a procedure C with mean $\bar{M}$ and positive-definite covariance matrix S, the Mahalonobis distance between the procedure to be classified, x, and a cluster, C_i, is

$$d(x, C_i) = \sqrt{(x - \bar{M})^T S^{-1} (x - \bar{M})}$$

The procedure k is assigned as

$$k = \underset{i}{\operatorname{argmin}}\{d(x, C_i)\}$$

Shortcuts usually contain less repeating steps and mistakes are the procedures that do not belong to any existing clusters. To detect whether a procedure to be classified is a shortcut or a mistake, we compute two Mahalanobis distances based on two representations of the procedure. In the first one, we merge all repeating edges of the procedure to create a simplified metrical representation of the procedure and compute the corresponding Mahalonobis distance. In the second one, repeating edges are not merged and a regular TON representation of the procedure is used. If a procedure belongs to a cluster in the merged representation but does not belong to any cluster in the multi-edge representation, it is considered as a shortcut. If a procedure does not belong to any cluster in both the merged and multi-edge representations, it is considered as a mistake.

2.5   New Procedure Generation (by Combination)

Medical procedures are typically structured into small, individual units to address one issue at a time, requiring multiple procedures for individuals with multiple injuries, which can be costly and time-consuming. The five structured life-saving procedures in the Trauma THOMPSON project generally have low variability between different physicians. If they can be combined to produce a new resuscitation procedure, the successful creation of such a procedure could represent a cost-effective approach to generating new medical knowledge. The TON has the potential to create such new procedures that can address multiple injuries in a single combined manner, thereby reducing delays in life-saving treatments. A method to generate a merged procedure through TON is Algorithm 2. With input and parameter changes, this method can essentially take a union of two task graphs. Replace all instances of the same procedure by two different procedures and then adjust the threshold value α to a small number. The resultant merged procedure is evaluated through the questionnaire discussed below.

2.6   Subjective Evaluation

In order to assess the effectiveness of all created procedures, a well-designed assessment survey was disseminated among medical experts. The questionnaire comprises baseline queries to determine their familiarity with the procedure under review. Subsequently, each procedure was evaluated based on its correctness, completeness, and clarity. The rating of each response is presented as Likert Scale (1–5). The complete evaluation survey is provided in the supplementary materials.

3   Experiments and Results

This section shows the result of stereotypical procedures that are generated based on TON, activity recognition accuracy, the recognition of shortcuts and mistakes, and new procedure generation by combinations.

3.1   Stereotypical Procedures (Average)

Table 1 shows the evaluated scores that each average procedure receives from medical experts.

Table 1. Average procedure evaluation result.

Procedure Name	Doctor 1	Doctor 2	Doctor 3	Doctor 4	Average Score
Interosseous insertion	3.1	3.1	4.3	3.9	3.6
Tube thoracostomy	1.8	3.5	2.3	3.5	2.8
Cricothyroidotomy	2.7	3.1	1.9	3.1	2.7
Needle thoracostomy	4.3	5.0	5.0	5.0	4.8
Tourniquet	4.3	5.0	5.0	5.0	4.8

The evaluation indicates satisfactory performance for needle thoracostomy and tourniquet, while difficulties are encountered in executing tube thoracostomy and cricothyroidotomy. Specifically, the challenges stem from concurrent and clustered actions, as well as varied approaches. In terms of concurrent actions, it can be arduous to monitor the progress of the Kelly clamps maneuver during tube thoracostomy insertion when multiple actions are occurring simultaneously. With regards to clustered actions, some actions require a specific sequence once initiated, such as swab usage or scalpel handling. Finally, different approaches may pose challenges in subsequent actions, for example, the use of a tracheal hook versus an obturator. Decisions concerning equipment selection may limit future actions in the same manner that a fork in the road can constrain subsequent paths. Figure 2 shows two examples of generated average procedures, tourniquet and tube thoracostomy.

The medical community demonstrates a lack of uniformity in executing the same medical procedures, and the manual annotations of procedure sequences in the Trauma THOMPSON project result in data imperfections. To enhance the quality of the generated average procedures, a weighting system can be implemented to individual procedure instances, where a higher weight should be assigned to superior practices.

3.2   Procedure Recognition Accuracy

For the activity recognition of TON graph trained Support Vector Machine (SVM), the Leave-One-Out cross-validation has an average accuracy of 100%. Figure 3 illustrates the confusion matrix of the classification result and all procedures are correctly classified. This finding suggests that TON can enhance the activity recognition performance of ViT.

3.3   Variations Between Procedures

Figure 4a illustrates the employment of normalized distance metrics to compare all instances of all procedures. The graph highlights certain threshold values that can effectively differentiate within-class variation from between-class variation. Specifically, the maximum within-class difference is 0.396 and the minimum between-class difference is 0.574. Consequently, any value falling within the range of 0.396 to 0.574 can serve as a suitable threshold for ascertaining whether two procedure instances belong to the same procedure or not.

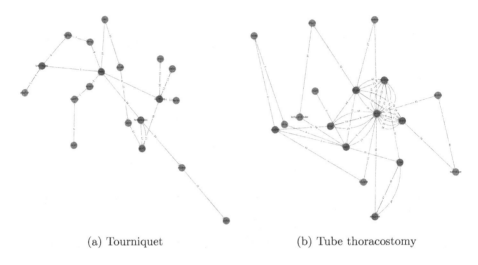

(a) Tourniquet   (b) Tube thoracostomy

Fig. 2. Examples of average procedures

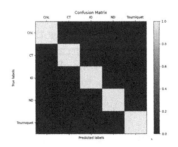

Fig. 3. Confusion matrix of TON trained SVM

3.4   Recognition of Mistakes and Shortcuts

Figure 5a shows the TON graph of an example of a shortcut tube thoracostomy procedure. It contains less repetitions. Figure 5b shows the TON graph of a mistake procedure example.

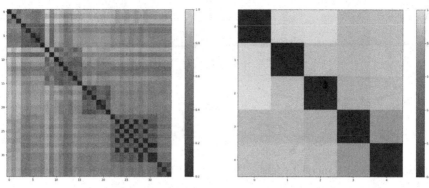

(a) Comparison between different medical procedure instances. CT: 0-7; Cric: 8-15; IO: 16-21; ND: 22-31; Tourniquet: 32-34

(b) Comparison between difference average procedures. CT: 0; Cric: 1; IO: 2; ND: 3; Tourniquet: 4

Fig. 4. Comparison between procedures

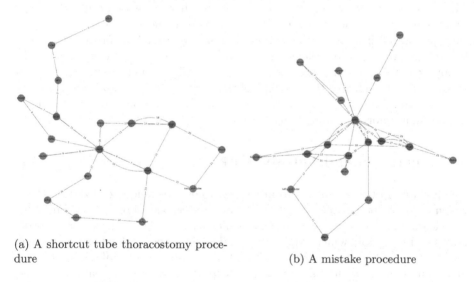

(a) A shortcut tube thoracostomy procedure

(b) A mistake procedure

Fig. 5. Example Procedures of Shortcuts and Mistakes

Figure 6a and Fig. 6b show the clustering result of the shortcut and mistake procedures after applying PCA and computing their Mahalanobis distance. As the figures indicates, a shortcut procedure belongs to a cluster in the merged edge representation and does not belong to any cluster in the multi-edge representation. In contrast, a mistake procedure does not belong to any cluster in both merged edge and multi-edge representations.

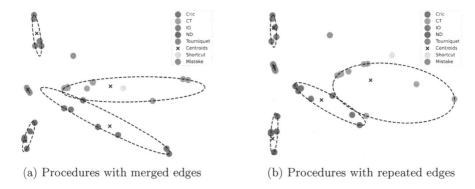

(a) Procedures with merged edges   (b) Procedures with repeated edges

Fig. 6. Graph clustering result

3.5 New Procedure Generation (by Combination)

The average score for the merged procedure that involves needle thoracostomy and tube thoracostomy is 2.4, as the two medical procedures are typically performed in a particular sequence. The way a medical procedure is performed involves numerous elements, including anatomy, equipment, safety, indications/contraindications, complications, microskills, economy of movement, and others. Thus, to enhance the outcome of the merging process, additional considerations should be incorporated, such as the implementation of different weighting techniques for the input data.

4 Conclusion and Future Work

The main aim of this paper is to introduce a novel technique called TON-ViT, which is an AI-based approach that combines neuro-symbolic processing with deep learning models to represent field-medicine knowledge. This approach incorporates a task graph, which explicitly describes the steps involved in each trauma resuscitation procedure through verb and noun pairs. The symbolic processing and manipulation of the task graph can help identify stereotypical procedures, irrespective of the performer's style. Additionally, the TON-ViT technique can also detect differences in styles, errors, shortcuts, and even generate procedures that have never been observed before.

When integrated with a transformer, the TON-ViT technique can be utilized to recognize actions in egocentric vision datasets. Furthermore, through symbolic manipulations on the graph, the TON-ViT model can generate new medical knowledge that was not previously seen. The preliminary results obtained by testing the TON-ViT technique with the Trauma Thompson Dataset have shown its effectiveness in representing field-medicine knowledge. As part of future work, the integration of medical language models can be explored to further enhance the ability to generate new medical knowledge.

References

1. Fensel, D., et al.: Introduction: what is a knowledge graph? In: Knowledge Graphs, pp. 1–10. Springer, Cham (2020). https://doi.org/10.1007/978-3-030-37439-6_1
2. Zou, X.: A survey on application of knowledge graph. J. Phys. Conf. Ser. **1487**(1), 012016 (2020). https://doi.org/10.1088/1742-6596/1487/1/012016
3. Yang, S., Zou, L., Wang, Z., Yan, J., Wen, J.-R.: Efficiently answering technical questions—a knowledge graph approach. Proc. AAAI Conf. Artif. Intell. **31**(1),(2017). https://doi.org/10.1609/aaai.v31i1.10956
4. Liew, C.Y., Labadin, J., Kok, W.C., Eze, M.O.: A methodology framework for bipartite network modeling. Appl. Netw. Sci. **8**(1), 6 (2023). https://doi.org/10.1007/s41109-023-00533-y
5. Li, Z., et al.: Temporal knowledge graph reasoning based on evolutional representation learning. In: Proceedings of the 44th International ACM SIGIR Conference on Research and Development in Information Retrieval, pp. 408–417. ACM (2021). ISBN 978-1-4503-8037-9. https://doi.org/10.1145/3404835.3462963.
6. Guo, Q., Zhuang, F., Qin, C., Zhu, H., Xie, X., Xiong, H., He, Q.: A survey on knowledge graph-based recommender systems. IEEE Trans. Knowl. Data Eng. **34**(8), 3549–3568 (2022). https://doi.org/10.1109/TKDE.2020.3028705
7. Wang, X., et al.: Learning intents behind interactions with knowledge graph for recommendation. In: Proceedings of the Web Conference 2021, pp. 878–887. ACM (2021). ISBN 978-1-4503-8312-7. https://doi.org/10.1145/3442381.3450133
8. Chen, H., Luo, X.: An automatic literature knowledge graph and reasoning network modeling framework based on ontology and natural language processing. Adv. Eng. Informatics **42**, 100959 (2019). https://doi.org/10.1016/j.aei.2019.100959
9. Peña, J., Rochat, Y.: Bipartite graphs as models of population structures in evolutionary multiplayer games. PLoS ONE **7**(9), e44514 (2012). https://doi.org/10.1371/journal.pone.0044514
10. Paulius, D., Huang, Y., Milton, R., Buchanan, W.D., Sam, J., Sun, Y.: Functional object-oriented network for manipulation learning. In: 2016 IEEE/RSJ International Conference on Intelligent Robots and Systems (IROS), pp. 2655–2662. https://doi.org/10.1109/IROS.2016.7759413.
11. Tiddi, I., Schlobach, S.: Knowledge graphs as tools for explainable machine learning: a survey. Artif. Intell. **302**, 103627 (2022). https://doi.org/10.1016/j.artint.2021.103627
12. Marino, K., Salakhutdinov, R., Gupta, A.: The more you know: using knowledge graphs for image classification. arXiv preprint arXiv:1612.04844 (2017)
13. Paulheim, H.: Knowledge graph refinement: a survey of approaches and evaluation methods. Semantic Web **8**(3), 489–508 (2016). https://doi.org/10.3233/SW-160218
14. Chaudhri, V.K., et al.: Knowledge graphs: introduction, history, and perspectives. AI Magazine **43**(1), 17–29 (2022). https://doi.org/10.1002/aaai.12033
15. Lecue, F.: On the role of knowledge graphs in explainable AI. Semantic Web **11**(1), 41–51 (2020). https://doi.org/10.3233/SW-190374
16. Zhang, W., Paudel, B., Zhang, W., Bernstein, A., Chen, H.: Interaction embeddings for prediction and explanation in knowledge graphs. In: Proceedings of the Twelfth ACM International Conference on Web Search and Data Mining, pp. 96–104. ACM (2019). ISBN 978-1-4503-5940-5. https://doi.org/10.1145/3289600.3291014
17. Manghani, K.: Quality assurance: importance of systems and standard operating procedures. Perspect. Clin. Res. **2**(1), 34 (2011). https://doi.org/10.4103/2229-3485.76288

18. Hitzler, P., Eberhart, A., Ebrahimi, M., Sarker, M.K., Zhou, L.: Neuro-symbolic approaches in artificial intelligence. Natl. Sci. Rev. **9**(6), nwac035 (2022). https://doi.org/10.1093/nsr/nwac035

19. Hitzler, P.: Some advances regarding ontologies and neuro-symbolic artificial intelligence. In: Brazdil, P., van Rijn, J.N., Gouk, H., Mohr, F. (eds.) ECMLPKDD Workshop on Meta-Knowledge Transfer, volume 191 of Proceedings of Machine Learning Research, pp. 8–10. PMLR (2022). www.proceedings.mlr.press/v191/hitzler22a.html

20. Xie, X., Kersting, K., Neider, D.: Neuro-symbolic verification of deep neural networks. arXiv preprint arXiv:2203.00938 (2022)

21. Hamilton, K., Nayak, A., Božić, B., Longo, L.: Is Neuro-symbolic AI Meeting Its Promises in Natural Language Processing? A Structured Review, pp. 1–42 (2022). https://doi.org/10.3233/SW-223228. www.medra.org/servlet/aliasResolver?alias=iospress&doi=10.3233/SW-223228

22. Oltramari, A., Francis, J., Henson, C., Ma, K., Wickramarachchi, R.: Neuro-symbolic architectures for context understanding. arXiv preprint arxiv.org/abs/2003.04707 (2020). https://doi.org/10.48550/ARXIV.2003.04707.Publisher: arXiv Version Number: 1

23. Dosovitskiy, A., et al.: An image is worth 16x16 words: transformers for image recognition at scale. arXiv preprint arXiv:2010.11929 (2021)

24. Pan, X., Ye, T., Han, D., Song, S., Huang, G.: Contrastive language-image pre-training with knowledge graphs. arXiv preprint arXiv:2210.08901 (2022)

25. Zhao, H., Torralba, A., Torresani, L., Yan, Z.: HACS: human action clips and segments dataset for recognition and temporal localization. arXiv preprint arXiv:1712.09374 (2019)

26. Rao, T.S., Radhakrishnan, R., Andrade, C.: Standard operating procedures for clinical practice. Ind. J. Psychiatry **53**(1), 1–3 (2011). https://doi.org/10.4103/0019-5545.75542

A New Similarity Metric for Deformable Registration of MALDI–MS and MRI Images

Florent Grélard[1]([⊠])(iD), Michael Tuck[1](iD), Elise Cosenza[2], David Legland[4,5](iD),
Marléne Durand[3](iD), Sylvain Miraux[2](iD), and Nicolas Desbenoit[1]([⊠])(iD)

[1] Univ. Bordeaux, CNRS, Bordeaux INP, CBMN, UMR 5248, 33600 Pessac, France
`florent.grelard@gmail.com, nicolas.desbenoit@u-bordeaux.fr`
[2] Univ. Bordeaux, CNRS, Résonance Magnétique des Systémes Biologiques
(UMR 5536), Bordeaux, France
[3] Univ. Bordeaux, CHU de Bordeaux, CIC-IT, INSERM, Institut Bergonié,
CIC 1401, 33000 Bordeaux, France
[4] UR BIA, INRAE, 44300 Nantes, France
[5] PROBE Research Infrastructure, BIBS Facility, INRAE, 44300 Nantes, France

Abstract. Multimodal imaging is a prominent strategy for biomedical research. For instance, Mass Spectrometry Imaging (MSI) can reveal the chemical composition of tissues with high specificity, helping to elucidate their metabolic pathways. However, this technique is not necessarily informative about the structural organization of a tissue. Other modalities, such as Magnetic Resonance Imaging (MRI), reveal functional areas in tissue. Images are analyzed jointly using several computational methods. Registration is a pivotal step that estimates a transformation to spatially align two images. Automatic methods usually rely on similarity metrics. Similarity metrics are used as optimization functions to find the transformation parameters. MALDI–MS and MR images have different intensity distributions that cannot be accounted for by traditional similarity metrics. In this article, we propose a novel similarity metric for deformable registration, based on the update of distance transformation values. We show that our method limits the intensity distortions while providing precisely registered images, on both synthetic and mouse brain images.

Keywords: Multimodal imaging · Image registration · Mass Spectrometry Imaging · Magnetic Resonance Imaging

1 Introduction

Multimodal imaging, or multiplexed imaging, is the combination of several imaging modalities. It supplements and enriches the data from one imaging modality, by using the information from a complementary modality. For instance, a relationship between chemical and cellular information can be established to evidence metabolic mechanisms in tissues or biomarkers of diseases [22].

G. Waiter et al. (Eds.): MIUA 2023, LNCS 14122, pp. 171–181, 2024.
https://doi.org/10.1007/978-3-031-48593-0_13

Matrix-Assisted Laser Desorption/Ionization – Mass Spectrometry Imaging (**MALDI–MSI**) is an acquisition technique which produces images of ionized molecules from a sample section based on their mass-to-charge ratio (m/z). This technique has gained traction in the recent years, with applications ranging from tissue metabolism [21], forensics [8], pharmacokinetics [4] to plant science [7]. A single MALDI–MS image encompasses thousands of molecules, without any prior tagging. The resulting image is hyperspectral: it has two spatial dimensions (image height and width) and a spectral dimension. Each pixel contains a spectrum which represents the molecular distribution at this position with high chemical specificity. This imaging technique can be combined with other techniques revealing the structural organization of a sample, such as Magnetic Resonance Imaging (**MRI**). The intensity in the images is linked to the proton density, which is mostly correlated to the amount of water. MALDI–MS and MR images are different in terms of signal quality, and dimensionality: MALDI–MSI encompasses the molecular distribution in two dimensions (2D) while MRI highlights the tissue structure in three dimensions (3D) at a glance.

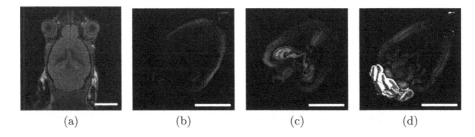

(a)   (b)   (c)   (d)

Fig. 1. Images of the mouse brain. Scale bars on the bottom-right corresponds to 5 mm. (a) MRI, *in vivo*, (b-d) MALDI–MSI, *ex vivo* for various ions: (b) m/z 764.520, (c) m/z 772.525 and (d) m/z 835.601 (all lipids). The intensity distributions of images from the two modalities are independent

Images from both modalities are processed and analyzed jointly in a multimodal computational workflow. A typical workflow involving MALDI–MSI and MRI has been described in [3]. First, a two-dimensional (2D) MR image which matches the MALDI–MS image is extracted from the 3D MR volume. Then, the images are *pre-processed* independently. This step involves various methods, such as denoising, normalization, and segmentation. The goal of this step is to obtain comparable images. *Registration* aims at spatially aligning the segmented images. Finally, images are analyzed jointly using *statistical analysis* methods, such as univariate statistical methods, clustering or machine learning approaches. Usually, the aim is to establish molecular fingerprints in MALDI–MSI that are spatially correlated to regions revealed by MRI.

Registration is the central step of multimodal workflows. It affects the quality of subsequent statistical analyses. MRI is performed on the whole object, in three dimensions, while MALDI–MS samples have two spatial dimensions and undergo various preparation steps which alter the tissue, both globally and locally. Deformable registration methods are necessary to compensate for local deformations. Moreover, the content of MALDI–MS and MR images is different in nature, resulting in images with discrepant intensity distributions (see Fig. 1). This complicates the usage of automatic registration methods which spatially align images based on intensity similarities.

In this paper, we focus on the registration of two-dimensional images. We propose a novel and precise registration method well-suited for images that have discrepant intensity distributions. Our method is based on the shape of the objects. More precisely, it is a similarity metric that uses distance transformation values. Distance transformation values are updated across registration iterations. Our metric is applicable to deformable registration methods, and allows for fine control over the transformation. It avoids spurious local deformations, and provides a precise spatial alignment of the images.

In Sect. 2, we present previous works related to the registration of 2D biomedical images. In Sect. 3, we detail our proposed method. Finally, we present the results obtained on both synthetic and real images, by comparison to state-of-the-art deformable registration methods in Sect. 4. We show our method is better able to find region-specific ions.

2   Related Works

Registration methods aim at finding the best transformation ϕ which aligns a target image T onto a reference image I. This is achieved through the optimization of a similarity metric $\mathcal{S}$, which quantifies the correspondences between the reference and deformed target images. The variational framework of [18] formulates a unique setting for deformable registration, in which the transformation is a deformation field $\phi(x)$. Deformable registration is an ill-posed problem because the number of parameters for the deformation field is greater than the information given by both images [19]. The regularization term $\mathcal{R}$ accounts for this problem by regularizing the deformation field. The deformable registration can be formulated as the following optimization problem:

$$\arg\min_{\phi} \mathcal{S}(I, T \circ \phi) + \mathcal{R}(\phi)$$

Similarity metrics can be classified into two categories: (a) intensity metrics, and (b) shape metrics.

Intensity metrics assess the correlation between pixel intensities in two images. Typical similarity metrics are the sum of squared intensity differences over both images (SSD), normalized cross-correlation, or mutual information, which estimates the statistical dependence of the intensity distributions of the target and reference images. For such metrics, intensity distributions are assumed to follow a joint probability density function. Mutual information and its normalized variant is widely used for biomedical image registration [10]. They are successfully applied to multimodal image registration involving modalities with similar intensity distributions, like computed tomography (CT) and positron emission tomography (PET) images [16]. However, mutual information does not accurately account for local deformations, because it considers the global relationship between intensity distributions. Local estimation of mutual information was done in [20] using square regions. Despite all these efforts, mutual information is not suited to images where the intensity distributions do not have any relationship, as is often the case for multimodal registration involving MALDI–MS.

Shape-based metrics compare the geometrical structure between two images. They usually involve fiducial markers, that is to say reference points which are found in both images. However, this either requires manual selections, or an automatic landmark detection methods, such as SIFT [11]. These methods are usually not suited to multimodal tasks involving MALDI–MSI because they rely on local intensity similarities.

In a more generic way, the authors of [5] register MR images by computing the distance transformation (DT) on a segmented image. The distance transformation is a representation of an object which maps each point to the distance to its nearest neighbor in the complement of the object. Let O be an object, $\bar{O}$, be its complement, the Euclidean DT maps the following value to each point $p \in O$:

$$DT(p) = \min_{q \in \bar{O}} \|p - q\|_2$$

For variational registration, distance transformations are inappropriate because the deformations induced at each step of the optimization process change the shape of the object. As a result, the interpolated values in the deformed distance maps do not correspond to the actual distances to the boundary (see Fig. 2). The resulting deformations can induce strong intensity distortions in the target image.

(a)   (b)   (c)   (d)   (e)

Fig. 2. Variational registration using SSD of distance transform values as a distance metric. (a) Reference image, (b) target image, and distance transformations of (c) the reference image, (d) the target after variational registration (e) overlay of figures c) in cyan, and d) in red, with similar values being gray. The deformations in the top-right part of the object in d) do not respect the theoretical shape from c). (Color figure online)

3   A New Registration Metric

In this section, we introduce a new similarity metric for the precise variational registration of images with different intensity distributions. This metric preserves the intensities of the target image.

Intensity-based methods are inadequate for the registration of images with different intensity distributions. We propose a new shape-based metric which makes use of distance transformations. Our method updates the distance transformation of the deformed target image at each optimization step (see Fig. 3). This metric, called Updated Distance Transformation – Sum of Squared Differences (UDT–SSD), compares the distance transformation of the reference image I, and the distance transformation of the target image T deformed by the transformation ϕ:

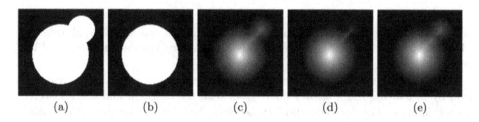

| (a) | (b) | (c) | (d) | (e) |

Fig. 3. Updated distance transform for variational registration. (a) Reference image, (b) target image, (c–e) distance transformations of (c) the reference image and (d–e) the target after variational registration, at different stages of the optimization process: (d) beginning, (e) end. The registered and the reference image have similar distance transformation values

$$\text{UDT–SSD}(I,T) = \sum_{i=0} \left(DT\big(I_i\big) - DT\big((\phi \circ T)_i\big) \right)^2$$

The update of distance transformation values ensures that the object is deformed continuously across iterations. This favors small deformations which better preserve the original intensities of the target image.

The definition for distance transformations is independent on the distance metric used. In this article, we use the Euclidean distance transformation, for which a linear-time algorithm was proposed [17]. This makes the computation of our metric time-efficient.

We opt for the variational framework proposed by [18] for deformable registration. It facilitates finding the right balance between shape preservation and intensity fidelity [9]. In the following, we use elastic regularization as it provides a fine control over the deformation field.

4   Results

In this section, the efficiency of our workflow is shown by comparison to state-of-the-art methods, using synthetic and mouse brain images. Mouse brain is used

Table 1. Evaluation of the variational registration on synthetic images by comparison of our metric, the Updated Distance Transformation (UDT), to the regular distance transformation (DT), and an intensity-based metric (Int), computed for different parameters (DT$_1$, DT$_2$, Int$_1$ and Int$_2$). This comparison is done quantitatively through four measures: the precision p, the recall r, the F-measure, and the mutual information m_I

	p	r	F	m_I
UDT	0.999	0.994	0.996	**0.572**
DT$_1$	0.967	0.995	0.980	0.569
DT$_2$	**1.00**	**0.996**	**0.998**	0.561
Int$_1$	0.947	0.988	0.967	0.463
Int$_2$	0.987	0.977	0.982	0.457

as a reference organ because regions can be identified quickly through existing MRI atlases (see e.g. [15]). These regions are used for the determination of ion co-localization in MALDI–MSI.

For both the synthetic and real dataset, we evaluate the quality of registration by measuring the intensity fidelity between the original, and the registered target images. The intensity fidelity is estimated by the mutual information m_I. Moreover, the deformation is assessed by quantifying the shape similarity. Objects are extracted by binarization, and three measures are used: (a) the precision p, i.e. the ratio between the number of common pixels and the number of pixels in the target image, (b) the recall r, i.e. the ratio between the number of common pixels and the number of pixels in the reference image and (c) the F-measure $F = 2 \cdot \frac{p \times r}{p+r}$.

The source code, with examples and documentation, is available online at https://github.com/fgrelard/RegistrationUDT, distributed under the Apache License 2.0 [1].

Synthetic Data. We generated synthetic images for the evaluation of our registration approach (see Fig. 4). The reference image encloses an ellipsoidal shape with various structures, such as a line segment and two disks. The target image is obtained by adding intensity variations and applying local deformations to the reference image.

We compare our metric, the Updated Distance Transformation (UDT), to an intensity-based metric (Int), as well as using the DT without update (DT). For all metrics, pointwise differences are evaluated by the Sum of Squared Differences (SSD). The resulting registered images are computed by using two sets of parameters: one set yields the best m_I values (Int$_1$, DT$_1$), and the other yields the best F values (Int$_2$ and DT$_2$).

The results for the variational registration are presented in Table 1 and Fig. 4. Our method yields the best mutual information value (0.572 bits) by comparison to the reference methods (see Fig. 4c). The shapes are also precisely aligned,

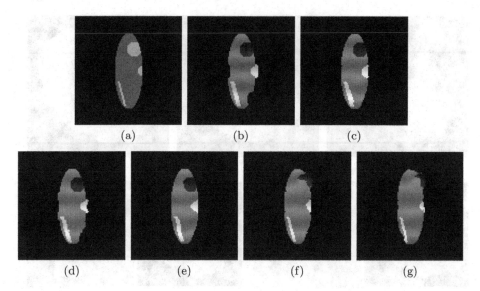

Fig. 4. Variational registration on synthetic images. (a) Reference image, (b) target image after registration using known rigid transformation parameters, (c–g) registered images using (c) our method, the updated distance transformation (UDT) metric, (d–e) the regular distance transformation (DT_1 and DT_2, respectively), and (f–g) the intensity-based metric (Int_1 and Int_2, respectively). Our method offers the best compromise between intensity fidelity and shape similarity

with a F-measure of 0.996. The registered images obtained by the other methods are less faithful to the original intensities. The best F-measure is obtained with parameter DT_2, but the structures have local distortions (see Fig. 4e). The intensities in the registered images obtained with Int_1 and Int_2 are not faithful. This is because the target image contains non-linear intensity variations (see Figs. 4f and 4g). Overall, the state-of-the-art methods make it difficult to find the right parameters to obtain both an accurate shape alignment and the preservation of original intensities. Our method yields the best compromise between shape similarity and intensity fidelity.

Real Data. The variational approach with the UDT is also evaluated on 2D mouse brain images. 3D MR images of *in vivo* mouse brains (C57Bl6) are acquired, then, the same mouse brain is sectioned and imaged in MALDI–MSI. More details about MRI and MALDI–MSI acquisition can be found in Supplementary information.

The sectioning and sample preparation steps induce global and local tissue deformations. First, images are pre-processed and rigidly aligned following the workflow of [9]. Then, MR images are mapped onto NeAt atlases [14,15], using the method of the authors in [12,13]. The resulting images contain twenty identified regions, such as hippocampus, ventricles or neocortex.

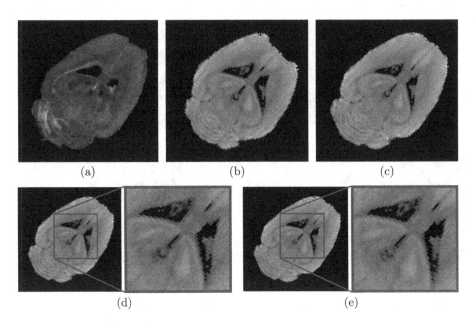

(a)   (b)   (c)

(d)   (e)

Fig. 5. Variational registration on mouse brain images. (a) MALDI–MS image segmentation (reference), (b) MR image after rigid registration (target) and (c) registered MR image after variational registration using our UDT, and (d–e) close-ups of overlays of registered images. Common pixels appear in grey. (d) Overlay of UDT (in cyan) and DT (in red). (e) Overlay of UDT (in cyan) and Int (in red). (Color figure online)

Table 2. Evaluation of the variational registration on real images by comparison of our metric, the Updated Distance Transformation (UDT), to the regular distance transformation (DT), and an intensity-based metric (Int), evaluating shape similarity (p, r, F) and intensity fidelity (m_I)

	p	r	F	m_I
UDT	**0.984**	**0.979**	**0.982**	**0.638**
DT	0.982	0.977	0.980	0.631
Int	0.984	0.976	0.980	0.635

After these initial steps, we evaluate our new metric during deformable registration. We compare our method to the regular distance transformation metric (DT), and to SSD of intensities (Int, see "Synthetic data" in Sect. 4). For each method, we choose the best trade-off between the F-measures and mutual information values (see Supplementary information for the used parameters).

The results are presented in Fig. 5 and Table 2. Our metric offers the best compromise between shape similarity ($F = 0.982$) and intensity fidelity ($m_I = 0.638$ bits). Overall, results are consistent with state-of-the-art methods. How-

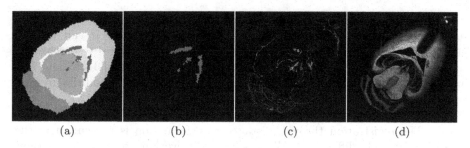

Fig. 6. Evidencing region-specific ions through Receiver-Operating Characteristic (ROC) analysis. (a) Regions found by MRI atlas, using the method of [13]. Each color corresponds to a region in the brain. (b) Ventricles (in red) from the MR image taken as reference for ROC computation and MALDI–MS images of (c) ion m/z 713.352 and (d) ion m/z 905.622. Ion m/z 713.352 is specific of ventricles while m/z 905.622 is not. The registered image from our metric confirms this statement, while the registered images from state-of-the-art methods do not

ever, differences that are larger than one pixel are visible in fine structures (see close-ups in Figs. 5d and 5e).

Even though those differences are small, they can impact the subsequent statistical analyses. We seek to identify ions that are specific of MRI atlas regions, registered by the previously computed transformations from the UDT, DT and Int similarity metrics. The area under the receiver-operating characteristic (AUC–ROC) curve gives an estimate of ion specificity for a given region. In the following, we restrict ROC analysis to ventricles because this region has the largest pixel-wise differences between previously registered images. AUC–ROC values are computed for each ion image. Then, they are sorted by descending order, and we compare the ion rankings across the various registration metrics.

The results are presented in Fig. 6. In the top 5% of AUC–ROC values, the average ranking difference is 4.64 between UDT and DT, and 4.23 between UDT and Int. Ion m/z 713.352 is specific of ventricles and ranked at the 28th position with our approach, while it is ranked 45th and 44th for the DT and Int approaches, respectively. Conversely, ion m/z 905.622 is not particularly specific of ventricles: it is ranked 42nd for our approach, while it is at the 32nd position for both DT and Int approaches. This means a region-specific ion is downgraded for a non-specific ion with the state-of-the-art methods. Thus, relevant and specific ions can be missed using these methods. Meanwhile, the registered image from UDT can reveal ions that are specific, while downgrading non-specific ions.

5   Conclusion and Perspectives

We proposed a novel similarity metric suited for deformable registration, based on updated distance transformation values. Our method represents the best compromise between shape preservation and intensity fidelity, by monitoring the deformation continuously across iterations.

Our method is fully automatic, and particularly well-suited in a multimodal setting where images have discrepant intensity distributions. It can lead to numerous benefits in image fusion in the context of biomedical research. Our results focus on the mouse brain, and the registration of MALDI–MS and MR images. However, our technique is not dependent on the sample and the imaging techniques.

The values of the distance transformation depend on the breadth of the object. The registration through distance transformation is not suited in the case of local scale differences. We are currently working on a scale-invariant distance transform. Our idea is to normalize the distance transformation values by the local scale of the object.

In this paper, we focused on the registration of 2D images. Several consecutive 2D sections can be acquired in MALDI–MSI. Those sections can be registered onto the 3D MR volume [6]. We plan on adapting the metric to three-dimensional (3D) images. Euclidean distance transformations have been extended to the 3D case [2]. The linear-time algorithm of [17] is independent on image dimensionality, which should keep the computation time tractable.

Acknowledgments. The authors would like to thank Bastien Arnaud, Mathieu Fanuel, Loïc Foucat and Héléne Rogniaux (UR BIA, BIBS facility, INRAE Nantes, France) for preliminary data acquisition and project supervision which allowed this work to be published.

Funding Information. This work was financially supported by the Agence National de la Recherche (France, Grants ANR-19-CE29-0010601 "MultiRaMaS"), the CNRS (Groupement de Recherche "GdR-MSI", GDR2125), and the STS Department of the University of Bordeaux.

References

1. Apache license. www.apache.org/licenses/LICENSE-2.0
2. Borgefors, G.: On digital distance transforms in three dimensions. Comput. Vision Image Underst. **64**(3), 368–376 (1996). https://doi.org/10.1006/cviu.1996.0065
3. Buchberger, A.R., DeLaney, K., Johnson, J., Li, L.: Mass spectrometry imaging: a review of emerging advancements and future insights. Anal. Chem. **90**(1), 240–265 (2017). https://doi.org/10.1021/acs.analchem.7b04733
4. Castellanos-Garcia, L.J., Sikora, K.N., Doungchawee, J., Vachet, R.W.: LA-ICP-MS and MALDI-MS image registration for correlating nanomaterial biodistributions and their biochemical effects. Analyst **146**(24), 7720–7729 (2021). https://doi.org/10.1039/d1an01783g
5. Chen, M., Carass, A., Bogovic, J., Bazin, P.L., Prince, J.L.: Distance transforms in multi channel MR image registration. In: Dawant, B.M., Haynor, D.R. (eds.) Medical Imaging 2011: Image Processing. SPIE (2011). https://doi.org/10.1117/12.878367
6. Dreisewerd, K., Yew, J.Y.: Mass spectrometry imaging goes three dimensional. Nat. Methods **14**(12), 1139–1140 (2017). https://doi.org/10.1038/nmeth.4513

7. Fanuel, M., et al.: Spatial correlation of water distribution and fine structure of arabinoxylans in the developing wheat grain. Carbohydrate Polym. **294**, 119738 (2022). https://doi.org/10.1016/j.carbpol.2022.119738
8. Francese, S.: Criminal profiling through MALDI MS based technologies – breaking barriers towards border-free forensic science. Aust. J. Forensic Sci. **51**(6), 623–635 (2019). https://doi.org/10.1080/00450618.2018.1561949
9. Grélard, F., Legland, D., Fanuel, M., Arnaud, B., Foucat, L., Rogniaux, H.: Esmraldi: efficient methods for the fusion of mass spectrometry and magnetic resonance images. BMC Bioinf. **22**(1) (2021). https://doi.org/10.1186/s12859-020-03954-z
10. Hill, D.L.G., Batchelor, P.G., Holden, M., Hawkes, D.J.: Medical image registration. Phys. Med. Biol. **46**(3), R1–R45 (2001). https://doi.org/10.1088/0031-9155/46/3/201
11. Lowe, D.: Object recognition from local scale-invariant features. In: Proceedings of the Seventh IEEE International Conference on Computer Vision. IEEE (1999). https://doi.org/10.1109/iccv.1999.790410
12. Ma, D., et al.: Automatic structural parcellation of mouse brain MRI using multi-atlas label fusion. PLoS ONE **9**(1), e86576 (2014). https://doi.org/10.1371/journal.pone.0086576
13. Ma, D., et al.: Study the longitudinal in vivo and cross-sectional ex vivo brain volume difference for disease progression and treatment effect on mouse model of tauopathy using automated MRI structural parcellation. Front. Neurosci. **13** (2019). https://doi.org/10.3389/fnins.2019.00011
14. Ma, Y., et al.: A three-dimensional digital atlas database of the adult c57bl/6j mouse brain by magnetic resonance microscopy. Neuroscience **135**(4), 1203–1215 (2005). https://doi.org/10.1016/j.neuroscience.2005.07.014
15. Ma, Y.: In vivo 3d digital atlas database of the adult c57bl/6j mouse brain by magnetic resonance microscopy. Front. Neuroanat. **2** (2008). https://doi.org/10.3389/neuro.05.001.2008
16. Mattes, D., Haynor, D., Vesselle, H., Lewellen, T., Eubank, W.: PET-CT image registration in the chest using free-form deformations. IEEE Trans. Med. Imaging **22**(1), 120–128 (2003). https://doi.org/10.1109/tmi.2003.809072
17. Maurer, C., Qi, R., Raghavan, V.: A linear time algorithm for computing exact euclidean distance transforms of binary images in arbitrary dimensions. IEEE Trans. Pattern Anal. Mach. Intell. **25**(2), 265–270 (2003). https://doi.org/10.1109/tpami.2003.1177156
18. Modersitzki, J.: Fair: Flexible Algorithms for Image Registration. Society for Industrial and Applied Mathematics, Philadelphia (2009)
19. Sotiras, A., Davatzikos, C., Paragios, N.: Deformable medical image registration: a survey. IEEE Trans. Med. Imaging **32**(7), 1153–1190 (2013). https://doi.org/10.1109/tmi.2013.2265603
20. Studholme, C., Drapaca, C., Iordanova, B., Cardenas, V.: Deformation-based mapping of volume change from serial brain MRI in the presence of local tissue contrast change. IEEE Trans. Med. Imaging **25**(5), 626–639 (2006). https://doi.org/10.1109/tmi.2006,872745
21. Trede, D., et al.: Exploring three-dimensional matrix-assisted laser desorption/ionization imaging mass spectrometry data: three-dimensional spatial segmentation of mouse kidney. Anal. Chem. **84**(14), 6079–6087 (2012). https://doi.org/10.1021/ac300673y
22. Tuck, M., Grélard, F., Blanc, L., Desbenoit, N.: MALDI-MSI towards multimodal imaging: challenges and perspectives. Front. Chem. **10** (2022). https://doi.org/10.3389/fchem.2022.904688

Decoding Individual and Shared Experiences of Media Perception Using CNN Architectures

Riddhi Johri[1], Pankaj Pandey[1], Krishna Prasad Miyapuram[1(✉)],
and James Derek Lomas[2]

[1] IIT Gandhinagar, Gandhinagar, India
kprasad@iitgn.ac.in
[2] TU Delft, Delft, The Netherlands

Abstract. The brain is an incredibly complex organ capable of perceiving and interpreting a wide range of stimuli. Depending on individual brain chemistry and wiring, different people decipher the same stimuli differently, conditioned by their life experiences and environment. This study's objective is to decode how the CNN models capture and learn these differences and similarities in brain waves using three publicly available EEG datasets. While being exposed to a variety of media stimuli, each brain produces unique brain waves with some similarity to other neural signals to the same stimuli. However, to figure out whether our neural models are able to interpret and distinguish the common and unique signals correctly, we employed three widely used CNN architectures to interpret brain signals. We extracted the pre-processed versions of the EEG data and identified the dependency of time windows on feature learning for song and movie classification tasks, along with analyzing the performance of models on each dataset. While the minimum length snippet of 5 s was enough for the personalized model, the maximum length snippet of 30 s proved to be the most efficient in the case of the generalized model. The usage of a deeper architecture, i.e., DeepConvNet was found to be the best for extracting personalized and generalized features with the NMED-T and SEED datasets. However, EEGNet gave a better performance on the NMED-H dataset. Maximum accuracy of 69%, 100%, and 56% was achieved in the case of the personalized model on NMED-T, NMED-H, and SEED datasets, respectively. However, the maximum accuracies dropped to 18%, 37%, and 14% on NMED-T, NMED-H, and SEED datasets, respectively, in the generalized model. We achieved a 5% improvement over the state of the art while examining shared experiences on NMED-T. This marked the out-of-distribution generalization problem and signified the role of individual differences in media perception, thus emphasizing the development of personalized models along with generalized models with shared features at a certain level.

Keywords: EEG · Neural responses · Music and Movie perception · Subjective differences

G. Waiter et al. (Eds.): MIUA 2023, LNCS 14122, pp. 182–196, 2024.
https://doi.org/10.1007/978-3-031-48593-0_14

Table 1. Synonyms for Evaluation and Experience Terminology

Experience	Evaluation Key Term	Model
Individual	Within-Subject	Personalized
Shared	Cross-Subject	Generalized

1   Introduction

The words "digital transformation", "innovation", and "media experience" have been a lot common in recent years, and companies aim to translate these concepts into tangible results. Creating an environment that provides customers with the most incredible user experience by allowing them to receive information customized to their needs and preferences is essential. One such method is to regulate the media experience using the user's brainwaves by detecting the person's concentration and excitement levels using electroencephalography (EEG) in both educational and entertainment applications [8,13,14,18] (Table 1).

Neurotechnology is a branch of neuroscience that has already made significant contributions to our knowledge of the brain and nervous system. It entails the creation of novel sensors and wearable gadgets for measuring, stimulating, and modulating brain activity. An EEG measures and records electrical activity in the brain using non-invasive sensors. EEG headsets are being enhanced and developed for wearable consumer applications, such as cognitive state detection, mental and neurological disorder detection, consumer choice prediction [19], and understanding and predicting people's responses to different media experiences [1,2], emotion prediction [20] and practicing meditation [4,15–17]. There is a lot of potential for EEG data to be used to improve media experiences. For example, EEG could be used to track how engaging a particular piece of media is and make recommendations accordingly to maximize the impact of the content [22,25].

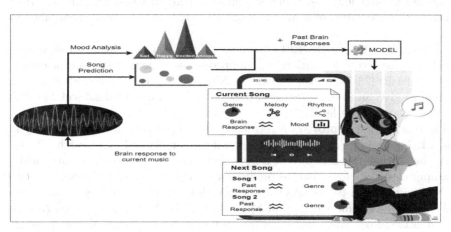

Fig. 1. Media Experience Brain Space: Each user's brain response and mood will be analyzed along with the prediction of the current song and its features. The model will then generate a playlist unique to the user, curated to evoke the emotion he/she is in

EEG is the most significant research area in the coming decade because EEG headsets can be pervasive as the fit-bit in our daily lives to monitor brain health. With the significant rise in the EEG domain, computation techniques should advance to capture intentional learning and reduce the unintentional factor of the learning. During the last decade, there has been a growing need for deep learning networks to be able to interpret learning in this field [3,21,23]. But to correctly elucidate these brain waves using neural architectures is the challenge. In this work, we predominantly discuss the disguise in feature learning that happens during the EEG classification task and the gap in the interpretations needed and interpretations done by these models.

The primary objective of this study is to evaluate Individual and Shared experiences in two different settings:

1. Song Classification
2. Movie Classification

and analyze the type and extent of feature learning done by deep learning networks. This is the first work which essentially presents the problem definition and proposes a direction for solving this problem.

2   Related Studies

Listening to music is a hobby, a tradition and a passion. The need for song recommendation systems arises as a result of system requirement that can recommend new songs to the users while being able to learn the user's past listening history, preferences, current listening habits, and mood. There are a few different ways that song recommender systems can operate. Some systems use collaborative filtering [6], which looks at the listening habits of a user's friends and followers to make recommendations. Other systems use content-based filtering [7], which looks at the attributes of a song (e.g. genre, artist, etc.) to make recommendations.

Yet another type is the EEG-based song identification which can be used to create playlist recommendations and improve song retrieval systems. The current state of the art [23] CNN-based model on NMED-T dataset for song identification is able to give an average test accuracy of 92.83% for within-subject but only 9%(less than chance level) for cross-subject classification. However, the cross-subject validation improved on retraining the model conditioned to classifying the EEG encodings into high and low-enjoyment classes. A research [18] has also shown that the first 20 s of a song segment can be used to train machine learning classifiers for accurate prediction from subsequent segments and that only β and γ band power spectra are enough to classify songs optimally. They achieved a maximum accuracy of 88% and 65% on Musin-G [12], and NMED-T datasets, respectively, using just one power band spectrum. However, as the testing shifted to cross-subject, the accuracy dropped to as low as 12% showing the out-of-generalization problem.

Listening to a song creates specific patterns in the brain, and these patterns are unique to each individual, which has been observed in the paper [24]. This is one reason why generalization is hard to get with EEG learning. While using a CNN architecture, the authors were able to predict the song using only 1 sec of EEG data and 10% of the data for training with an accuracy of 84.96% for within-participant. The results dropped to a 9.44% for cross-participant.

3   Data Description

3.1   NMED-T

This study analyzed NMED-T [11], a publicly available dataset containing behavioural responses and EEG from twenty participants engaged in a naturalistic song-listening. The EEG recordings were made with the electrode net attached, and behavioural ratings were obtained afterwards. Songs were presented in random order during the acquisition of the dataset. Each trial was followed by participants rating their familiarity and enjoyment of the music on a scale of 1–9. The EEG experiment was divided into two consecutive recording blocks to minimize participant tiredness and facilitate electrode impedance testing between the recording blocks. The preprocessed version of this dataset, which contains 125 channels of EEG data captured at 125 Hz, was primarily used in our study.

3.2   NMED-H

This publicly available naturalistic music EEG dataset [9] contains recorded brain signals of 48 adults listening to full-length Hindi pop songs. A total of sixteen stimuli, four songs, and four stimulus conditions per song are included in the dataset. A different version of the song was played twice to each group of twelve participants assigned to each stimulus. Each piece has four versions: Original, Reversed, Phase-scrambled, and Measure-shuffled, each lasting around 4.5 min. The results of this study are based on clean EEG Matlab files that had been cleaned, preprocessed, anonymized, and aggregated across participants into various stimulus matrices. Additionally, the dataset contains data structures listing the participants, stimulus and their behavioural ratings. Participants were only provided with one version of a song to listen to in the experiment.

3.3   SEED

SEED [28] contains EEG signals collected from fifteen subjects, seven males and eight females, who watched fifteen excerpts from Chinese films. The film clips lasted approximately four minutes and were well-edited to create coherent emotions that evoked and maximized three emotions: positive, negative, and neutral. Fifteen trials were conducted per subject, lasting 305 s, including a hint of starting for 5 s, a movie clip for 4 min, a self-assessment for 45 s, and a rest for

15 s. The data collected from the 62-electrode EEG cap was then downsampled to 200 Hz and processed using a bandpass filter ranging from 0 to 75 Hz [5]. The extracted differential entropy (DE) features of the EEG signals were smoothed further with conventional moving average and linear dynamic systems (LDS) methods.

4   Methodology

(See Fig. 2).

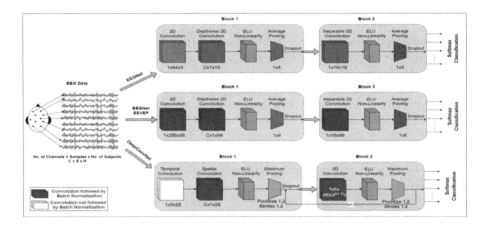

Fig. 2. The different architectures used for classification from EEG data

4.1   EEGNet

EEG data consisting of C channels, S time samples, and N subjects are passed for 150 epochs to the EEGNet model [10] composed of three convolutions in sequence. The input is routed through eight 2D convolution filters of size (1,64) to generate feature maps at various bandpass frequencies in the first block to obtain temporal information. Then, D*8 depthwise convolutions of size (C,1) are employed to learn spatial information within each temporal filter. The depth parameter D determines how many spatial filters should be learned for each feature map. With a dropout rate of 0.5, the model is regularized after applying exponential linear unit (ELU) Non-Linearity and Average Pooling layer of size (1, 4).

This is followed by a Separable 2D Convolution layer consisting of sixteen filters of size (1,16) in Block 2. This helps to combine spatial filters across temporal bands optimally. After applying ELU Non-Linearity, dimensionality reduction is achieved using an Average Pooling layer of size (1, 8). All the convolutions are followed by Batch Normalization. Finally, features after dropout are passed to the Softmax Classification layer.

4.2   EEGNet_SSVEP

The SSVEP variant of EEGNet [27] was explicitly designed for classifying Asynchronous Steady-State Visual Evoked Potentials signals. This differs from the above network in terms of size and number of kernels used in each convolution layer, as shown in Fig. 1. In block 1, ninety-six 2D Convolution and Depthwise 2D Convolution ($D = 1$) of size (1,256) and (C,1), respectively, are used to obtain frequency-specific spatial filters. Furthermore, depthwise convolutions reduce the number of free parameters to fit when compared to fully-connected convolutions.

In block 2, ninety-six separable convolutions of size (1,16) are used, which benefits by reducing the number of parameters to fit as well as explicitly decoupling the relationship between feature maps within and across them. In turn, a kernel summarizing each feature map is learned, followed by the optimal merging of the outputs. Each Convolution layer is followed by Batch Normalization. After convolutions in both blocks, the input passes through ELU non-linear activation, 2D average pooling, and dropout layers. Lastly, a dense layer and a softmax activation function are connected to the final layer.

4.3   DeepConvNet

The deep ConvNet architecture [26] to extract features and decode EEG signals is inspired by computer vision architectures. This architecture has four blocks, each consisting of a 2D convolution layer with max_norm constraint, batch normalization, ELU non-linearity activation, max pooling of size (1,2) with strides (1,2), and a dropout layer with a dropout rate of 0.5.

The convolution of the first block is split into two convolution layers of 25 filters each, one temporal layer (1,5) and one spatial layer (C,1). By using two layers, a linear transformation is forced into a combination of a temporal and a spatial filter, which implicitly regularizes the overall convolution. Finally, the fifth layer is a dense layer with a softmax activation function for classification.

4.4   Evaluation Strategy

To carry out this study, the datasets were divided into 5 s, 10 s, 20 s, and 30 s windows and an analysis of the performance of architectures was made to find the most efficient time window.

To decode how the models capture and comprehend the individuality and commonality in the perception of media by every individual, the experiments were performed in two settings. In the within-subject analysis, the data was split into the train, validate, and test datasets, thus having leakage of subjects' information. However, in the cross-subject analysis, the data was split such that the data of subjects present in the test dataset were not shown to the model while training. This resulted in a visible difference in the architecture's performance in both settings. They were inadequate to extract generalized features in the case of distribution shift. To verify these results, further experiments were done to plot t-SNE graphs showing song and subject classification. The t-SNE plots displayed the groups formed in training and testing data according to songs in the case of the personalized model and subjects in the case of the generalized model.

5   Experimental Results

Table 2. Classification Accuracy

Within Subject			
Chunks	NMED-T	NMED-H	SEED-M
5	**0.69 ± 0.01**	1.0 ± 0.001	**0.56 ± 0.01**
10	0.68 ± 0.02	**1.0 ± 0.0**	0.51 ± 0.01
20	0.57 ± 0.04	**1.0 ± 0.0**	0.48 ± 0.02
30	0.46 ± 0.02	0.97 ± 0.04	0.45 ± 0.02
Cross Subject			
Chunks	NMED-T	NMED-H	SEED-M
5	0.11 ± 0.04	0.35 ± 0.15	**0.14 ± 0.04**
10	0.13 ± 0.04	0.35 ± 0.16	0.13 ± 0.04
20	0.13 ± 0.03	0.33 ± 0.16	**0.14 ± 0.04**
30	**0.18 ± 0.07**	**0.37 ± 0.18**	0.14 ± 0.05

5.1   Time Window for Personalized Model

With different-sized time windows of the same dataset, Fig. 3(a) compares the best performance of all the architectures. The results of NMED-T, NMED-H, and SEED datasets, also evident in Table 2, indicate that 5 s window size is adequate and give the best results with 69%, 100%, and 56% accuracy, respectively. This shows that the architectures are capable of identifying and classifying even from the smallest snippet of brain signals. Thus, the features can be learnt and identified accurately and efficiently from tiniest fragments of brain activity by the networks in the case of within-subject evaluation.

5.2   Time Window for Generalized Model

Figure 3(b) compares the best performance for cross-subject evaluation from all the architectures when fed with different-sized time windows of the same dataset. Here the results are contrary to that of the personalized model as the 30 s window size gives the best results. Moreover, the architectures achieved a maximum accuracy of 18%, 37%, and 14%, respectively, on NMED-T, NMED-H, and SEED datasets as shown in Table 2. This shows the disguise in feature learning done by the models resulting in such low accuracies even when fed by large chunks of data at a time. However, if the dataset size is increased, the model might be able to learn features from a higher number of 30-second windows, leading to better results.

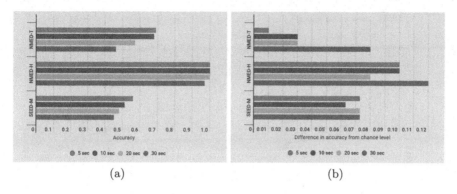

(a)   (b)

Fig. 3. (a) Within-Subject and (b) Cross-Subject Classification accuracies using different architectures on different-sized time windows

Table 3. Best accuracies achieved for each dataset

Within Subject

Dataset	EEGNET	EEGNet_SSVEP	DeepConvNet
NMED-T	0.61 ± 0.05	0.61 ± 0.04	**0.69 ± 0.01**
NMED-H	**1.0 ± 0.001**	**1.0 ± 0.001**	1.0 ± 0.002
SEED-M	0.32 ± 0.02	0.32 ± 0.01	**0.56 ± 0.01**

Cross Subject

Dataset	EEGNET	EEGNet_SSVEP	DeepConvNet
NMED-T	0.14 ± 0.04	0.16 ± 0.05	**0.18 ± 0.07**
NMED-H	**0.37 ± 0.18**	0.28 ± 0.09	0.27 ± 0.12
SEED-M	0.13 ± 0.05	**0.14 ± 0.05**	0.13 ± 0.05

5.3 Within-Subject Evaluation

We can see from the Table 3 and Fig. 4(a) that DeepConvNet outperformed the other two architectures on every dataset, whereas EEGNET and EEG-Net_SSVEP performed similarly on the three datasets. Although the difference in accuracy on NMED-T is only 8%, the accuracy on SEED movie classification has increased significantly from 32% to 56%. In view of this, a deeper neural network can be said to do a better training on brain signal data for classifying individual experiences since it has more layers in its architecture, allowing it to learn the features more accurately.

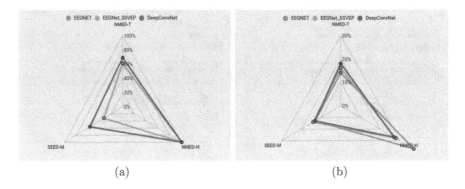

(a)   (b)

Fig. 4. Best (a) Within-Subject and (b) Cross-Subject Classification accuracies using different architectures on the datasets

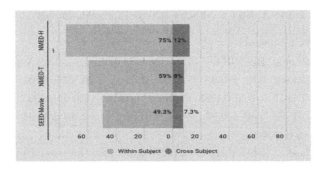

Fig. 5. The maximum increase in accuracy achieved from chance level on the datasets

5.4 Cross-Subject Evaluation

In contrast to the results of the personalized model, for the generalized model, no particular architecture performed well on all the datasets, as evident from Fig. 4(b). The architectures have showcased their inability to learn the appropriate feature to classify the data of unseen subjects. Table 3 shows that, on NMED-T data, DeepConvNet outperformed the other neural networks with a performance of 18%, EEGNet claimed a performance of 37% on NMED-H data, and EEGNet_SSVEP performed well on SEED data with a performance of 14%. However, recognizing the general characteristic of different brain signals for the same stimuli was not decoded by any architecture.

Table 4. Maximum classification accuracies achieved in the datasets considering all the architectures and various window sizes

Dataset	Chance Level	Within Subject	Increase	Cross Subject	Increase
NMED-T	0.1	**0.69**	0.59	**0.18**	0.08
NMED-H	0.25	**1**	0.75	**0.37**	0.12
SEED-Movie	0.067	**0.56**	0.493	**0.14**	0.073

5.5   Difference in Accuracies from Chance Level

A considerable difference between the increase in accuracies from chance level for within-subject and cross-subject classification can be seen in Fig. 5 and Table 4. In the personalized model, there was at least a 49% increase over chance levels, whereas, in the generalized model, it was barely 12%. This is due to the fact that the models did not learn relevant features of the song/movie to be able to classify when getting tested on distinct subjects indicating the distribution shift problem. However, the models performed exceedingly well when there was data leakage of the subjects. This suggests that models are influenced by the subjects' features and are learning properties specific to the song/movie as well as the subjects.

5.6   Network

Each of the experiments demonstrated that the DeepConvNet had shown consistent performance, with the model either providing the best accuracy or a similar level of precision to others. This is mainly due to its ability to capture the complex features of the data by increasing the depth of the network. The network uses a large number of filters and ReLU activations to increase the depth of the network and improve its performance. It also utilizes batch normalization and dropout layers to prevent overfitting, and the use of pooling layers has enabled the network to reduce the number of parameters used and thus reduce the computational complexity.

6   Discussion

6.1   Same Brain Perceives Different Stimuli Differently

The brain has a lot to say about who we are as individuals. Different parts of the brain may be activated depending on the type of stimuli and the person's individual response to it. Thus this difference is reflected in the EEG signals that help the models to identify the class of a new signal. Depending on the unique features found in EEG signals of different stimuli, the model gets trained to identify those features during testing to classify the song or movie. The t-SNE plots in Fig. 6(a) and (b) show how accurately the models were able to group the different songs of the NMED-T dataset.

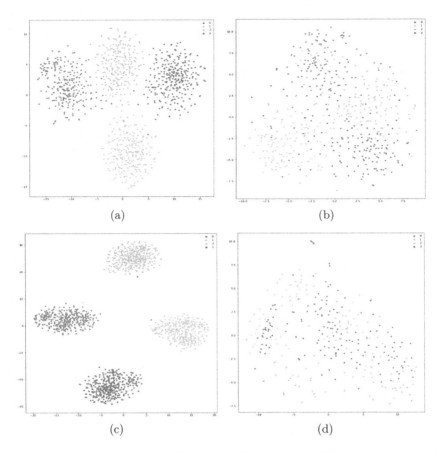

Fig. 6. t-SNE plots for Song classification on (a) train data, (b) test data after train-test split, and (c) train data, (d) test data after cross subject

6.2   Different Brains Perceive the Same Stimuli Differently

Variation in every individual's brain perception results in the models learning irrelevant features that might be specific to the subject rather than the media. This fact is visible while doing the cross-subject evaluation. The t-SNE plots in Fig. 6(c) and (d) show clear groups being formed for every song with the training dataset; however, on the test dataset, no such categorization is visible. On the other hand, for subject classification in Fig. 7(c) and (d), there is a clear group formation of subjects on the test dataset but not on the training dataset. Thus, the model has learnt features specific to the subjects and is categorizing based on that resulting in a low classification accuracy of only 18% for the generalized model. This is called the distribution shift or the out-of-distribution (OOD) generalization problem where the models are not able to accurately make predictions on data from the new, unknown distribution, which are new subjects here.

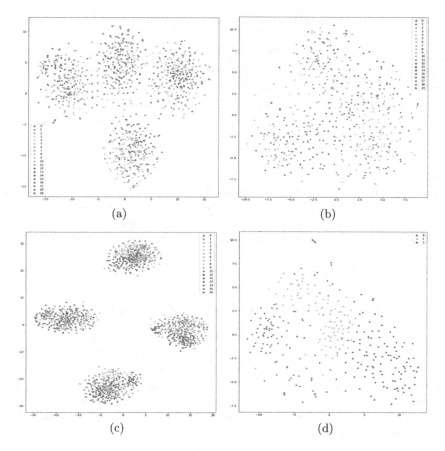

Fig. 7. t-SNE plots for Subject classification on (a) train data, (b) test data after train-test split, and (c) train data, (d) test data after cross subject

Since the same brain perceives the same stimuli invariably, the t-SNE plots in Fig. 6(a) and (b) of song classification in case of within-subject evaluation showcase precise group formation of tunes on both train and test datasets. This categorization is not achieved for subject classification, apparent in Fig. 7(a) and (b). Thus, when there was no distribution shift and the model was aware of all the subjects during the training and validation phase, it learned useful features of the media stimuli and achieved an accuracy of 69%.

6.3 Outperform State of the Art on NMED-T Music Identification

Recent works echoed in Table 5 have examined the song identification task using EEG signals to exhibit the subjective differences of neural responses in music perception. One of the studies [23] used a CNN architecture to identify songs of the NMED-T dataset with leave one subject out cross-validation and achieved a maximum precision of 9.9%. Another study [18] experimented with the relation

Table 5. Comparision with other works on Cross-Subject Song Classification

Article	Initial Input	Classifier	Accuracy
Dhananjay Sonawane et al. [24]	Time Frequency Plots	CNN architecture	9.44%
Gulshan Sharma et al. [23]	Topoplots	CNN architecture	9.9%
Pankaj Pandey et al. [18]	Frequency Bandpower	Random Forest	12.9%
Ours	Time Series	DeepConvNet	**18%**

between brain signals and unique and repetitive patterns present in the songs to classify the stimulus achieving a comparable maximum accuracy of 12.9% and 12.5% on NMED-T and MUSIN-G datasets respectively. Similarly, the authors of [24] showed the use of CNN architectures on the frequency domain dataset of MUSIN-G to gain a cross-subject song classification accuracy of 9.44%. Our work surpass these state-of-the-art works by achieving a maximum accuracy of 18% on the generalized model.

6.4   Why NMED-H Reflects the Highest Accuracy?

The models used in this research learned features not only unique to the media stimuli but to the subjects too. In the NMED-H dataset, a subject did not listen to a different version of the same song or an other song of the same version. Hence, no two songs considered for classification have the same subject. This results in corresponding groups of songs and subjects. This becomes more evident from the contrasting accuracies for within-subject and cross-subject evaluation of 100% and 37%, respectively.

7   Conclusion

In the future, EEG will empower the creation of large datasets that link people's brain activity to their responses to different media experiences. This could potentially be used to create personalized media experiences that are tailored to each individual's preferences and brain activity.

Our results show that different brains have different responses to the same experiences, and the same brain has different reactions to different experiences. The demand for new models that can adapt to this distribution shift between subjects is also evident from the t-SNE plots of song and subject classification for within-subject and cross-subject categories. It can be seen that while the test data t-SNE plot has clear and distinct groups of different songs in the case of the personalized model, the generalized model has the same for various subjects rather than the songs. The development of generalized models will elevate the media experiences of every individual using EEG wearables technology. The users will be able to better understand their cognitive and emotional processes, and thus make better decisions regarding their media consumption.

References

1. Building the world's most valuable brain data models. www.kernel.com/
2. Transforming music into medicine. www.lucidtherapeutics.com/
3. Bedmutha, P., Pandey, P., Ahmed, N., Miyapuram, K.P., Lomas, D.: Canonical correlation analysis (CCA) reveal neural entrainment for each song and similarity among genres (2022)
4. Chaudhary, S., Pandey, P., Miyapuram, K.P., Lomas, D.: Classifying EEG signals of mind-wandering across different styles of meditation. In: Brain Informatics: 15th International Conference, BI 2022, Padua, Italy, 15–17 July 2022, Proceedings, pp. 152–163. Springer, Heidelberg (2022). https://doi.org/10.1007/978-3-031-15037-1_13
5. Duan, R.N., Zhu, J.Y., Lu, B.L.: Differential entropy feature for EEG-based emotion classification. In: 6th International IEEE/EMBS Conference on Neural Engineering (NER), pp. 81–84. IEEE (2013)
6. Elahi, M., Ricci, F., Rubens, N.: A survey of active learning in collaborative filtering recommender systems. Comput. Sci. Rev. **20**, 29–50 (2016). https://doi.org/10.1016/j.cosrev.2016.05.002
7. Geetha, G., Safa, M., Fancy, C., Saranya, D.: A hybrid approach using collaborative filtering and content based filtering for recommender system. J. Phys. Conf. Ser. **1000**(1), 012101 (2018). https://doi.org/10.1088/1742-6596/1000/1/012101
8. Johri, R., Pandey, P., Miyapuram, K.P., Lomas, D.: Brain activity recognition using deep electroencephalography representation. In: 2023 IEEE Applied Sensing Conference (APSCON), pp. 1–3. IEEE (2023)
9. Kaneshiro, B., Nguyen, D.T., Dmochowski, J.P., Norcia, A.M., Berger, J.: Naturalistic music EEG dataset - hindi (nmed-h) (2014–2016). www.exhibits.stanford.edu/data/catalog/sd922db3535
10. Lawhern, V.J., Solon, A.J., Waytowich, N.R., Gordon, S.M., Hung, C.P., Lance, B.J.: EEGNET: a compact convolutional neural network for EEG-based brain-computer interfaces. J. Neural Eng. **15**(5), 056013 (2018). www.stacks.iop.org/1741-2552/15/i=5/a=056013
11. Losorelli, S., Nguyen, D.T.T., Dmochowski, J.P., Kaneshiro, B.: Naturalistic music EEG dataset - tempo (nmed-t) (2017). www.exhibits.stanford.edu/data/catalog/jn859kj8079
12. Miyapuram, K.P., Ahmad, N., Pandey, P., Lomas, J.D.: Electroencephalography (EEG) dataset during naturalistic music listening comprising different genres with familiarity and enjoyment ratings. Data Brief **45**, 108663 (2022). https://doi.org/10.1016/j.dib.2022.108663. www.sciencedirect.com/science/article/pii/S235234092200868X
13. Pandey, P., Ahmad, N., Miyapuram, K.P., Lomas, D.: Predicting dominant beat frequency from brain responses while listening to music. In: 2021 IEEE International Conference on Bioinformatics and Biomedicine (BIBM), pp. 3058–3064 (2021). https://doi.org/10.1109/BIBM52615.2021.9669750
14. Pandey, P., Bedmutha, P.S., Miyapuram, K.P., Lomas, D.: Stronger correlation of music features with brain signals predicts increased levels of enjoyment. In: 2023 IEEE Applied Sensing Conference (APSCON), pp. 1–3. IEEE (2023)
15. Pandey, P., Gupta, P., Miyapuram, K.P.: Brain connectivity based classification of meditation expertise. In: Mahmud, M., Kaiser, M.S., Vassanelli, S., Dai, Q., Zhong, N. (eds.) BI 2021. LNCS (LNAI), vol. 12960, pp. 89–98. Springer, Cham (2021). https://doi.org/10.1007/978-3-030-86993-9_9

16. Pandey, P., Miyapuram, K.P.: Nonlinear EEG analysis of mindfulness training using interpretable machine learning. In: 2021 IEEE International Conference on Bioinformatics and Biomedicine (BIBM), pp. 3051–3057. IEEE (2021)

17. Pandey, P., Rodriguez-Larios, J., Miyapuram, K.P., Lomas, D.: Detecting moments of distraction during meditation practice based on changes in the EEG signal. In: 2023 IEEE Applied Sensing Conference (APSCON), pp. 1–3. IEEE (2023)

18. Pandey, P., Sharma, G., Miyapuram, K.P., Subramanian, R., Lomas, D.: Music identification using brain responses to initial snippets. In: ICASSP 2022–2022 IEEE International Conference on Acoustics, Speech and Signal Processing (ICASSP), pp. 1246–1250 (2022). https://doi.org/10.1109/ICASSP43922.2022.9747332

19. Pandey, P., Swarnkar, R., Kakaria, S., Miyapuram, K.P.: Understanding consumer preferences for movie trailers from eeg using machine learning. arXiv preprint arXiv:2007.10756 (2020)

20. Pandey, P., Tripathi, R., Miyapuram, K.P.: Classifying oscillatory brain activity associated with Indian rasas using network metrics. Brain Inf. 9(1), 1–20 (2022)

21. Roy, Y., Banville, H., Albuquerque, I., Gramfort, A., Falk, T.H., Faubert, J.: Deep learning-based electroencephalography analysis: a systematic review - iopscience (2019). www.iopscience.iop.org/article/10.1088/1741-2552/ab260c

22. Salehzadeh, A., Calitz, A.P., Greyling, J.: Human activity recognition using deep electroencephalography learning. Biomed. Signal Process. Control 62, 102094 (2020). https://doi.org/10.1016/j.bspc.2020.102094. www.sciencedirect.com/science/article/pii/S1746809420302500

23. Sharma, G., Pandey, P., Subramanian, R., Miyapuram, K.P., Dhall, A.: Neural encoding of songs is modulated by their enjoyment (2022). https://doi.org/10.48550/ARXIV.2208.06679

24. Sonawane, D., Miyapuram, K.P., Rs, B., Lomas, D.J.: Guessthemusic: song identification from electroencephalography response (2020). https://doi.org/10.48550/ARXIV.2009.08793

25. Su, J., Wen, Z., Lin, T., Guan, Y.: Learning disentangled behaviour patterns for wearable-based human activity recognition. In: Proceedings of the ACM on Interactive, Mobile, Wearable and Ubiquitous Technologies, vol. 6, no. 1, pp. 1–19 (2022). https://doi.org/10.1145/3517252

26. Tibor, S.R., et al.: Deep learning with convolutional neural networks for EEG decoding and visualization. Human Brain Mapp. 38(11), 5391–5420 (2017). https://doi.org/10.1002/hbm.23730. www.onlinelibrary.wiley.com/doi/abs/10.1002/hbm.23730

27. Waytowich, N., et al.: Compact convolutional neural networks for classification of asynchronous steady-state visual evoked potentials. J. Neural Eng. 15(6), 066031 (2018). www.stacks.iop.org/1741-2552/15/i=6/a=066031

28. Zheng, W.L., Lu, B.L.: Investigating critical frequency bands and channels for EEG-based emotion recognition with deep neural networks. IEEE Trans. Auton. Ment. Dev. 7(3), 162–175 (2015). https://doi.org/10.1109/TAMD.2015.2431497

Revolutionizing Cancer Diagnosis Through Hybrid Self-supervised Deep Learning: EfficientNet with Denoising Autoencoder for Semantic Segmentation of Histopathological Images

Mostafa A. Hammouda[1], Marwan Khaled[1], Hesham Ali[1,2], Sahar Selim[1,2], and Mustafa Elattar[1,2(✉)]

[1] School of Information Technology and Computer Science, Nile University, Giza 12677, Egypt
mos.ahmed@nu.edu.eg
[2] Medical Imaging and Image Processing Research Group, Center for Informatics Science, Nile University, Giza 12677, Egypt

Abstract. Machine Learning technologies are being developed day after day, especially in the medical field. New approaches, algorithms and architectures are implemented to increase the efficiency and accuracy of diagnosis and segmentation. Deep learning approaches have proven their efficiency; these approaches include architectures like EfficientNet and Denoising Autoencoder. Accurate segmentation of nuclei in histopathological images is essential for the diagnosis and prognosis of diseases like cancer. In this paper, we propose a novel method for semantic segmentation of nuclei using EfficientNet and Denoising Auto-encoder on the PanNuke dataset. The denoising auto-encoder pre-processing step is used to enhance the feature representations of input images, and EfficientNet is the model that has been used as the semantic segmentation model. Our proposed method achieved state-of-the-art results, outperforming most of the previously proposed methods by a significant margin, as our proposed method achieved a higher Dice score of 83.33 compared to the previous related work methods, which achieved scores varying from 69.3 to 80.28. The efficiency of our proposed approach will be demonstrated in this paper by discussing, exploring, and comparing it with previously proposed methods and their results.

Keywords: semantic segmentation · histopathological images · self-supervised deep learning · EfficientNet · Denoising Autoencoder

1 Introduction

Medical imaging is a powerful tool in healthcare that allows physicians and healthcare professionals to see inside the body and diagnose diseases. Histopathology images are a type of medical imaging that is used to diagnose and study cancer. Cancer is a complex disease that can be difficult to diagnose accurately. In many cases, cancer diagnosis

G. Waiter et al. (Eds.): MIUA 2023, LNCS 14122, pp. 197–214, 2024.
https://doi.org/10.1007/978-3-031-48593-0_15

requires the examination of tissue samples, which are obtained through a biopsy or surgery. These tissue samples are then examined under a microscope by a pathologist, who looks for signs of cancer cells which is time consuming process.

Histopathological images are critical for cancer diagnosis and treatment for several reasons. First, they allow pathologists to examine tissue samples in detail and identify specific cellular and molecular characteristics that can help diagnose cancer [1]. Second, histopathology images provide a visual representation of the tumor, allowing physicians to determine its size, shape, and location. In addition to cancer diagnosis, histopathological images are also important for cancer treatment. They can help physicians determine the stage and grade of the cancer, which can guide treatment decisions [2]. In case the cancer is detected at an early stage and confined to a specific region, surgery might be considered as the optimal course of treatment. However, if the cancer has metastasized to other areas of the body, chemotherapy or radiation therapy could be required.

Advances in deep learning and computer vision have also made it possible to automate the analysis of histopathological images. These algorithms can help pathologists identify cancer cells more accurately and efficiently, leading to faster and more accurate diagnoses [3]. Additionally, these algorithms can be used to identify new biomarkers and potential drug targets for cancer treatment. Deep learning has revolutionized the field of computer vision, including semantic segmentation. Semantic segmentation is the task of assigning a class label to each pixel in an image, resulting in a dense labeling of the image. Deep learning models, particularly Convolutional Neural Networks (CNNs), have been successful in solving this task due to their ability to learn complex features and patterns from the input image. These models consist of several layers that process the input image in a hierarchical manner, gradually learning to extract higher-level features [4]. In recent years, several deep learning models have been proposed for semantic segmentation, including Fully Convolutional Networks (FCNs) [5], U-Net [6], and DeepLab [7].

One of the main problems of histopathological images used for segmentation is the lack of labeled data, because labeling data is a time-consuming process, especially for segmentation as it takes a long time to segment the cells particularly. So, to overcome this problem we are using a self- supervised method. Self-supervised learning is a training technique that allows a model to learn from a large amount of unlabeled data without relying on explicit supervision [8]. It does this by training the model to predict certain properties or relationships within the data itself, rather than being given explicit labels or targets. It can help address the limitation of the lack of labeled data by using a large amount of unlabeled data to pre-train a model, which can then be fine-tuned on a smaller labeled dataset for a specific task. By training the model on unlabeled data and defining a surrogate task that does not require explicit labels, the model can learn useful representations that can be transferred to other downstream tasks, such as semantic segmentation. Semantic segmentation involves the process of labeling every pixel within an image with a specific category or class based on the corresponding object or region it represents. In addition, Self-supervised methods have shown promising results in several medical image analysis tasks, including nuclear segmentation, by leveraging the large amount of available unlabeled data to learn useful representations. Nuclear segmentation refers to the process of identifying and segmenting the nuclei in an image. Nuclei are the most prominent structures in most biological images, and their segmentation is an essential task in many applications, including medical diagnosis and research, bioimaging, and cell biology. Accurate segmentation of nuclei can provide valuable information on the number, size, shape, and spatial distribution of cells, which can help in various tasks such as quantification of cell features, classification of cell types, and detection of abnormalities.

Denoising auto-encoder is one of the techniques that has been used for self-supervised learning. It is a type of neural network that can be used for image denoising, which is the process of removing noise from an image. This technique can be helpful in improving the accuracy of semantic segmentation, by removing noise and enhancing the features of the input images. In this research, we demonstrate that efficient denoising of medical images may be achieved using denoising autoencoders built using convolutional layers. EfficientNet [9] has gained significant attention in the computer vision community due to its superior performance and efficiency in various computer vision tasks, including semantic segmentation. We used EfficientNet as an encoder for our self-supervision approach, which is based on a compound scaling method that optimizes the depth, width, and resolution of the network simultaneously [9].

EfficientNet models are designed to be efficient in terms of computational resources. Despite their efficiency, EfficientNet models have shown remarkable performance on various benchmarks, outperforming larger and more complex models. In semantic segmentation, EfficientNet models have been used as an encoder in encoder-decoder architectures, where the encoder extracts features from the input image and the decoder upsamples these features to obtain a dense output. The advantages of using EfficientNet models in semantic segmentation include their high accuracy, efficiency, and scalability. Overall, EfficientNet has proven to be a powerful tool for semantic segmentation, achieving competitive results while being efficient in terms of computational resources.

The PanNuke dataset [10] used in this research is a popular dataset for segmentation, which contains histopathology images of breast cancer tissues, annotated with pixel-level labels for various tissue types. One of the challenges in semantic segmentation is dealing with class imbalance and class- specific features. We can overcome this challenge by using different loss functions, such as the Dice loss or cross-entropy loss, which can handle class imbalance and encourage the model to focus on hard- to-segment regions. Overall, self-supervised learning has shown promising results in segmentation, achieving high performance on several benchmark datasets. The ability of self-supervised learning models to learn complex features and patterns from the input image even when labeled data is limited has opened new possibilities for applications in fields such as medical imaging, where accurate and efficient segmentation of images is critical for diagnosis and treatment.

The objective of this research is to utilize self-supervised deep learning methods for semantic segmentation. The study provides valuable insights into the field of medical imaging, specifically regarding the segmentation of histopathology images. This research is expected to make significant contributions towards advancing the field of medical image analysis. First it proves that self-supervised deep learning can efficiently and accurately perform segmentation on histopathological images. Secondly, the combination of EfficientNet and Denoising Autoencoder approaches that are proposed in this paper can improve the accuracy and efficiency of semantic segmentation on histopathological images. Thirdly, the results achieved by using the powerful feature extraction capability of EfficientNet and the denoising capability of denoising auto-encoder were compared to the previous related works to prove the efficiency and accuracy of our proposed method.

2 Related Work

The research on semantic segmentation in computer vision has been ongoing for a considerable period of time. More recently, self-supervised deep learning techniques have emerged as a dominant force, delivering exceptional results across diverse semantic segmentation tasks. This section provides an overview of related research in semantic segmentation and explores the use of EfficientNet and denoising auto-encoder for image processing tasks.

The aim of semantic segmentation is to class label each pixel in an image [1]. Fully convolutional neural networks (FCNs) have been widely used for semantic segmentation tasks due to their ability to learn hierarchical feature representations. Shelhamer et al. [5] proposed the first FCN for semantic segmentation, which achieved competitive results on the PASCAL VOC dataset. Following this work, a variety of FCN architectures and optimization techniques have been proposed to improve the accuracy and efficiency of semantic segmentation like the Graph-FCN [11]. In the medical image analysis domain, semantic segmentation has been widely used for various applications, such as tumor segmentation, organ segmentation, and cell segmentation. For example, Ronneberger et al. proposed the U-Net architecture for medical image segmentation, which achieved significant results on various datasets [6]. Similarly, Çiçek et al. proposed a 3D U-Net architecture for volumetric medical image segmentation, which achieved competitive results on the MICCAI 2012 dataset [12].

Semantic segmentation can be used to automatically identify and segment specific structures or regions of interest in these images, such as cancerous cells or regions. This can help clinicians make more accurate diagnoses and treatment plans. Deep learning models, such as convolutional neural networks (CNNs), have shown promising results in segmenting histopathology images. For example, a study published in the Cancers Journal [13] used a U-Net architecture [6], a type of CNN, to segment breast cancer histopathology images into various tissue structures, including adipose, ductal carcinoma, and stroma. The study achieved a high accuracy in segmenting the tissue structures. Another example is the use of deep learning models in segmenting brain tumors from magnetic resonance imaging (MRI) scans. Another study [14] used a 3D U-Net architecture to segment brain tumors from MRI scans. The study achieved high accuracy in segmenting the tumors. These are just few examples of the many applications of deep learning approaches in semantic segmentation, which shows the ability of deep learning models to learn complex features and patterns from the input image has opened up new possibilities for accurate and efficient segmentation of various types of medical imaging, leading to improved diagnosis and treatment of diseases such as cancer.

Nuclear segmentation is an important task in medical image analysis, particularly in the field of pathology. Many previous studies have proposed different approaches for nuclear segmentation using various image processing techniques and deep learning methods. For example, one study [15] introduced a nuclei segmentation method based on two-stage learning framework consisting of two connected Stacked U-Nets (SUNets), and another study [16] presented a systematic survey on nucleus segmentation using deep learning in the years 2017–2021, highlighting various segmentation models (U-Net, SCPP-Net, Sharp U-Net, and LiverNet) and exploring their similarities, strengths, datasets utilized, and unfolding research areas. A study [17] proposed rotation and translation-equivariant U- Net architecture for nuclear segmentation in histopathology images on a dataset from TCGA samples for various types of tissue [18]. Another study proposed a self-supervised learning approach for nuclear segmentation in histopathology images, named NormToRaw [19], which is designed specifically for the task of nuclei segmentation and can extract semantic information from various stain types through style transfer, by utilizing a generative adversarial network, the method is capable of transferring normalized images to raw images, it is pre-trained on a dataset of over 8,000 unlabeled images and further trained on 16 labeled images, their results showed that the method is effective in improving the performance of nuclei segmentation. A similar study [20] proposed a self-supervised training approach to train a fully convolutional neural network (FCNN) that can identify features in input images and generate confidence maps which capture the network's belief about the objects belonging to the same class, they evaluated the proposed method for the task of segmenting nuclei from two histopathology datasets and showed comparable performance with relevant self-supervised and supervised methods. These studies demonstrate the potential of self-supervised learning approaches for nuclear segmentation in histopathology images, particularly in cases where labeled data is limited or difficult to obtain.

Denoising auto-encoder has been used in medical imaging before, for example, one of the previous similar works [21] proposed a model based on classification, denoising autoencoder, and transfer learning. The medical images are first pre-processed in order to optimize the probability of recovering inputs and clearly demonstrating their features, it relies on a denoising autoencoder (DAE) algorithm with a modified loss function based on a Gaussian distribution for decoder output. This helps to improve the diagnosis process. Then, classification is carried out to identify pneumonia using a transfer learning model and a four-layer convolution neural network (FCNN). Their suggested approach covers multi-class classification of chest X-ray images as well as binary classification of chest computed tomography (CT) images. In the context of semantic segmentation, denoising auto-encoder can be used to enhance the features of the input images by removing noise and improving the image quality.

The deep learning-based models have demonstrated considerable promise after outperforming all classic machine learning methods. However, the need for a large training sample size and significant computing costs limits the use of these methods [22]. Another limitation that deep learning-based models suffer from is the lack of consistency as the output of the models can be inconsistent and unpredictable, especially when dealing with noisy or ambiguous data. These limitations can make it challenging to apply deep learning models to segmentation tasks, especially in the medical domain where accuracy and interpretability are critical. Our study highlights the need for more investigation into addressing these limitations in deep learning models for segmentation.

3   Methods

The proposed approach in this paper aims to use a Denoising autoencoder model with EfficientNet-B3 backbone for preprocessing where the input images with noise are pre-processed. More preprocessing techniques like Data augmentation are performed to increase the size and diversity of the training dataset and to learn more robust and generalized features that are less dependent on the specifics of the training data, and more applicable to new, unseen data and avoid overfitting. Finally, preprocessed input is trained using the EfficientNet-B3 architecture which finally leads to the segmented mask as shown in Figure 1. In order to provide a clearer understanding of our proposed DAE with EfficientNet-B3 approach, we will provide a detailed description of the dataset used, elaborate on the preprocessing techniques employed, and discuss the Denoising Autoencoder and EfficientNet-B3 architectures utilized. Additionally, we will outline our training and evaluation methodologies.

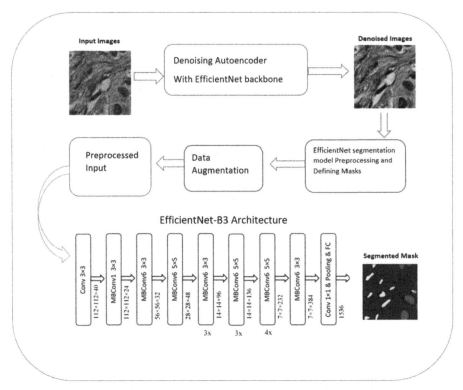

Fig. 1. Proposed DAE with EfficientNet-B3 Approach. Our proposed method involves prepro-cessing the input images with noise using a Denoising Autoencoder model. These processed images are then further preprocessed using the EfficientNet-B3 model, with six masks being introduced to the model. To prepare the input images for training, we perform data augmentation using the EfficientNet-B3 architecture. Finally, the model is trained to obtain the segmented mask.

3.1 Dataset Description

The PanNuke dataset is a publicly available dataset that contains 205,343 labeled nuclei categorized into 5 clinically important classes for the challenging tasks of segmenting and classifying nuclei from 19 different tissue types, including breast, lung, prostate, and colon [10]. The dataset used in this study comprises 481 visual fields, selected randomly from a pool of over 20,000 whole slide images captured at various magnifications. The dataset provides ground-truth annotations for each image, which include the following categories: Connective, Neoplastic, Inflammatory, epithelial, and Dead masks. In this study, we used the PanNuke dataset for semantic segmentation of nuclei.

3.2 Preprocessing

The exact size of the original images in the PanNuke dataset may vary from image to image, as they are derived from various sources and have different dimensions. Therefore, to comply with the input requirements of EfficientNet-B3, the original images in the Pannuke dataset were resized or cropped to a square shape with dimensions of 300x300 pixels. This resizing process ensures that the images are compatible with the EfficientNet-B3 architecture and can be effectively processed for semantic segmentation. Prior to training the models, we preprocessed the dataset by reading the folds and defining the masks. We also shuffled the training data and applied data augmentation techniques, including random horizontal flip, shift with limit equals 0.0625, rotation with 45 limit, and scale limit equals to 0.5, to increase the diversity of the training data and prevent overfitting.

3.3 Denoising Autoencoder Architecture

To enhance the feature representations of the input images, we also used a denoising auto-encoder [23] as a pre-processing step. The Denoising Autoencoder (DAE) employed in our study consists of multiple hidden layers to capture and encode the salient features of the input data. Each hidden layer is equipped with its own set of learnable parameters, including weights and biases, allowing the model to progressively learn hierarchical representations of the input. The DAE architecture is designed to balance sensitivity to input data and insensitivity to prevent overfitting. Ideally, an autoencoder model should be sensitive enough to accurately reconstruct the original input from the encoded representation, capturing the essential information and structural characteristics of the data. At the same time, it should be insensitive enough to avoid memorizing or overfitting the training data, ensuring generalization to unseen examples. This balance is crucial for achieving robust and effective representations, enabling the autoencoder to learn meaningful and compact representations of the input data.

The Denoising Autoencoder consisted of an encoder network, a decoder network, and a noise function. The encoder network mapped the input image to a latent space representation, while the decoder network mapped the latent space representation back to the image space. The noise function added random Gaussian noise to the input image, and the auto-encoder was trained to reconstruct the original image from the noisy input, to provide more specific details, the Gaussian noise is generated using a function that follows a Gaussian distribution. The mean and standard deviation values of the Gaussian distribution are utilized to control the magnitude and characteristics of the noise added to the input image. The particular parameters used for the mean and standard deviation are determined based on considerations for achieving an appropriate level of noise perturbation while ensuring effective denoising during the autoencoder training process. Figure 2 shows the architecture of the Denoising Autoencoder and a sample image through the Denoising Autoencoder process. The original images are corrupted by adding noise, then the corrupted images are trained using the layers of the EfficientNet-B3 architecture, which is considered the backbone of the Denoising Autoencoder model, which finally leads to a better preprocessed denoised image.

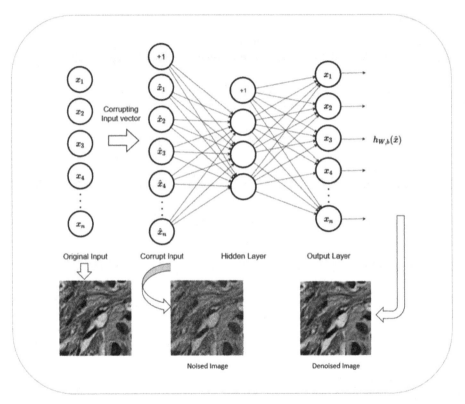

Fig. 2. Denoising Autoencoder (DAE) Architecture [23]. The first column displays the initial input embeddings, while the second column illustrates the input and hidden layer of the Denoising Auto-Encoder. Finally, the last column exhibits the output layer, which presents the denoised image.

3.4 EfficientNet-B3 Architecture

We used EfficientNet-B3 [9] as the backbone architecture for our semantic segmentation model. EfficientNet-B3 is a family of neural network architectures that have been shown

to achieve competitive results on various computer vision tasks, including image classification, object detection, and semantic segmentation. In this study, we used ImageNet
for encoder weights for the EfficientNet-B3 model. As for the number of classes, it was
set to 6 classes. EfficientNet-B3 was selected for the semantic segmentation task in this
research based on several considerations. While EfficientNet-B7 offers a higher level of
model complexity and potentially higher performance, there are specific reasons why
EfficientNet-B3 was chosen:

1. Trade-off between complexity and resource requirements: EfficientNet-B7 is a deeper
   and more complex model compared to EfficientNet-B3. This increased complexity
   comes at the cost of higher computational resources, such as memory and processing
   power. Considering the limitations of available computational resources, EfficientNet-
   B3 strikes a balance between model complexity and resource requirements, making
   it more suitable for practical implementation.
2. Dataset size and complexity: The choice of the model architecture may depend on
   the size and complexity of the dataset. EfficientNet-B7 might be more suitable for
   extremely large and complex datasets where a higher level of model capacity is
   required to capture intricate patterns. However, if the dataset, such as the PanNuke
   dataset, is relatively smaller or has specific characteristics that make it less complex, EfficientNet-B3 can still achieve satisfactory results without being excessively
   complex.

In summary, the selection of EfficientNet-B3 was driven by the considerations of
balancing model complexity, computational resources, dataset size, and previous performance. EfficientNet-B3 offers a reasonable trade-off between performance and resource
requirements, making it a suitable choice for the semantic segmentation task on the
PanNuke dataset.

3.5 Training

We trained the semantic segmentation model using the denoising auto-encoder preprocessed images. The model was trained on Colab Pro using its GPU as hardware
accelerator and using the PyTorch deep learning framework. Adam optimizer is used
with a learning rate of 1e-3 and a batch size of 4. We also used Cross Entropy Loss as a
loss function to train the model. To train the Denoising Autoencoder, Adam optimizer
was used with learning rate 3e-4 and Mean Squared Error (MSE) as loss function, in
addition it was trained on 65 epochs. Then the EfficientNet-B3 model was trained on
another 65 epochs. Finally, to evaluate the effectiveness of Denoising Auto-Encoder we
trained an EfficientNet-B3 model on 130 epochs without combining it to the denoising
autoencoder.

3.6 Evaluation and Performance Metrics

The performance of our model is evaluated using Dice evaluation metric, which measures
the overlap between the predicted segmentation masks and the ground-truth masks by
Multiplying 2 to the Area of Overlap divided by the total number of pixels in both
images. For Validation we performed per class Cross Validation on three folds, where

the number of epochs in each fold is set to 50 epochs. Furthermore, we compared our model's performance with the results of the previously proposed approaches to validate its effectiveness.

The Dice coefficient is used as an evaluation metric for measuring the performance of image segmentation algorithms. It is often preferred due to its simplicity and sensitivity to small differences between segmentation masks. The Dice coefficient measures the overlap between the predicted segmentation mask and the ground truth segmentation mask, with values ranging from 0 to 1. A Dice coefficient of 1 indicates perfect overlap between the predicted and ground truth segmentation masks, while a coefficient of 0 indicates no overlap [24].

The Dice coefficient is calculated as follows:

$$Dice = (2 * |A \cap B|) / (|A| + |B|) * 100$$

where A is the predicted segmentation mask, B is the ground truth segmentation mask, $|A|$ is the total number of pixels in A, and $|B|$ is the total number of pixels in B. $|A \cap B|$ represents the number of pixels where the predicted segmentation mask and the ground truth segmentation mask overlap.

4 Results

Initially, we used the ResNet-18 model as the backbone architecture for our semantic segmentation network. The key innovation of ResNet-18 is the introduction of residual connections, or skip connections, which allow the network to learn residual mappings. These connections bypass a few stacked layers by adding the input of those layers directly to the output. This enables the network to learn the residual information, or the difference between the input and the desired output, making it easier to optimize the network's weights. However, we found that the EfficientNet-B3 model achieved better results in terms of dice score, with a score of 82.6% compared to 78.6% for ResNet-18. In addition, the segmentation results with EfficientNet-B3 showed better visual quality compared to ResNet-18. Therefore, we chose EfficientNet-B3 as the backbone architecture for our final model. After that to further enhance the results, we used the Denoising Autoencoder for preprocessing and then performed the segmentation using the EfficientNet-B3 model. We compared the results of the EfficientNet-B3 model without using the denoising autoencoder with the results of the proposed DAE with EfficientNet-B3 method, to prove the efficiency of the denoising autoencoder. Furthermore, we conducted a comparative analysis between our results and the results from related work. In this section, we will go through the results of the EfficientNet-B3 model with and without the combination of Denoising Autoencoder, and the results of the comparison of our DAE with EfficientNet-B3 method with the results of the related work methods that used the same dataset and evaluation metric. Finally, the visual results of sample images are presented which show the predicted segmentation masks generated by our method compared to the ground truth masks.

4.1 EfficientNet-B3 with and Without the Combination of Denoising Autoencoder

We started our experiments by testing ResNet-18 as a backbone in which we obtained 78.6% dice score, and the results were not competitive. Then we used the EfficientNet-B3 approach which achieved better results than the ResNet-18 model, and 82.6% dice score. The improvement of the score by this margin (from 78.6 to 82.6) shows that the performance of EfficientNet-B3 in segmentation is better, but still the dice score and the visual results needed further enhancement. That is why we targeted the limitations that causes the inconsistency of the results which are image noise and artifacts, variability in staining and tissue characteristics and ambiguity in boundary delineation, therefore, the choice of DAE with EfficientNet-B3 was motivated by the aim to mitigate the mentioned limitations and enhance the consistency of the segmentation results. After Combining the Denoising Autoencoder, the dice score of the testing results and the visual segmentation results were clearly enhanced, as the Dice score increased from **82.6%** to **83.2%**. Table 1 shows a comparison of the dice score of the testing results between the ResNet-18, EfficientNet-B3 and the proposed DAE with EfficientNet-B3 approaches.

Table 1. Dice score results of the ResNet-18, EfficientNet-B3, and DAE with proposed EfficientNet-B3 method.

Model	Dice score
ResNet-18	78.6%
EfficientNet-B3	82.6%
Proposed DAE with EfficientNet-B3	**83.2%**

Enhancing the dice score and the visualization results (as shown in Fig. 3) of the EfficientNet-B3 model by this margin highlight's the significance of our approach and shows that it can contribute to addressing the limitation of inconsistency of self-supervised deep learning approaches in medical image segmentation by providing a way to improve the quality of input images, which can lead to more consistent segmentation results. These results prove that our novel DAE with EfficientNet-B3 approach is efficient and successful, and that it can help to improve the overall accuracy of segmentation and potentially lead to better diagnosis and treatment outcomes.

4.2 Comparison with Related Work Results

We evaluated the performance of our proposed method using Dice evaluation metric. Then our results were compared with the known self-supervised deep learning approaches that used the same dataset (PanNuke dataset) to perform the segmentation using the same evaluation metric that we used which is Dice evaluation metric. The research [25] compared the performance of different models which are Mask-RCNN [26], DIST [27] and Micronet [28]. Table 2 shows the cross-validation evaluation results of our proposed DAE with EfficientNet-B3 method compared to the approaches that were mentioned in their research.

EfficientNet

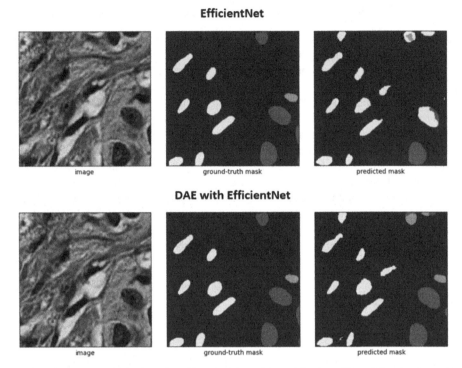

Fig. 3. Visual comparison of results with and without the combination of Denoising Autoencoder. The images on the top show the original image and ground truth mask compared to the predicted mask of the EfficientNet-B3 model without combining the Denoising Autoencoder, while the bottom images show the original image and ground truth mask compared to the predicted mask of the combination of the DAE with EfficientNet-B3 proposed model.

Table 2. Evaluation results of the DAE with the proposed EfficientNet-B3 method compared to the methods of the research [25] mentioned above.

Model	Dice score
Kaiming He [26]	69.36%
DIST [27]	75.23%
Micronet [28]	80.28%
Proposed DAE with EfficientNet-B3	**83.33%**

As shown in Table 2, the proposed method achieved a higher Dice score of 83.33 compared to the Mask-RCNN, DIST and Micronet approaches, which achieved scores varying from 69.3 to 80.28. The high score of our proposed method indicates that it performed better in accurately segmenting nuclei in histopathological images with high efficiency.

4.3 Visual Results

We also evaluated the visual results of our proposed method by comparing the segmentation masks generated by our method to the ground truth masks. The following figure (Fig. 4) shows the difference between the ground truth mask and the predicted mask of our proposed DAE with EfficientNet-B3 model.

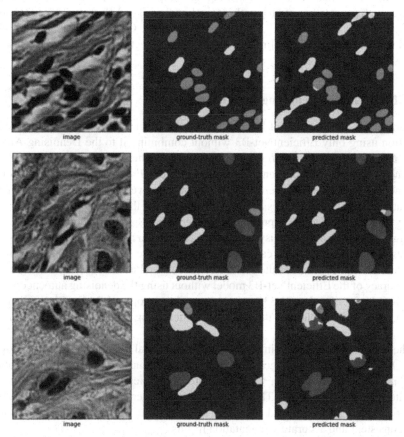

Fig. 4. Sample Segmentation Results of the Proposed Method. The images on the left show the original histopathology image, while the images in the middle show the ground truth segmentation mask. The image on the right shows the segmentation mask predicted by our proposed method.

5 Discussion

This research paper proposes a novel approach for semantic segmentation of histopathology images using a combination of denoising autoencoder and EfficientNet-B3 models. The Denoising Auto-Encoder enhanced the accuracy of the EfficientNet-B3 model. In addition, the proposed DAE with EfficientNet-B3 method achieved an 83.33% dice score on the PanNuke dataset, outperforming previous work and demonstrating the effectiveness of denoising autoencoder for preprocessing. The limitations of the study include the limited scope of the segmentation task and the potential for further investigation of alternative preprocessing techniques and self-supervised deep learning architectures. Future work could explore the use of instance segmentation approaches and other architectures such as HoVerNet and SpaNet. Overall, this paper contributes to the advancement of semantic segmentation in medical imaging and presents a promising approach for the accurate segmentation of histopathological images. The Effectiveness of Denoising Auto-Encoder on EfficientNet-B3, and the insights of the Comparison with the previous related work, in addition to the limitations and future directions will be further presented in this section.

5.1 Effectiveness of Denoising Auto-Encoder

To evaluate the effectiveness of Denoising Auto-Encoder, we performed semantic segmentation using only EfficientNet-B3 without combining it to the Denoising Autoencoder, and then compared its results to the results of the proposed method where we combined the denoising auto-encoder and the EfficientNet-B3 model. The denoising auto-encoder pre-processing step was found to be effective in enhancing the feature representations of the input images as well as the scores of the Dice evaluation metric. By adding random noise to the input images and training the auto-encoder to reconstruct the original image from the noisy input, we were able to reduce the effect of noise and artifacts in the images, which can negatively impact the performance of the semantic segmentation model. This resulted in a more robust and accurate segmentation model. The accuracy of the EfficientNet-B3 model without using the denoising autoencoder was 79.19% and it increased to 83.33% when we used the proposed DAE with EfficientNet-B3 approach, which highlights the efficiency of denoising autoencoder in enhancing the results.

The impact of using denoising autoencoder in medical image segmentation is significant. One of the main challenges in medical image segmentation is the high variability and complexity of the images, which leads to difficulties in obtaining consistent and accurate segmentation results. The denoising autoencoder can help overcome this limitation by reducing the noise and enhancing the features of the input images, resulting in more consistent and accurate segmentation results.

Furthermore, the combination of denoising autoencoder and EfficientNet-B3 models proposed in this paper has the potential to contribute to solving the limitation of inconsistency in self-supervised deep learning models. By enhancing the features of the input images using denoising autoencoder and using a model like EfficientNet-B3 for segmentation, the proposed approach can achieve more consistent and accurate results compared to previous related work. Overall, the proposed approach has the potential to significantly improve the accuracy and consistency of medical image segmentation, which can lead to better diagnosis and treatment of diseases. It can also pave the way for further research in the field of self-supervised deep learning for medical image segmentation, with potential applications in various medical fields.

5.2 Comparison with Previous Related Work

The DAE with EfficientNet-B3 approach proposed in this research for semantic segmentation of nuclei achieved 83.33% dice score, which outperformed most of the previous methods. The results of the DAE with EfficientNet-B3 proposed method are compared with the results of different methods that were proposed by a thesis [25] submitted for the Degree of PhD at the University of Warwick in which a comparative performance study of different models that also performed semantic segmentation on the PanNuke dataset using Dice as an evaluation metric. Our proposed method demonstrated a higher degree of effectiveness in accurately segmenting nuclei in histopathology images compared to the majority of other methods, as reflected by its superior scores. However, two more approaches were mentioned in the research [25] that we are comparing our results with, one of these approaches is HoVerNet [29] which achieved a similar result to ours (83.68%), but the other approach which is SpaNet [25] used instance segmentation and achieved 84.12% dice score which is better than our dice score. The fact that their approach achieved higher dice score suggests that we consider instance segmentation in the future work.

5.3 Limitations and Future Directions

Despite achieving competitive results, our proposed method has some limitations. First, the PanNuke dataset only contains annotations for nuclear, epithelial, and lymphocyte masks, which limits the scope of the segmentation task. Future studies could focus on extending the dataset to include annotations for additional tissue structures, such as blood vessels and connective tissue.

Second, while the denoising auto-encoder was effective in reducing the effect of noise and artifacts in the images, it may not be optimal for all datasets and segmentation tasks. Future studies could investigate alternative pre-processing techniques that are better suited for different datasets and segmentation tasks.

Finally, while EfficientNet-B3 achieved significant results in our study, there may be other architectures that are better suited for semantic segmentation tasks in histopathology images. Future studies could explore the use of other neural network architectures, such as HoVerNet [29] and SpaNet [25] which was used to perform instance segmentation in the thesis [25] and achieved higher scores than our proposed method. Which means that instance segmentation might lead to better segmentation, that is why future studies might also include performing instance segmentation instead semantic segmentation.

In conclusion, our proposed method of using EfficientNet-B3 and denoising auto-encoder achieved state-of-the-art results in semantic segmentation of nuclei. The denoising auto-encoder was found to be effective in enhancing the feature representations of the input images, resulting in a more robust and accurate segmentation model. Future studies could extend the dataset and explore alternative pre- processing techniques and neural network architectures to further improve the performance of semantic segmentation on histopathology images.

6 Conclusions

This paper aimed to address the importance of semantic segmentation in histopathology images using self-supervised deep learning approaches. The proposed method combined denoising autoencoder and EfficientNet-B3 models to perform semantic segmentation on histopathology images. The results achieved a dice score of 83.33%, which outperformed the previous works, proving that the usage of denoising autoencoder for preprocessing enhances the results and increases the accuracy. The proposed DAE with EfficientNet-B3 approach has a significant impact compared to the previously used approaches in histopathology image segmentation, contributing to the field of medical imaging and assisting in cancer diagnosis and treatment.

However, there are some limitations in the used dataset where it only contains annotations for nuclear, epithelial, and lymphocyte masks, which limits the scope of the segmentation task. Additionally, while our proposed approach achieved good results for semantic segmentation, instance segmentation approaches such as SpaNet achieved better results. Therefore, future studies could investigate instance segmentation using our proposed methods and other methods such as SpaNet. Additionally, alternative preprocessing techniques that are better suited for different datasets and segmentation tasks could be incorporated in the future work.

In summary, this paper proposed a novel approach to semantic segmentation in histopathology images using a combination of denoising autoencoder and EfficientNet-B3 models. The achieved results demonstrated the effectiveness of the proposed approach and its potential to improve cancer diagnosis and treatment. However, there is still room for further research and improvement in this field.

References

1. Wang, S., Yang, D.M., Rong, R., Zhan, X., Xiao, G.: Pathology image analysis using segmentation deep learning algorithms. Am. J. Pathol. **189**(9), 1686–1698 (2019). https://doi.org/10.1016/j.ajpath.2019.05.007
2. Banerji, S., Mitra, S.: Deep learning in histopathology: a review. Wiley Interdisc. Rev. Data Mining Knowl. Disc. **12**(1), 1439 (2022). https://doi.org/10.1002/widm.1439
3. Veta, M., Pluim, J.P.W., Van Diest, P.J., Viergever, M.A.: Breast cancer histopathology image analysis: a review. IEEE Trans. Biomed. Eng. **61**(5), 1400–1411 (2014). https://doi.org/10.1109/TBME.2014.2303852
4. Chen, L.-C., Papandreou, G., Kokkinos, I., Murphy, K., Yuille, A.L.: Semantic image segmentation with deep convolutional nets and fully connected CRFs (2014). http://arxiv.org/abs/1412.7062
5. Shelhamer, E., Long, J., Darrell, T.: Fully convolutional networks for semantic segmentation (2016). http://arxiv.org/abs/1605.06211
6. Ronneberger, O., Fischer, P., Brox, T.: U-Net: convolutional networks for biomedical image segmentation (2015). http://arxiv.org/abs/1505.04597
7. Chen, L.C., Papandreou, G., Kokkinos, I., Murphy, K., Yuille, A.L.: DeepLab: semantic image segmentation with deep convolutional nets, atrous convolution, and fully connected CRFs. IEEE Trans. Pattern Anal. Mach. Intell. **40**(4), 834–848 (2018). https://doi.org/10.1109/TPAMI.2017.2699184

8. Yuan, Y., Hou, J., Nüchter, A., Schwertfeger, S.: Self-supervised point set local descriptors for point cloud registration (2020). http://arxiv.org/abs/2003.05199

9. Tan, M., Le, Q.V.: EfficientNet: rethinking model scaling for convolutional neural networks (2019). https://proceedings.mlr.press/v97/tan19a/tan19a.pdf

10. Gamper, J., et al.: PanNuke dataset extension, insights and baselines (2020). http://arxiv.org/abs/2003.10778

11. Lu, Y., Chen, Y., Zhao, D., Chen, J.: Graph-FCN for image semantic segmentation. In: Lu, H., Tang, H., Wang, Z. (eds.) ISNN 2019. LNCS, vol. 11554, pp. 97–105. Springer, Cham (2019). https://doi.org/10.1007/978-3-030-22796-8_11

12. Ourselin, S., Joskowicz, L., Sabuncu, M.R., Unal, G., Wells, W. (eds.): MICCAI 2016. LNCS, vol. 9901. Springer, Cham (2016). https://doi.org/10.1007/978-3-319-46723-8

13. Jin, Y.W., Jia, S., Ashraf, A.B., Hu, P.: Integrative data augmentation with u-net segmentation masks improves detection of lymph node metastases in breast cancer patients. Cancers (Basel) **12**(10), 1–13 (2020). https://doi.org/10.3390/cancers12102934

14. Wang, F., Jiang, R., Zheng, L., Meng, C., Biswal, B.: 3d u-net based brain tumor segmentation and survival days prediction. In: Crimi, A., Bakas, S. (eds.) BrainLes 2019. LNCS, vol. 11992, pp. 131–141. Springer, Cham (2020). https://doi.org/10.1007/978-3-030-46640-4_13

15. Kong, Y., Genchev, G.Z., Wang, X., Zhao, H., Lu, H.: Nuclear segmentation in histopathological images using two-stage stacked U-Nets with attention mechanism. Front. Bioeng. Biotechnol. **8**, 573866 (2020). https://doi.org/10.3389/fbioe.2020.573866

16. Basu, A., Senapati, P., Deb, M., Rai, R., Dhal, K.G.: A survey on recent trends in deep learning for nucleus segmentation from histopathology images. Evol. Syst. (2023). https://doi.org/10.1007/s12530-023-09491-3

17. Chidester, B., Ton, T.-V., Tran, M.-T., Ma, J., Do, M. N.: Enhanced rotation-equivariant U-Net for nuclear segmentation. https://github.com/thatvinhton/G-U-Net

18. Kumar, N., Verma, R., Sharma, S., Bhargava, S., Vahadane, A., Sethi, A.: A dataset and a technique for generalized nuclear segmentation for computational pathology. IEEE Trans. Med. Imaging **36**(7), 1550–1560 (2017). https://doi.org/10.1109/TMI.2017.2677499

19. Chen, X., Zhong, X., Li, T., An, Y., Mo, L.: NormToRaw: a style transfer based self-supervised learning approach for nuclei segmentation. In: 2022 International Joint Conference on Neural Networks (IJCNN) , pp. 1–7. IEEE (2022). https://doi.org/10.1109/IJCNN55064.2022.9892957

20. Boserup, N., Selvan, R.: Efficient self-supervision using patch-based contrastive learning for histopathology image segmentation. In: Proceedings of the Northern Lights Deep Learning Workshop, vol. 4 (2023). https://doi.org/10.7557/18.6798

21. El-Shafai, W., et al.: Efficient deep-learning-based autoencoder denoising approach for medical image diagnosis. Comput. Mater. Continua **70**(3), 6107–6125 (2022). https://doi.org/10.32604/cmc.2022.020698

22. Gondara, L.: Medical image denoising using convolutional denoising autoencoders (2016). https://doi.org/10.1109/ICDMW.2016.102

23. Parashar, M.: Jaypee Institute of Information Technology University, University of Florida. College of Engineering, Institute of Electrical and Electronics Engineers. Delhi Section, and Institute of Electrical and Electronics Engineers, 2014 Seventh International Conference on Contemporary Computing (IC3): 7–9 August 2014, Jaypee Institute of Information Technology, Noida, India (2014)

24. Raudonis, V., et al.: Automatic detection of microaneurysms in fundus images using an ensemble-based segmentation method. Sensors **23**(7), 3431 (2023). https://doi.org/10.3390/s23073431

25. Koohbanani, N.A.: Working with scarce annotations in computational pathology (2020). http://wrap.warwick.ac.uk/153064

26. He, K., Gkioxari, G., Girshick, R.: Mask r-cnn. In: Proceedings of the IEEE International Conference on Computer Vision, pp. 2961–2969 (2017)
27. Naylor, P., Laé, M., Reyal, F., Walter, T.: Segmentation of nuclei in histopathology images by deep regression of the distance map. IEEE Trans. Med. Imaging **38**(2), 448–459 (2019). https://doi.org/10.1109/TMI.2018.2865709
28. Raza, S.E.A., et al.: Micro-Net: a unified model for segmentation of various objects in microscopy images. Med. Image Anal. **52**, 160–173 (2019). https://doi.org/10.1016/j.media.2018.12.003
29. Graham, S., et al.: Hover-Net: simultaneous segmentation and classification of nuclei in multi-tissue histology images. Med. Image Anal. **58**, 101563 (2019). https://doi.org/10.1016/j.media.2019.101563

Baseline Models for Action Recognition of Unscripted Casualty Care Dataset

Nina Jiang[1], Yupeng Zhuo[1], Andrew W. Kirkpatrick[2], Kyle Couperus[3,5], Oanh Tran[3,5], Jonah Beck[3,5], DeAnna DeVane[3,5], Ross Candelore[3,4], Jessica McKee[2], Chad Gorbatkin[3], Eleanor Birch[3], Christopher Colombo[3,5], Bradley Duerstock[1], and Juan Wachs[1(✉)]

[1] School of Industrial Engineering, Purdue University, West Lafayette, USA
jpwachs@purdue.edu
[2] University of Calgary, Calgary, Canada
[3] Madigan Army Medical Center, Tacoma, USA
[4] William Beaumont Army Medical Center, Fort Bliss, USA
[5] The Geneva Foundation, Tacoma, USA

Abstract. This paper presents a comprehensive framework of datasets and algorithms for action recognition in scenarios where data is scarce, unstructured, and unscripted. The long-term objective of this work is an intelligent assistant to the medic, a surrogate buddy, that can tell the medic what needs to get done in every step of trauma resuscitation. As an essential part of this objective, we collected datasets and developed algorithms suitable for emergent contexts, such as casualty care in the field, disaster response and recovery scenarios, and other related high-risks/high-stakes scenarios where real-time decision-making is crucial. The proposed framework enables the development of new algorithms by providing a standardized set of evaluation metrics and test cases for assessing their performance. Ultimately, this research seeks to enhance the capabilities of practitioners and emergency responders by enabling them to better anticipate and recognize actions in challenging and unpredictable situations. Our dataset, referred to as Trauma Thompson, includes Tourniquet Application, Tracheostomy, Tube Thoracostomy, Needle Thoracostomy, and Interosseous Insertion procedures. The proposed algorithms based on the relative position embedding for the Vision Transformer referred as to ReVit, can achieve competitive performance with the state-of-art algorithms on our dataset.

Keywords: Egocentric datasets · Action recognition · Action anticipation · Combat Casualty Care · Life-saving interventions · Surgical simulation

---

Disclaimers: The views expressed are those of the author(s) and do not reflect the official policy of the Department of the Army, the Department of Defense, or the U.S. Government. The investigators have adhered to the policies for the protection of human subjects as prescribed in 45 CFR 46.

1 Introduction

The current state-of-the-art in action recognition algorithms falls short in scenarios where the activities being performed exhibit a high degree of variability and where the available data is scarce. The scarcity of data compounds the challenges, as algorithms require substantial amounts of data to effectively learn patterns and make predictions. Consequently, the paucity of algorithmic frameworks capable of addressing the aforementioned challenges calls for the development of new, more robust frameworks capable of addressing these complex scenarios. Deep Learning-based approaches rely on large amounts of data with well-defined priors about the underlying data distributions [1]. These techniques make the implicit assumption that the tasks being performed consist of a sequence of more or less known actions, with little variation amongst them. This assumption may not hold in scenarios where the activities being performed exhibit significant variability or where the sequence of actions is not well-defined, such as in open-world settings. In such cases, Deep Learning-based techniques may struggle to accurately recognize and predict the actions, leading to decreased performance and increased uncertainty. The key concept of this paper is that by collecting few examples from very diverse procedures and using innovative Machine Learning algorithms, we can capture the variability existing in the aforementioned settings. We propose a Trans-former based approach that allows generalization due to the Vision Transformer has shown promising results in action recognition [2], the framework of ViT is more generalization and is a robust method to solve computer vision problems. These algorithms are tested in datasets that are unscripted and correspond to procedures in the field, Tourniquet Application (TOUR), Tracheostomy (CRIC), Tube Thoracostomy (CT), Needle Thoracostomy (ND), and Interosseous Insertion (IO) procedures where medics are required to conduct lifesaving procedures. Next, we will discuss the state of the art as it relates to dataset development and action recognition.

1.1 Egocentric Dataset

In recent years, significant advancements have been made in various domains such as image classification, object detection, captioning, and visual question answering. These developments have been largely attributed to the advancements in deep learning techniques and the availability of large-scale image benchmarks. However, progress in the field of video understanding has been slow, primarily due to the lack of annotated datasets. Nonetheless, this scenario has started to change with the release of action classification benchmarks like UCF101 [3], HMDB51 [4], Kinetics [5], Something-Something [6], and AVA [7]. Most of these datasets contain short videos, typically only a few seconds in duration, focusing on a single action. The Charades dataset, on the other hand, aims to recognize daily activities by collecting 10K videos of humans performing various tasks in their homes [8]. Nevertheless, the scripted nature of these videos recorded by AMT workers often makes them less natural and lacks the progression and multi-tasking that occur in real-life situations. To address this, research has shifted focus to first-person vision, which provides a unique viewpoint of people's daily activities. Egocentric data, in particular, has been found to be valuable for human-to-robot imitation learning and has a direct impact on Human-Computer Interaction (HCI) applications [9]. However, datasets to evaluate first-person vision algorithms are significantly smaller than their third-person counterparts and are often captured in a single environment, which limits the diversity of data available [10]. Daily interactions from wearable cameras are also scarcely available online, making it a largely untapped source of information. Therefore, egocentric datasets are getting more popular since they provide a novel perspective of actions, the first-person view, in the computer vision field. Most of them record the activities of daily living (ADL). For example, Pirsiavash et al. created a dataset that contains 20 participants and collected 10 h of ADL videos [11]. Damen et al. produced a dataset of egocentric ADL collected by wearable cameras. EGTEA Gaze+ contains 28 h (de-identified) of egocentric-view cooking activities from 86 unique sessions of 32 subjects [12]. The Epic-Kitchen dataset is a large collection of egocentric videos that contain 100 h of cooking actions of 32 participants [13]. Most of those datasets provide challenges in action recognition, and Epic-kitchen also holds challenges for action anticipation.

1.2   Action Recognition

In the realm of computer vision, the identification of human actions involves the categorization of actions depicted in a given video, while the localization or extraction of specific instances of actions within the video is the focus of action detection and segmentation. To achieve accurate representations of actions, deep learning models typically require access to vast quantities of video data, which presents a challenge due to the high dimensionality and volume of video data. It is crucial for these models to possess the capability to extract spatial and temporal complexity within video representations to facilitate action recognition. The study of action recognition has been a subject of extensive research over the past decade, with numerous survey papers and review articles available for reference [14–16]. Its relevance remains high due to its wide-ranging practical applications such as video retrieval and visual surveillance. Despite notable progress, creating an intelligent system that can recognize complex human actions in complex environments has proven challenging. The primary challenge in developing effective machine learning models for scenarios in the wild, particularly those associated with high-risk and high-stakes contexts such as crisis and recovery, natural disasters, and civilian casualties, is the requirement for large quantities of data. However, gathering such volumes of data in these contexts is often challenging. As a result, there is a need for machine learning algorithms that can generalize well from a limited number of observations that are representative of the diverse styles and scenarios encountered in the wild. Addressing this gap is crucial to developing models that can operate effectively and reliably in high-risk and high-stakes settings, where the consequences of errors or inaccuracies can be severe. In this paper, we represent the first egocentric unscripted dataset of life-saving intervention (LSI) procedures with detailed annotations by medical professionals. Unlike most of egocentric datasets, in which actions follow a pre-defined script, we have compiled 99 unscripted procedure videos with variability in terms of environment, simulator, and type. Based on this dataset, we propose multiple tasks to design impactful and LSI algorithms, including action recognition, and action anticipation. Moreover, it has the potential for multiple tasks including action detection, and vision questioning answering (VQA).

2   Methods

2.1   Dataset

Data Collection. We focus on capturing natural, unscripted LSI procedures from the first-person perspective, which involves operating a medical tool, searching for an item, changing one's mind, and encountering unexpected problems. We introduce the Trauma Thompson dataset, which is the first egocentric view dataset of LSI procedures to date. By focusing on first-person vision, this dataset offers a unique viewpoint on physicians' procedures, reflecting their goals, motivations, and ability to multitask while performing a variety of important tasks. It is expected that this dataset can facilitate the development and evaluation of first-person vision algorithms, which have significant implications for human-computer interaction (HCI) applications and human-to-robot imitation learning in the field of Tactical Combat Casualty Care (TCCC) [17].

Fig. 1. Examples of procedure video clips

Data Sources. The egocentric video capture method using a GoPro 7 (or higher) was utilized to obtain videos of medical procedures. These videos were captured by Subject Matter Expert in the field of trauma management and resuscitation (such as Emergency Medicine, Trauma Surgery, Critical Care Medicine), including representatives from University of Calgary, Madigan Army Medical Center, and William Beaumont Army Medical Center, to increase the diversity of the dataset. They have over 10 years of clinical experience and ensure that the camera angle was adjusted to capture the hands in the center of the image during the procedure. The videos were stored as MP4 files with 1080p resolution, and all actions in the videos were annotated by timestamps and extracted into frames. Guidelines for annotation were provided to the physicians, including using the present tense, verb-object pairs, and avoiding pronouns to ensure privacy. If an action took a long time, the medical experts could add extra lines with different timesteps for the same action in the same video clip. The physicians' extensive clinical experience guarantees the accuracy of the annotations. The dataset includes various procedures captured under different environments, including simulators, to enhance the diversity of the dataset. The procedure videos are expected to contribute to the development of new medical training and assessment tools. The procedure videos are captured with egocentric video capture using a GoPro 7 (minimum). As illustrated in Fig. 1, we have collected various procedures under different environments to increase the diversity of our dataset.

Data Pre-processing Methods. Once the procedure videos and annotations have been collected from the physicians, the data is undergoing pre-processing. Raw videos are processed to extract frames. Annotations are parsed into verb and noun pairs, and then classes are clustered for similar verbs and nouns. For instance, verbs such as "pick" and "take" are grouped into the same verb class.

Action Annotation. The dataset's annotation consists of the beginning timestamp, ending timestamp, and the action executed on a target, expressed as a verb-noun pair, for the corresponding video clip. The expected output for testing is the labels for the action, verb, and noun. Medical professionals were responsible for annotating the data, and providing the timestamps and actions for each procedural step. To reduce the possibility of errors in timestamping and video segmentation, the annotations underwent peer review. The detailed annotation process is described in Fig. 2.

2.2   Action Recognition Models

Inspired by the Vision Transformer introduced in the above section and relative position encoding methods, we present our Relative Vision Transformer (ReViT). As illustrated in Fig. 3, the ReViT is composed of three main components: the patches embedding layer, the Transformer encoder, and the classifier.

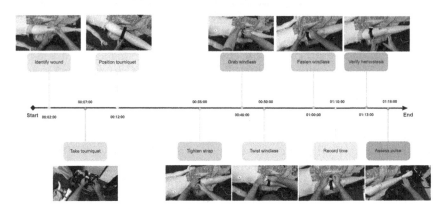

Fig. 2. Action annotations for the tourniquet application

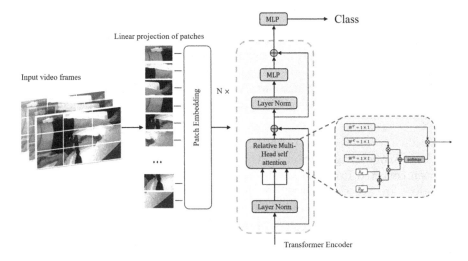

Fig. 3. Structure of Relative Vision Transformer.

First, extract the input video $V \in R^{(T \times H \times W \times C)}$ into separate frames, The extracted frames are sampled uniformly, which is similar to ViT [2]. After sampling n_t input frame, all the resulting patches were rasterized and concatenated into a single sequence of tokens. Therefore, the input video $V \in R^{(T \times H \times W \times C)}$ was processed into non-overlapping patches, $x_i \in R^{(h \times w)}$. Then, the patches were embedded, and the tokens passed into the Transformer encoder. For the Transformer algorithm, the self-attention block was the key part of the model. This self-attention block generated output by mapping a query and a group of key-value pairs. In the conventional model [18], self-attention computed the output sequence $z_i \in R^{d_z}$ as defined in the followings:

$$Z = (z_1, z_2, ..., z_n) \tag{1}$$

$$z_i = \sum_{i=1}^{n} \alpha_{ij}(x_i W_V) \qquad (2)$$

$$a_{ij} = softmax(\frac{(x_i W_Q)(x_j W_k)^T}{\sqrt{d_Z}}) \qquad (3)$$

The W_V, W_Q, and $W_K \in R^{(d_x \times d_z)}$ are layer-uniquely parameter matrices [18].

In transformer-based models, there are two primary methods to encode positional representations: absolute and relative. Absolute methods assign unique encoding vectors to each input token based on their absolute positions in the sequence, ranging from 1 to the maximum sequence length. These encoding vectors are then combined with the input tokens to convey positional information to the model. In contrast, relative position methods encode the pairwise relations of tokens by calculating the relative distance between input elements. Relative position encoding (RPE) is often achieved using a lookup table with learnable parameters that interact with queries and keys in self-attention modules. This approach enables the modules to capture long dependencies between tokens effectively. Relative position encoding has been proven to be effective in natural language processing [19]. Therefore, relative relations are presumably to be crucial for tasks that depend on the relative ordering or distance of elements. In order to improve the effectiveness of the model, compared to the standard Trans-former encoder, the relative position encoding was introduced into the muti-head self-attention modular. The self-attention that added the relative position embedding is: where the r_{ij}^W and r_{ij}^H are learnable relative width and height position weights between x_i and x_j, respectively [20].

$$z_i = \sum_{i=1}^{n} \alpha_{ij}(x_i W_V + r_{ij}^W + r_{ij}^H) \qquad (4)$$

3 Experiments and Results

3.1 Dataset Results

At present, the dataset comprises 62 videos demonstrating 5 medical procedures and contains 1020 fully annotated video clips. The procedures are (1) tube thoracostomy, (2) needle thoracostomy, (3) interosseous insertion, (4) tracheostomy, and (5) tourniquet application. We are still in the process of collecting data for the dataset. For classification tasks, we have selected class distribution based on the frequency of occurrence in real-world scenarios. We have extracted 323,565 separate interventional frames from 1080p videos (1920×1080) and the dataset includes 37 verb, 41 noun, and 127 action classes, as illustrated in Fig. 4 and Fig. 5.

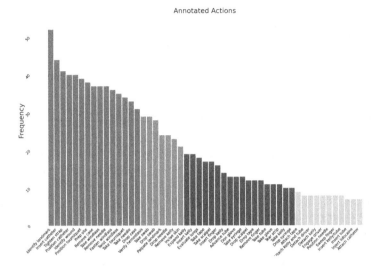

Fig. 4. Action classes of THOMPSON Dataset

Dataset Analysis and Insights. Figure 4 and Fig. 5 illustrate the imbalanced nature of the current dataset. This is due to the actions being collected from an unscripted dataset, which is distributed according to the real-life frequency of procedures. However, it is important to note that the performance of an algorithm is adversely affected by an increasing degree of class imbalance in the training data [21]. Therefore, the presence of an imbalanced dataset renders algorithm development more challenging compared to a balanced dataset. Figure 6 depicts the frequency of verb-noun combinations within the dataset. Evidently, it can be concluded that the verbs'take','remove', and'drop' exhibit a higher frequency of co-occurring with nouns. This phenomenon is indeed consistent with the frequency of actions observed during real-life life-saving procedures.

3.2   Action Recognition Results and Benchmarks

Train and Test Split. The video clips in both the training and testing datasets represent the designated procedure, but the training set contains full annotations for the action (verb + noun) and timestamp of the corresponding video clip, whereas the testing set only includes the timestamp. The total number of video clips is split into 70% for the training, 10% for the validation, and 20% for the test. Resulting in 714, 102, and 204 video clips for the training, validation, and test case, respectively.

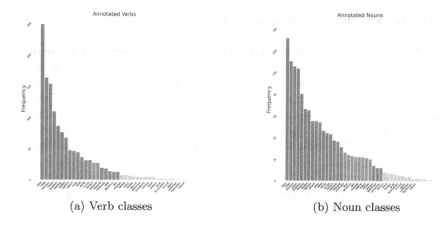

(a) Verb classes   (b) Noun classes

Fig. 5. Verb and noun classes

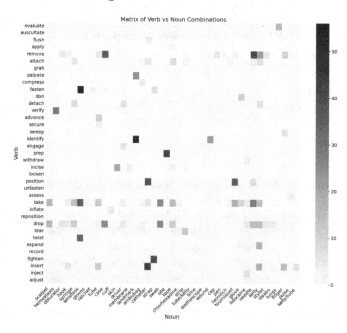

Fig. 6. Frequently of the co-occurring verb and nouns

Implementation Details. Each model was trained on a GeForce RTX™ 3090 Ti for 80 epochs with a mini-batch size of 512. We set the learning rate to 0.001 for spatial and temporal streams. Similar to [22], after averaging the 25 samples within the action segment, each with 10 spatial croppings. We fused both streams by averaging output predicted classes with the same weights. All unspecified parameters were set to the same values as [22].

Evaluation Metrics and Results Analysis. For the validation and test sets, a class-agnostic approach is used to assess accuracy [23]. A sequence of frames at the beginning of each clip is sampled for each action. Accuracy is calculated by dividing the number of correctly predicted action clips by the total number of action clips in the validation and test sets. The Top 1 and Top 5 accuracies for verb, noun, and action (verb + noun) are evaluated. In Table 1, the ReVit model produces better performance when compared to other models on the Thompson dataset, as evidenced by its accuracy of 18.18%. This is particularly noteworthy as the other models evaluated in the study, including Ego-rnn, SlowFast [24], and TSN [25], achieved 10.02%, 15.76%, and 12.27%, respectively. The ReVit is 2.42% higher than the TSN. However, note that this does not necessarily mean that the ReVit model is "better" than the other models in a general sense, but it was seen that performed better in the Thompson dataset.

Table 1. Action Recognition Benchmarks.

Model Name	EGTEA Gaze+	Epic-Kitchen 100	Epic-Kitchen 55	THOMPSON
Ego-rnn	60.13%	33.16%	14.47%	10.02%
TSN	55.26%	27.40%	19.85%	15.76%
Slowfast	54.72%	36.81%	23.96%	12.27%
ReVit	39.27%	35.33%	22.16%	18.18%

Table 2. Action Recognition Results for Each Procedure.

Procedure	Top-1 Verb	Top-1 Noun	Top-1 Act	Top-5 Verb	Top-5 Noun	Top-5 Act
Tour	20.00%	46.67%	24.44%	84.44%	95.56%	48.89%
CRIC	6.67%	15.39%	6.67%	30.77%	23.08%	15.39%
CT	35.63%	33.33%	13.79%	81.61%	78.16%	20.69%
ND	33.90%	38.98%	27.12%	83.05%	91.53%	54.24%
IO	23.81%	19.04%	14.29%	66.67%	42.86%	28.57%
Average	**27.69%**	**33.73%**	**18.18%**	**74.89%**	**74.76%**	**34.14%**

The overall Top-1 action recognition and Top-5 action recognition accuracy of the ReVit on our dataset are 18.18% and 34.14%, respectively. The limited availability of data and the imbalanced nature of the dataset make it challenging for algorithms to perform effectively. Table 2 presents the performance of the ReVit model for predicting the Top-1 and Top-5 verbs, nouns, and actions for five medical procedures: Tourniquet Application, Tracheostomy, Tube Thoracostomy, Needle Thoracostomy, and Interosseous Insertion. The model's accuracy is measured by comparing its predictions to the ground truth labels for each procedure. The results indicate that the model's performance varies depending on the procedure being performed since the dataset has more videos on the Needle Thoracostomy procedures. Needle Thoracostomy has the highest accuracy for Top-1 action recognition, with 27.12% accuracy. Tourniquet Application has the highest accuracy for top-5 verb and noun predictions, with 84.44% and 95.56% accuracy, respectively. The visualization of the results is illustrated in Fig. 7.

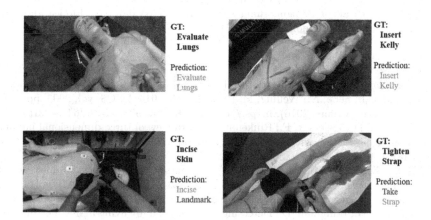

Fig. 7. Results visualization for action recognition

4   Conclusion

The authors present the Trauma Thompson dataset, which is the first unscripted egocentric view dataset for casualty care composed of 99 videos of 5 critical LSI procedures annotated by medical professionals. The dataset includes 127 action classes, 37 verb classes, and 41 noun classes. Moreover, the author proposed a Relative Vision Transformer (ReVit) to add relative position embedding in the multi-head self-attention for action recognition. Experimental results proved that the ReVit achieved competitive performance with the state-of-art algorithms on the Trauma Thompson dataset. In future endeavors, we plan to collect more videos with procedures, simulators, and environment variabilities, and develop the algorithm for action anticipation based on the relative position encoding.

References

1. Zhang, Y., et al.: Neural network-based approaches for biomedical relation classification: a review. J. Biomed. Inform. **99**, 103294 (2019). https://doi.org/10.1016/j.jbi.2019.103294
2. Dosovitskiy, A., et al.: An image is worth 16 × 16 words: transformers for image recognition at scale (2020). https://doi.org/10.48550/ARXIV.2010.11929

3. Soomro, K., Zamir, A.R., Shah, M.: UCF101: a dataset of 101 human actions classes from videos in the wild (2012). https://doi.org/10.48550/ARXIV.1212.0402

4. Kuehne, H., Jhuang, H., Garrote, E., Poggio, T., Serre, T.: HMDB: a large video database for human motion recognition. In: 2011 International Conference on Computer Vision, pp. 2556–2563. IEEE, Barcelona, Spain (2011). https://doi.org/10.1109/ICCV.2011.6126543

5. Kay, W., et al.: The kinetics human action video dataset (2017)

6. Goyal, R., et al.: The "something something" video database for learning and evaluating visual common sense (2017)

7. Gu, C., et al.: AVA: a video dataset of spatio-temporally localized atomic visual actions (2018)

8. Sigurdsson, G.A., Varol, G., Wang, X., Farhadi, A., Laptev, I., Gupta, A.: Hollywood in homes: crowdsourcing data collection for activity understanding. In: Leibe, B., Matas, J., Sebe, N., Welling, M. (eds.) ECCV 2016. LNCS, vol. 9905, pp. 510–526. Springer, Cham (2016). https://doi.org/10.1007/978-3-319-46448-0_31

9. Bachmann, D., Weichert, F., Rinkenauer, G.: Review of three-dimensional human-computer interaction with focus on the leap motion controller. Sensors 18(7), 2194 (2018). https://doi.org/10.3390/s18072194

10. Abebe, G., Catala, A., Cavallaro, A.: A first-person vision dataset of office activities. In: Schwenker, F., Scherer, S. (eds.) MPRSS 2018. LNCS (LNAI), vol. 11377, pp. 27–37. Springer, Cham (2019). https://doi.org/10.1007/978-3-030-20984-1_3

11. Pirsiavash, H., Ramanan, D.: Detecting activities of daily living in first-person camera views. In: 2012 IEEE Conference on Computer Vision and Pattern Recognition, pp. 2847–2854. IEEE, Providence, RI (2012). https://doi.org/10.1109/CVPR.2012.6248010

12. Li, Y., Liu, M., Rehg, J.M.: In the eye of beholder: joint learning of gaze and actions in first person video. In: Ferrari, V., Hebert, M., Sminchisescu, C., Weiss, Y. (eds.) ECCV 2018. LNCS, vol. 11209, pp. 639–655. Springer, Cham (2018). https://doi.org/10.1007/978-3-030-01228-1_38

13. Damen, D., et al.: The EPIC-KITCHENS dataset: collection, challenges and baselines (2020)

14. Sun, Z., Ke, Q., Rahmani, H., Bennamoun, M., Wang, G., Liu, J.: Human action recognition from various data modalities: a review. IEEE Trans. Pattern Anal. Mach. Intell. (2022). https://doi.org/10.1109/TPAMI.2022.3183112

15. Yao, G., Lei, T., Zhong, J.: A review of convolutional-neural-network-based action recognition. Pattern Recognit. Lett. 118, 14–22 (2019). https://doi.org/10.1016/j.patrec.2018.05.018

16. Abdulazeem, Y., Balaha, H.M., Bahgat, W.M., Badawy, M.: Human action recognition based on transfer learning approach. IEEE Access 9, 82058–82069 (2021). https://doi.org/10.1109/ACCESS.2021.3086668

17. Butler, F.K., Hagmann, J., Butler, E.G.: Tactical combat casualty care in special operations. Mil. Med. 161, 3–16 (1996). https://doi.org/10.1093/milmed/161.suppl_1.3

18. Wu, K., Peng, H., Chen, M., Fu, J., Chao, H.: Rethinking and improving relative position encoding for vision transformer (2021). https://doi.org/10.48550/ARXIV.2107.14222

19. Qu, A., Niu, J., Mo, S.: Explore better relative position embeddings from encoding perspective for transformer models. In: Proceedings of the 2021 Conference on Empirical Methods in Natural Language Processing, pp. 2989–2997. Association for Computational Linguistics, Online and Punta Cana, Dominican Republic (2021). https://doi.org/10.18653/v1/2021.emnlp-main.237

20. Bello, I., Zoph, B., Vaswani, A., Shlens, J., Le, Q.V.: Attention augmented convolutional networks (2020)
21. Mazurowski, M.A., Habas, P.A., Zurada, J.M., Lo, J.Y., Baker, J.A., Tourassi, G.D.: Training neural network classifiers for medical decision making: the effects of imbalanced datasets on classification performance. Neural Netw. **21**(2), 427–436 (2008). https://doi.org/10.1016/j.neunet.2007.12.031
22. Wang, L., et al.: Temporal segment networks: towards good practices for deep action recognition (2016)
23. Zhao, H., Torralba, A., Torresani, L., Yan, Z.: HACS: human action clips and segments dataset for recognition and temporal localization (2019)
24. Feichtenhofer, C., Fan, H., Malik, J., He, K.: SlowFast networks for video recognition (2019)
25. Wang, L., et al.: Temporal segment networks for action recognition in videos (2017)

Biomarker Detection

Web-Based AI System for Medical Image Segmentation

Hao Chen[1], Taowen Liu[1], Songyun Hu[1], Leyang Yu[1], Yiqi Li[1], Sihan Tao[1], Jacqueline Lee[1], and Ahmed E. Fetit[1,2(✉)]

[1] Department of Computing, Imperial College London, London, UK
`a.fetit@imperial.ac.uk`
[2] UKRI CDT in Artificial Intelligence for Healthcare, Imperial College London, London, UK

Abstract. Image segmentation is a crucial step in the diagnosis of brain tumours, and machine learning has emerged as a promising tool for tumour characterisation from medical imaging data. Despite their enormous potential in automatic segmentation of brain tumours from complex MRI scans, the implementation and use of machine learning algorithms can often present practical challenges to medical imaging researchers. This paper introduces a web-based GUI application designed to integrate all the components needed in deep learning workflows, allowing medical imaging researchers to seamlessly train and infer on data stored on in-house servers or on local machines. Our platform simplifies the process of training and inferring on MRI data using state-of-the-art models, supports integration with XNAT servers, and incorporates powerful tools for visualizing inference results.

Keywords: Deep learning systems · Magnetic resonance imaging · Image segmentation · XNAT · Image informatics

1 Introduction

Magnetic Resonance Imaging (MRI) is a non-invasive technique that can be used to detect, characterise, and monitor various diseases and conditions including brain tumours. Machine learning (ML) techniques, including deep learning-based algorithms, have demonstrated enormous potential in automatic segmentation of brain tumours from complex MRI scans. However, the way modern ML workflows are implemented in research settings presents a number of barriers to medical imaging researchers who may be novice to the computing aspects of the work. First, GPU acceleration is normally needed to provide the necessary compute power for training computationally intensive models, making GPU cards an invaluable component in most ML workflows. Furthermore, implementing the source code of a network architecture, as well as configuring GPU cards on research labs and hospital hardware may impede the adoption of powerful ML innovations by non-computer scientists. Moreover, researchers interested in

---

H. Chen and T. Liu—Equal contributions.

G. Waiter et al. (Eds.): MIUA 2023, LNCS 14122, pp. 231–241, 2024.
https://doi.org/10.1007/978-3-031-48593-0_17

training and deploying ML models should ideally be able to seamlessly use in-house servers and compute resources for getting access to the training data as well as storing any processed data, introducing further challenges in terms of writing data to and from the available servers.

To address these challenges, in this paper we introduce a web-based GUI application to the medical imaging community, integrating all the components typically needed in deep learning workflows into a single system. Our proposed web-based system allows users to seamlessly train and infer on MRI data stored on in-house servers or local machines, without the need for a programming background, using a variety of advanced ML models whose architectures and training parameters can be easily configured in the GUI.

Our contributions are as follows: *1)* we develop a web-based platform that enables medical imaging researchers to configure, train, and evaluate segmentation models using MRI data stored locally, *2)* we incorporate ML segmentation models (e.g. HyperDenseNet3D [3], ResNet3D [13], Gibbs ResUnet [2]) while ensuring that the model architectures and training hyper-parameters remain configurable by the user, and *3)* we integrate the eXtensible Neuroimaging Archive Toolkit (XNAT) [9] into our system, a popular medical image data management system, to allow users to easily use data stored on XNAT servers and carry out model training and/or inference.

Our web-based artificial intelligence (AI) system for medical image segmentation enables the following capabilities: *i)* tracking training progress through TensorBoard[1], *ii)* saving trained models for later use or further fine-tuning, *iii)* visualising inferred results using embedded third-party Neuroimaging Informatics Technology Initiative (NIfTI) [10] viewers, *iv)* allowing users to also evaluate their trained models on a selected test dataset and view the results in the provided evaluation history table, *v)* allowing users to log in to existing XNAT accounts, download data and upload inferred results from/to XNAT servers, and *vi)* evaluating on images from XNAT or uploaded locally. Whilst the main use-case discussed throughout this paper is brain tumour segmentation from MRI scans, the system can be easily adapted to support a variety of datasets and other anatomies or imaging modalities.

2   Related Work

Several frameworks have been developed that enable researchers to store, manage, and share medical imaging data, e.g. Dicoogle [14], an open source Picture Archiving and Communications System (PACS) archive, the Open Health Imaging Foundation (OHIF) Viewer [17], and the eXtensible Neuroimaging Archive Toolkit (XNAT) [9]. XNAT has been widely adopted as an infrastructure backbone for the organisation, management, and distribution of large imaging repositories [4]; examples include NeuroAI-HD [16], the Human Connectome Project (HCP) [15], the Developing Human Connectome Project (dHCP) [8], and the NITRC image repository [5]. Additionally, several projects have successfully

---

[1] https://www.tensorflow.org/tensorboard.

interfaced XNAT with programming languages and interoperability standards, e.g. RXNAT [4], PyXNAT [12], and FHIR on XNAT [6]. Whilst researchers should be able to directly use the data held on available XNAT servers to carry out model training and/or inference, little work was reported in the literature on providing no-code interfaces for incorporating deep learning workflows with XNAT.

3 Designing a Web-Based AI System for Medical Image Segmentation

In this section, we discuss the system architecture of our proposed web-based platform for medical image segmentation and the datasets used when deploying our system.

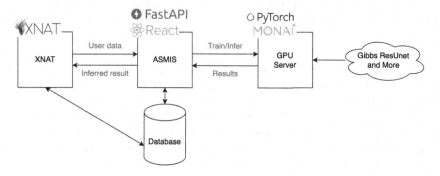

Fig. 1. System architecture of our web-based system for medical image segmentation.

3.1 System Architecture

Figure 1 gives an overview of the system's architecture. Our web-based platform can be used to either load MRI data stored locally or using XNAT [9]. XNAT provides a variety of tools for storing, organising, and exporting research imaging data and is widely used by medical imaging researchers worldwide across research labs, hospitals, and universities. We used FastAPI[2] and React[3] to implement the web interface that allows users to seamlessly communicate with XNAT and to upload data. Crucially, we used HTTPS in both the back-end and the front-end in order to ensure the safety of network traffic.

Regarding training and inference, we used PyTorch[4], a popular open-source machine learning library used for deep learning, for configuring, training deep

---

[2] https://fastapi.tiangolo.com/.
[3] https://react.dev/.
[4] https://pytorch.org/.

learning models, and ultimately using the trained models to infer on MRI images. The web-based system assumes that a GPU-enabled server is available for the users to carry out image training and inference. However, if no GPU resources are available, our system automatically falls back to available CPUs instead. The web server communicates with the GPU server using a combination of FastAPI HTTPS calls and WebSocket connections; FastAPI HTTPS calls are used to transmit the training configuration, dataset, and model architecture, while Web-Socket connections are used to provide real-time updates on the training process.

Figure 2 illustrates the workflow for training and inferring on medical images. The system uses data from XNAT or stored locally to create a dataset. With regards to data management in the context of MRI image segmentation, a 'dataset' is a collection of 3D scans and segmentation pairs used to train a model. To write a new dataset onto the system, the user can either upload local NIfTI images and segmentation labels, or retrieve and download an existing dataset from XNAT, which would require the XNAT Dataset Plugin. For XNAT access, the login page of our application enables users to log into their XNAT account using their access credentials, and identify the XNAT server they wish to use via a URL. This functionality enables users to easily switch between different XNAT servers, and ensures that they can only access data held on XNAT for which they have been granted permission. Crucially, XNAT offers fine-grained control of authorization which our system inherits so users could have great flexibility of authorization.

Once a dataset is written onto the system, it will be stored in the application's database in a ML-friendly storage format and could be readily used for model training and evaluation. The final step in the workflow is represented by the inference stage, where users can upload images from XNAT or from their local machines, and then view and save results.

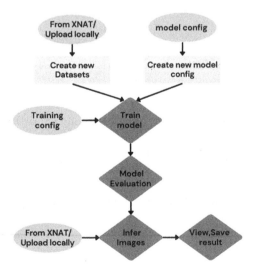

Fig. 2. Workflow for training and inferring on medical images.

To ensure that the application is robust and reliable, we employed static type checks and unit tests during our continuous iteration/continuous development pipeline. Specifically, we employed pytest for testing and mypy for performing static type checks; and we carried out a variety of validation checks for inputs to our application, e.g. training hyper-parameters (Fig. 3).

3.2 Datasets

Our application provides users with the option to upload their own MRI data onto XNAT, which can help adapt any developed models to the statistical distributions of a specific hospital or research centre. In our system, we made use of the publicly available brain tumour segmentation task data obtained from the Medical Segmentation Decathlon (MSD) challenge [1]. The MSD is a benchmark dataset that was designed to evaluate the performance of state-of-the-art segmentation algorithms. It consists of 10 different medical imaging segmentation tasks, covering several anatomical structures, including the liver, pancreas, and brain tumours. Our system was deployed with the brain tumour segmentation task data, which comprises two 3D NIfTI files for each data point, one for the MRI scan and the other for the label. The NIfTI files in the dataset are $240 \times 240 \times 155$ voxels, with a voxel size of 1 mm^3. The labeled data points include information about the location and size of the brain tumour, as well as the surrounding brain tissues.

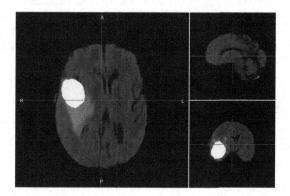

Fig. 3. Tumour label overlaid over a 3D MRI scan; scan obtained from the MSD dataset [1].

4   Training and Inference

In this section, we describe the key components regarding training and evaluation in our system: model configuration, training configuration, and inference and evaluation, as well as evaluate the usability of our system. Figure 4 shows the user view of the web application for the training step.

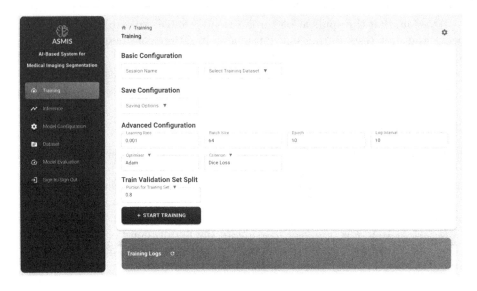

Fig. 4. User interface of the web application for the training step.

4.1   Model Configuration

To create a new segmentation model, users can use our system's model configuration page and choose their preferred network architecture. Several architectures are currently available, including a variety of models from the Medical Imaging Model Zoo [11] such as HyperDenseNet3D [3], HRNet3D [7], and ResNet3D [13]; as well as Gibbs ResUNet [2]. All models are pre-defined in PyTorch.

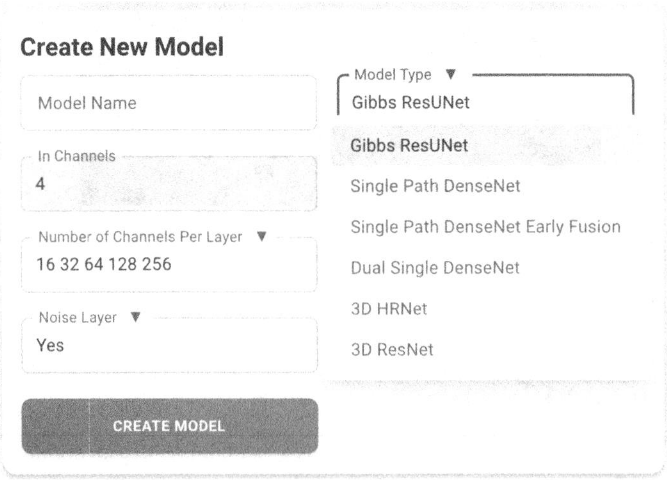

Fig. 5. User interface showing how a model can be configured.

Figure 5 shows a selection of the model types that can be selected and configured. Users can also specify exact parameters before training, such as the number of channels, number of layers, and whether to include a Gibbs Noise Layer in the case of Gibbs ResUnet. This level of customization allows users to tailor the model's architecture to suit the specific needs of their dataset and experiment, potentially improving the overall performance of the segmentation task. Once the user has chosen their desired model configuration, these settings are stored in the system's database for easy retrieval and management. This feature enables users to revisit and modify their configurations, experiment with different settings, and compare the performance of various model architectures and training parameters. By providing a user-friendly interface for selecting and customizing pre-defined models in PyTorch, our system aims to lower the barriers for medical professionals to access and apply advanced ML techniques for brain tumour segmentation tasks.

4.2 Training Configuration

Model training can be initiated using the GUI via a training page (see Fig. 6). Upon selecting the training set, the user can either train a new model or fine-tune an existing one. Hyperparameters can be specified in the advanced configuration section, including the learning rate, batch size, number of epochs, optimizer, as

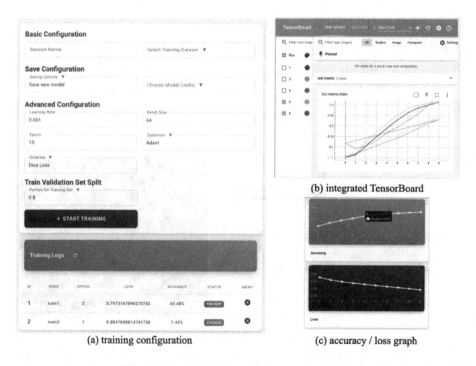

(a) training configuration

(b) integrated TensorBoard

(c) accuracy / loss graph

Fig. 6. User interface of the training page.

well as loss function. Users can then set the interval at which training logs are saved, and the split ratio determines the proportion of the training set and validation set. After the configuration step is complete, the training process can begin, and relevant metrics and the training status will be displayed in the 'training log' table. Moreover, our application has an integrated TensorBoard display which helps visualize training progress and results. A detailed training page will then be available for each training session created, plotting the computed training accuracy and loss in real-time.

The dataset is stored in server in the supported NIfTI format. Upon training, the dataset is firstly preprocessed, which involves cropping, normalization, and other necessary adjustments to ensure consistency. The preprocessed data is then converted into PyTorch tensors, which are compatible with the machine learning library. The network architecture is pre-stored in the dataset and is automatically loaded during the training process.

Once the GPU server receives the necessary information, it begins the training process, periodically sending updates to the web server with the latest training metrics, such as loss and accuracy. The web server then updates the training log table and the integrated TensorBoard display in real-time, allowing users to monitor the progress of the ongoing training session. This implementation design ensures seamless communication between the user interface, web server, and GPU server, enabling users to focus on configuring and monitoring the training process without worrying about the underlying complexities.

4.3   Inference and Evaluation

Our application enables users to evaluate their trained models on a selected test set, providing valuable insights into the model's performance. The Intersection

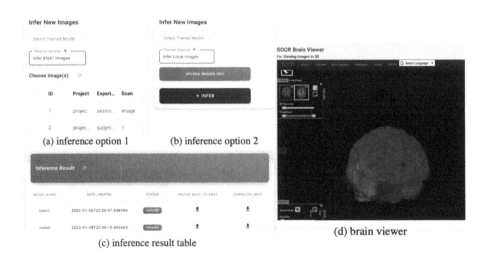

(a) inference option 1

(b) inference option 2

(c) inference result table

(d) brain viewer

Fig. 7. User interface of the inference page, including the third-party brain viewer.

over Union (IoU) metric is displayed and automatically updated as the evaluation progresses, offering users a clear understanding of the model's segmentation accuracy. Once a user has finalized the training process, they can employ the trained model to infer the brain tumour segmentation of an MRI image. The user has the option to choose between using images from XNAT or uploading local images for inference, and all the data can either be downloaded to the local machine or uploaded to the XNAT server as needed, providing ample flexibility in data management.

Furthermore, our application incorporates two third-party viewers, SOCR Brain Viewer[5] and Papaya[6], that allow users to rapidly inspect and visualize the segmentation output. These image viewers provide various visualization options, allowing users to assess the quality of the segmentation and identify potential areas for improvement. Figure 7 shows the user view of the inference page, including the brain viewer.

4.4   Evaluating Usability

To evaluate the usability of our system, we asked five users with extensive ML background to use our application, and assigned them tasks focused on model training, evaluation and inference. With an average of 2.2 questions asked per user, each task took an average of 341 s to complete. We also asked the users to estimate the amount of time they would take to perform the same process with Python code, based on their prior experience. This amounted to 35 min on average, suggesting that our application can substantially improve the efficiency of modern ML workflows (see Fig. 8).

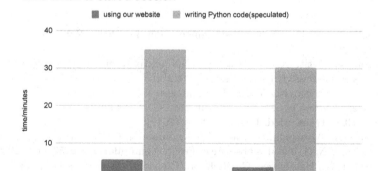

Fig. 8. Time taken to start an inference/evaluation session

---

[5] https://socr.umich.edu/HTML5/BrainViewer/.
[6] https://www.fmrib.ox.ac.uk/ukbiobank/group_means/index.html.

5   Conclusion and Future Work

In this paper, we presented an efficient web-based platform that integrates ML models with medical imaging tools through a user-friendly and intuitive interface which does not require any programming experience. The application greatly simplifies the process of training and inferring MRI data for brain tumour segmentation research, empowering users to interact with the system by integrating all the components typically needed in deep learning workflows. The platform also supports uploading and downloading data to/from to XNAT servers, tracking training progress through TensorBoard, and viewing inferred results using embedded third-party NIfTI viewers.

Our web-based AI system surpasses the individual components it integrates by carefully handling edge cases and validating all inputs before training, evaluation, and inference sessions, thus simplifying the user process for debugging ML algorithms. There are several avenues for future work; first, we aim to provide flexibility for users to be able to define, implement, and upload their own PyTorch model. Moreover, we plan to extend our system to incorporate other datasets and other anatomies or imaging modalities. Finally, for a more secure approach to handling sensitive medical data, implementing federated learning with XNAT servers represents a promising direction of future work.

Acknowledgments. The research of Dr Ahmed E. Fetit was supported by the UKRI CDT in Artificial Intelligence for Healthcare in his role as Senior Teaching Fellow (grant number EP/S023283/1). For the purpose of open access, the author has applied a Creative Commons Attribution (CC BY) licence to any Author Accepted Manuscript version arising.

References

1. Antonelli, M., et al.: The medical segmentation decathlon. Nat. Commun. **13**(1), 4128 (2022)
2. Cabrera, Y., Fetit, A.E.: Reducing CNN textural bias with k-space artifacts improves robustness. IEEE Access **10** (2022)
3. Dolz, J., Gopinath, K., Yuan, J., Lombaert, H., Desrosiers, C., Ben Ayed, I.: HyperDense-Net: a hyper-densely connected CNN for multi-modal image segmentation. IEEE Trans. Med. Imaging **38**(5) (2019)
4. Gherman, A., Muschelli, J., Caffo, B., Crainiceanu, C.: Rxnat: an open-source R package for XNAT-based repositories. Front. Neuroinform. **14**, 572068 (2020)
5. Kennedy, D.N., Haselgrove, C., Riehl, J., Preuss, N., Buccigrossi, R.: The NITRC image repository. Neuroimage **124**, 1069–1073 (2016)
6. Khvastova, M., Witt, M., Essenwanger, A., Sass, J., Thun, S., Krefting, D.: Towards interoperability in clinical research-enabling FHIR on the open-source research platform XNAT. J. Med. Syst. **44**, 1–5 (2020)
7. Li, S., Ke, L., Pratama, K., Tai, Y.W., Tang, C.K., Cheng, K.T.: Cascaded deep monocular 3D human pose estimation with evolutionary training data. In: 2020 IEEE/CVF CVPR, June 2020

8. Makropoulos, A., et al.: The developing human connectome project: a minimal processing pipeline for neonatal cortical surface reconstruction. Neuroimage **173**, 88–112 (2018)
9. Marcus, D.S., Olsen, T.R., Ramaratnam, M., Buckner, R.L.: The extensible neuroimaging archive toolkit: an informatics platform for managing, exploring, and sharing neuroimaging data. Neuroinformatics **5**(1) (2007)
10. Moore, C.M.: Nifti (File format) | radiology reference article | radiopaedia.org
11. Nikolaos, A.M.: Deep learning in medical image analysis : a comparative analysis of multi-modal brain-MRI segmentation with 3D deep neural networks, July 2019
12. Schwartz, Y., et al.: PyXNAT: XNAT in python. Front. Neuroinform. **6**, 12 (2012)
13. Tran, D., Wang, H., Torresani, L., Ray, J., LeCun, Y., Paluri, M.: A closer look at spatiotemporal convolutions for action recognition (2017)
14. Valente, F., Silva, L.A.B., Godinho, T.M., Costa, C.: Anatomy of an extensible open source PACS. J. Digit. Imaging **29**, 284–296 (2016)
15. Van Essen, D.C., et al.: The WU-Minn human connectome project: an overview. Neuroimage **80**, 62–79 (2013)
16. Vollmuth, P., et al.: Artificial intelligence (AI)-based decision support improves reproducibility of tumor response assessment in neuro-oncology: an international multi-reader study. Neuro Oncol. **25**(3), 533–543 (2023)
17. Ziegler, E., et al.: Open health imaging foundation viewer: an extensible open-source framework for building web-based imaging applications to support cancer research. JCO Clin. Cancer Inform. **4**, 336–345 (2020)

A New Approach for Identifying Skin Diseases from Dermatological RGB Images Using Source Separation

Mustapha Zokay[✉] and Hicham Saylani

Laboratoire de Matériaux, Signaux, Systèmes et Modélisation Physique Faculté des sciences, Université Ibnou Zohr, 8106 Cité Dakhla, Agadir, Morocco
mustapha.zokay@edu.uiz.ac.ma, h.saylani@uiz.ac.ma

Abstract. In this article, we propose a new BSS approach for identifying skin diseases from RGB images that proceeds in two steps. We begin by separating the three main chromophores (oxyhemoglobin, deoxyhemoglobin and melanin) using Non-negative Matrix Factorization (NMF). For this purpose, we propose a special initialization of the solution matrices based on the sparsity of the chromophores, instead of initializing them with random matrices as is the case for basic versions of NMF. We then propose a new disease identification criterion that exploits the three contributions of each chromophore on the three spectral bands of our RGB dermatological image. To validate our approach, we used an open access database containing RGB images of melanoma and neavus. The results obtained showed good performance for our approach in terms of chromophore separation, compared to the most commonly used method in the literature, as well as disease identification compared to identification based on the most popular criterion.

Keywords: RGB dermatological images · Melanoma · Neavus · Chromophores · Melanin · Oxyhemoglobin · Deoxyhemoglobin · Blind Source Separation · Non-negative Matrix Factorization · Sparse Component Analysis

1 Introduction

Nowadays, the identification of skin diseases has become a major concern in the field of dermatology. Indeed, in most cases, they are diagnosed by dermatologists with the naked eye, which can lead to a large margin of error in identification, since the quality of the diagnosis heavily depends on the stage of the disease and the quality of the diagnosis made by an expert. Thus, the development of invasive and non-invasive methods exploiting medical imaging for identification has always been a major challenge for the scientific community interested in this subject, with special attention to non-invasive methods due to their great advantages. Indeed, the latter are very interesting both for the patient and the

G. Waiter et al. (Eds.): MIUA 2023, LNCS 14122, pp. 242–256, 2024.
https://doi.org/10.1007/978-3-031-48593-0_18

practitioner since they are simpler in terms of medical procedures and less penalising for the patient compared to invasive methods[1] [11,12]. Using multispectral imaging, it is possible to differentiate between skin diseases by calculating the average reflectance from the area affected by the disease as a function of wavelength [14,18,26].

Although the use of multispectral imaging is preferable to RGB imaging in dermatology, their processing is complex and requires specific equipment and arrangements, making them inaccessible to most dermatologists. Therefore, most databases of dermatological images consist of RGB images, i.e. images with three spectral bands, namely Blue, Green and Red. Therefore, it is crucial to develop identification methods exploiting only RGB images [15,23]. Among these methods, we find the estimation of skin chromophore distributions, namely oxyhemoglobin, deoxyhemoglobin and melanin [15,23]. Indeed, we can model the RGB image obtained at wavelengths λ_i, $i \in \{1,2,3\}$ which correspond to the Blue, Green and Red bands as a function of the skin chromophores, as follows:

$$b_i(\boldsymbol{u}) = \sum_{j=1}^{j=3} m_{ij}.c_j(\boldsymbol{u}), \quad i \in \{1,2,3\}, \tag{1}$$

where:

- $b_i(\boldsymbol{u})$ is defined as the inverse logarithm of the reflectance of the pixel $\boldsymbol{u} = (x,y)$ at wavelength λ_i,
- $c_j(\boldsymbol{u})$ represent the concentrations of oxyhemoglobin, deoxyhemoglobin and melanin respectively,
- m_{ij} are the mixing coefficients that depend on the absorption of the chromophore and the penetration of the light into the skin.

In order to simplify the estimation of chromophores from this model, we are interested in transforming RGB images, which are 2D signals, into 1D signals using a matrix vectorization technique[2]. This is defined by the operator $f(.)$, as follows:

$$b_i(v) = f(b_i(\boldsymbol{u})) = \sum_{j=1}^{j=3} m_{ij} \cdot f(c_j(\boldsymbol{u})) = \sum_{j=1}^{j=3} m_{ij} \cdot c_j(v), \quad i \in \{1,2,3\}. \tag{2}$$

We can then write the Eq. (2) in a matrix form:

$$\mathbf{X} = \mathbf{M} \cdot \mathbf{C}, \tag{3}$$

where $\mathbf{X} = [b_1(v), b_2(v), b_3(v)]^T$, $\mathbf{C} = [c_1(v), c_2(v), c_3(v)]^T$ and $\mathbf{M} = (m_{ij})_{1 \leqslant i,j \leqslant 3}$.

---

[1] Indeed, for example a biopsy is painful, costly and time-consuming, especially considering that it may need to be repeated for further examination.

[2] The goal of this transformation defined by the operator $f(.)$ is to transform a matrix of dimensions $n_x \times n_y$ into a row matrix of dimension $1 \times n_x \cdot n_y$, by assembling its columns one after the other.

To estimate the distribution of the three chromophores from RGB images, two main approaches have been proposed [9,13,23,27,28]. The first identification approach is to exploit the chromophore absorption spectra [13,27,28]. In [13,27, 28], the authors proposed to directly estimate the three chromophores from an RGB image by exploiting the absorption spectra of these chromophores [10,29], i.e. the knowledge of the mixing matrix $\mathbf{M}$ to invert the system of Eq. (3). The main disadvantage of this approach lies in the use of special RGB imaging, in which the bands of the three colors are very selective and are represented by only one wavelength for each of the three bands.

The second approach to estimate chromophores [9,23] is based on *Blind Source Separation* (BSS) [6,22]. It aims to estimate the sources, represented by the chromophores from the mixtures, which are the RGB images, without any prior information. All chromophore identification methods using BSS aim to estimate only two particular chromophores which are melanin and hemoglobin [9,23]. Indeed, for this family, hemoglobin is a mixture of oxyhemoglobin and deoxyhemoglobin. In [23], the distributions of these two chromophores are estimated assuming that these two chromophores are independent using a BSS method based on *Independent Component Analysis*. In [9], these two chromophores are estimated using a BSS method based on *Non-negative Matrix Factorization* by initializing the source solution matrix with a particular version derived from mixtures that has been adopted in [30]. However, the major drawback of this approach is that most methods do not allow separation between oxyhemoglobin and deoxyhemoglobin.

In this article, we propose a new approach based on BSS to identify skin diseases from an RGB image. This approach proceeds in two steps. In the first step, we propose a new procedure to estimate the distributions of the three main chromophores, unlike existing methods that only allow estimation of melanin and hemoglobin [9,23] or rely on a priori information on chromophore absorption coefficients [13,27,28]. To achieve this, we begin by estimating an initial version of the three chromophores by exploiting their sparsity, which we then use to initialize an *NMF* algorithm to obtain the final version. For the second step of our approach, we propose a new numerical criterion to identify skin diseases that exploits the contributions of each chromophore in the three spectral bands of our RGB image, unlike the classical criterion that is calculated relative to each of these bands in the raw state [14,18,26]. The tests are carried out in two steps. In the first step, we evaluate the performance of chromophore estimation and in the second step, we test the performance of our new criterion for identifying skin diseases. The rest of this article is organized as follows. Section 2 presents our new approach for estimating chromophore distributions and identifying skin diseases. Section 3 presents the results of the tests performed, followed by a final section dedicated to the conclusion and perspectives of our work.

2   Proposed Approach

The new approach proposed in this paper aims to distinguish between different skin diseases and proceeds in two steps. The first step consists to separate the three chromophores from their mixtures provided by the three spectral bands of the RGB image, which allows obtaining an estimation of the distribution of each chromophore separately. In the second step, we use these estimations to identify the type of disease, using a new numerical criterion, unlike most existing methods that generally use a subjective criterion based on the visual analysis of the different estimated chromophore distributions in a qualitative manner. The two steps of our approach are discussed in the two Subsects. 2.1 and 2.2, respectively.

2.1   Chromophores Separation

During this first step, we are interested in estimating each of the distributions of the three chromophores separately from their mixtures provided by the three spectral bands of the treated RGB dermatological image, using *Blind Source Separation* (BSS) [6,22]. There are several BSS methods in the literature that differ only in the hypotheses made about the source signals and/or the mixing matrix. There are three categories of BSS methods. The first category includes methods based on Independent Component Analysis (ICA) which assume that the sources are independent. The second category includes methods based on Sparse Component Analysis (*SCA*) which assume that the source signals are sparse[3]. Finally, the third category includes methods based on Non-negative Matrix Factorization (*NMF*) which assume that both the source signals and the mixing coefficients are positive. In this paper, we propose a new BSS approach that exploits the last two hypotheses, sparsity and positivity of the source signals. Indeed, we propose to exploit both the sparsity and positivity of the chromophores, so our BSS approach proceeds in two steps. In the first step, we are looking for an estimation of the chromophores using a BSS method based on *SCA*. But since this estimation is not very satisfactory, due to the fact that the sparsity hypotheses of the chromophores is not perfectly verified, we use a BSS method based on *NMF* in the second step to improve this estimation.

Preliminary Chromophores Estimation Using *SCA*
There are several BSS methods based on *SCA*. Here we opted for the method proposed by Abrard et al. in [1], known for its effectiveness and simplicity of implementation. This method is based on exploiting the ratios between different mixtures after selecting a reference mixture. By taking the first mixture $b_1(v)$ as the reference mixture, we are interested on the following ratios:

$$\frac{b_2(v)}{b_1(v)} = \frac{m_{21}c_1(v) + m_{22}c_2(v) + m_{23}c_3(v)}{m_{11}c_1(v) + m_{12}c_2(v) + m_{13}c_3(v)} \tag{4}$$

---

[3] A signal is sparse if the majority of its elements are zero or very close to zero.

$$\frac{b_3(v)}{b_1(v)} = \frac{m_{31}c_1(v) + m_{32}c_2(v) + m_{33}c_3(v)}{m_{11}c_1(v) + m_{12}c_2(v) + m_{13}c_3(v)} \tag{5}$$

Then, we assume that there is at least one single-source zone for each chromophore, i.e. a zone where this chromophore is the only one present. In other words, we assume that [1]:

1. There is at least one zone Z_1 in the mixtures such that:

$$\forall v \in Z_1, \ c_1(v) \neq 0 \ \text{ and } \ c_2(v) = c_3(v) = 0 \tag{6}$$

2. There is at least one zone Z_2 in the mixtures such that:

$$\forall v \in Z_2, \ c_2(v) \neq 0 \ \text{ and } \ c_1(v) = c_3(v) = 0 \tag{7}$$

3. There is at least one zone Z_3 in the mixtures such that:

$$\forall v \in Z_3, \ c_3(v) \neq 0 \ \text{ and } \ c_1(v) = c_2(v) = 0 \tag{8}$$

By exploiting these hypotheses (6), (7) and (8) and noting $\frac{m_{ij}}{m_{1j}} = m'_{ij}$, we obtain:

$$\forall v \in Z_1, \quad \frac{b_2(v)}{b_1(v)} = \frac{m_{21}}{m_{11}} = m'_{21} \ \text{ and } \ \frac{b_3(v)}{b_1(v)} = \frac{m_{31}}{m_{11}} = m'_{31} \tag{9}$$

$$\forall v \in Z_2, \quad \frac{b_2(v)}{b_1(v)} = \frac{m_{22}}{m_{12}} = m'_{22} \ \text{ and } \ \frac{b_3(v)}{b_1(v)} = \frac{m_{32}}{m_{12}} = m'_{32} \tag{10}$$

$$\forall v \in Z_3, \quad \frac{b_2(v)}{b_1(v)} = \frac{m_{23}}{m_{13}} = m'_{23} \ \text{ and } \ \frac{b_3(v)}{b_1(v)} = \frac{m_{33}}{m_{13}} = m'_{33} \tag{11}$$

By taking up the matrix Eq. (3) which models our mixtures we can write:

$$\mathbf{X} = \mathbf{M} \cdot \mathbf{C} = \mathbf{M'} \cdot \mathbf{C'}, \tag{12}$$

where:

$$\mathbf{M'} = \begin{pmatrix} 1 & 1 & 1 \\ m'_{21} & m'_{22} & m'_{23} \\ m'_{31} & m'_{32} & m'_{33} \end{pmatrix} = \begin{pmatrix} 1 & 1 & 1 \\ \frac{m_{21}}{m_{11}} & \frac{m_{22}}{m_{12}} & \frac{m_{23}}{m_{13}} \\ \frac{m_{31}}{m_{11}} & \frac{m_{32}}{m_{12}} & \frac{m_{33}}{m_{13}} \end{pmatrix} \ \text{ and } \ \mathbf{C'} = \begin{pmatrix} m_{11}c_1(v) \\ m_{12}c_2(v) \\ m_{13}c_3(v) \end{pmatrix}. \tag{13}$$

Thus, Eqs. (9), (10) and (11) allow us to identify the elements m'_{ij} of the new mixing matrix $\mathbf{M'}$, knowing that $m'_{1j} = 1$, $\forall j$. As for the identification of the single-source zones Z_i, it is done by using a criterion based on the calculation of the variance of one of the ratios $\frac{b_i(v)}{b_1(v)}$ [1]. Indeed, according to the Eqs. (9), (10) and (11), the variance of these ratios is zero over any single-source zone Z_i. In practice, we divide the mixtures into slices of the same length L, on which we calculate these ratios and their variance. If, for a given slice, this variance is very low, it is considered as a single-source zone. Then, these slices are sorted in ascending order of the variance of the ratio $\frac{b_i(v)}{b_1(v)}$. To identify the first column of the mixing matrix $\mathbf{M'}$, we select the first value of the ratio mean $\frac{b_i(v)}{b_1(v)}$ with low variance. As for the last two columns of the mixing matrix, they are identified using an appropriate threshold (see [1] for more details). Thus, an estimate of the matrix $\mathbf{M}'$, denoted $\widehat{\mathbf{M}'}$, is given by the following relation:

$$\widehat{\mathbf{M}'} = \mathbf{M}' \cdot \boldsymbol{\Delta} + \mathbf{E}, \tag{14}$$

where $\boldsymbol{\Delta}$ is a permutation matrix and $\mathbf{E}$ is a matrix that models the estimation errors which are due to the fact that the sparsity hypothesis of each chromophore is not perfectly verified. The next phase for our separation step will therefore aim to improve this estimation by using a BSS method based this time on the *Non-negative Matrix Factorization (NMF)*.

Chromophores Estimation Using NMF

The principle of BSS methods based on *NMF* is to factorize the matrix $\mathbf{X}$ representing the mixtures into a product of two matrices $\mathbf{A}$ and $\mathbf{S}$ such that these two matrices are approximations of the two matrices $\mathbf{M}'$ and $\mathbf{C}'$ respectively, i.e.:

$$\mathbf{X} = \mathbf{M}' \cdot \mathbf{C}' = \mathbf{A} \cdot \mathbf{S}, \tag{15}$$

with:

$$\mathbf{A} \simeq \mathbf{M}' \quad \text{and} \quad \mathbf{S} \simeq \mathbf{C}'. \tag{16}$$

In this sense, in order to estimate the matrix $\mathbf{C}'$ of the chromophores, we must determine the matrices $\mathbf{A}$ and $\mathbf{S}$, which amounts to minimizing a criterion (or objective function) based on measuring the error between $\mathbf{X}$ and $\mathbf{AS}$. Since several criteria have been proposed in the literature to measure this error [7], we have opted here for the *Euclidean Distance* as a criterion, which is the simplest and most widely used by the community of *BSS* based on *NMF*. Denoted $\mathcal{D}_e(\mathbf{X}|\mathbf{AS})$, this criterion is then defined as follows:

$$\mathcal{D}_e(\mathbf{X}|\mathbf{AS}) = \frac{1}{2}||\mathbf{X} - \mathbf{AS}||^2. \tag{17}$$

To minimize this criterion, we can use one of the most commonly used optimization algorithms of *BSS* based on *NMF*, namely the *Multiplicative Update (MU)* algorithm [16], the *Alternating Least Squares (ALS)* algorithm [25] and the *Projected Gradient* (PG) algorithm [20]. In this paper, we opted for the *ALS* algorithm, known for its good performance, particularly in terms of fast convergence [4]. Below are the two update rules for the two solution matrices $\mathbf{S}$ and $\mathbf{A}$ using this algorithm [25]:

$$\mathbf{S} \longleftarrow (\mathbf{A}^T\mathbf{A})^{-1}\mathbf{A}^T\mathbf{X} \tag{18}$$

$$\mathbf{A} \longleftarrow \mathbf{X}\mathbf{S}^T(\mathbf{S}\mathbf{S}^T)^{-1} \tag{19}$$

However, it is well known that the factorization provided by *NMF* is not unique. Indeed, it is evident that if the couple $(\mathbf{A}, \mathbf{S})=(\mathbf{M}', \mathbf{C}')$ is a solution, then any couple $(\mathbf{A}', \mathbf{S}')=(\mathbf{M}'\mathbf{K}, \mathbf{K}^{-1}\mathbf{C}')$ is also a solution for any invertible matrix $\mathbf{K}$ since we have:

$$\mathbf{X} = \mathbf{AS} = \mathbf{M}'\mathbf{C}' = \mathbf{M}'\mathbf{K} \cdot \mathbf{K}^{-1}\mathbf{C}' = \mathbf{A}'\mathbf{S}'. \tag{20}$$

Therefore, instead of converging to the global minimum corresponding to the solution couple $(\mathbf{A},\mathbf{S})=(\mathbf{M}',\mathbf{C}')$ our *ALS* algorithm can very well converge to a local minimum corresponding to a solution couple $(\mathbf{A}',\mathbf{S}')=(\mathbf{M}'\mathbf{K},\mathbf{K}^{-1}\mathbf{C}')$, where $\mathbf{K}$ is any invertible matrix. To address this issue, one of the solutions that have been adopted in the literature is to replace the random initialization of the solution couple $(\mathbf{A},\mathbf{S})$ with a particular initialization usually chosen based on prior information available on the sought solution [5]. Similarly, here we proposed a solution to this problem based not on prior information but rather on the exploitation of the first version of the estimated matrix $\mathbf{M}'$ during the first phase of our chromophore separation step by exploiting their sparsity. Thus, we propose to initialize the matrix $\mathbf{A}$ with the matrix $\widehat{\mathbf{M}'}$ given by the relation (14).

After the convergence of our *ALS* algorithm, the solution matrices $\mathbf{A}$ and $\mathbf{S}$ are ideally given by the following two relationships:

$$\mathbf{A} = \mathbf{M}' \cdot \mathbf{DP} \quad \text{and} \quad \mathbf{S} = (\mathbf{DP})^{-1} \cdot \mathbf{C}', \tag{21}$$

where $\mathbf{P}$ is a permutation matrix and $\mathbf{D}$ is a diagonal matrix. By referring to the relations (21), we can see that the columns of the mixing matrix and the sources are estimated to a permutation and a scaling factor, which does not allow us to establish a direct correspondence between the estimated sources and the chromophores. First, by omitting the effect of the permutation matrix $\mathbf{P}$ (by assuming that $\mathbf{P} = \mathbb{I}_d$), we can clearly see that the scale factor problem can be solved by dividing each column of the estimated matrix $\mathbf{A}$ by its first coefficient. Indeed, if we denote $\alpha_j = \mathbf{D}(j,j)$ as the diagonal elements of the matrix $\mathbf{D}$, then each column of index "j" of the matrix $\mathbf{A}$ can be written as $[\alpha_j, \alpha_j m'_{2j}, \alpha_j m'_{3j}]^T$. By dividing this column by its first coefficient (α_j) we obtain the new normalised column $[1, m'_{2j}, m'_{3j}]^T$, which is no other than the column of index "j" of the mixing matrix $\mathbf{M}'$. Now, it is evident that taking into account the permutation matrix $\mathbf{P}$ in the relations (21) (by assuming that $\mathbf{P} \neq \mathbb{I}_d$), will result a permutation of the columns of the matrix $\mathbf{M}'$, which corresponds to a permutation of the rows of the matrix $\mathbf{C}'$. This means that we cannot determine the nature of each of the estimated sources. To overcome this problem and associate each estimated source to the correct chromophore, we have exploited the absorption spectra of the three chromophores provided in [10] and represented in Fig. 1. By exploiting these spectra (which indirectly provide us an estimation of the coefficients $m'_{2j} = \frac{m_{2j}}{m_{1j}}$ and $m'_{3j} = \frac{m_{3j}}{m_{1j}}$), we can clearly identify the permutation matrix $\mathbf{P}$ involved and subsequently establish the correspondence between each of the normalised columns of our estimated mixing matrix $\mathbf{A}$ and the correct chromophore that corresponds to it.

2.2   Disease Identification

After estimating the chromophores in the first step, our objective is now to exploit them to identify different skin diseases by proposing a new numerical criterion that allows for this identification. The originality of our new criterion

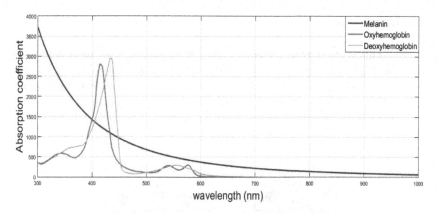

Fig. 1. Absorption coefficient of chromophores provided in [10].

lies in considering the three contributions of each chromophore $c_j(v)$ at the level of the three available spectral bands, unlike the classical criterion which is commonly used for both RGB images [18] and multispectral images [14,26]. Indeed, the latter consists of calculating the average reflectance from the region of the skin affected by the disease, for each spectral band $b_i(v)$ containing all three mixed chromophores. To implement our criterion, we begin by segmenting the region affected by the disease using one of the classical segmentation algorithms [2,17,24]. Denoted $\mathscr{C}_j$, our new criterion associated to each chromophore $c_j(v)$ is defined as follows:

$$\mathscr{C}_j = \frac{1}{N_r} \sum_{v \in R} \sum_{i=1}^{i=3} a_{ij} s_j(v), \quad j \in \{1, 2, 3\}, \tag{22}$$

where R is the region affected by the disease, N_r is the number of pixels contained in this region and a_{ij} are the coefficients of the matrix $\mathbf{A}$.

Consequently, the global algorithm of our approach, which proceeds in two steps (chromophores separation and disease identification) is presented below.

Algorithm

1: Estimation of a first version of the mixing matrix $\mathbf{M}'$, noted $\hat{\mathbf{M}}'$, using SCA
2: Initialization of $\mathbf{A}$ by $\hat{\mathbf{M}}'$
3: **While** $\mathcal{D}_e(\mathbf{X}|\mathbf{AS}) > threshold$, do:
   - $\mathbf{S} = (\mathbf{A}^T\mathbf{A})^{-1}\mathbf{A}^T\mathbf{X}$
   - Resetting all negative values of $\mathbf{S}$ to zero
   - $\mathbf{A} = \mathbf{XS}^T(\mathbf{SS}^T)^{-1}$
   - Resetting all negative values of $\mathbf{A}$ to zero
4: Solving the permutation problem of estimated chromophores
5: Computing the criterion $\mathscr{C}_j$, $j \in \{1, 2, 3\}$
6: Disease identification

---

3   Results

The goal of this section is to measure the performance of our approach, using a database of RGB dermatological images [8] which originally contains 100 nevus images and 70 melanoma images. However, our tests were carried out on a set of only 85 images, as some images containing a significant amount of hair and/or of poor quality were excluded from our tests. Figure 2 shows an example of such images.

Fig. 2. Example of images excluded from our tests.

At first, we evaluate the performance of our first step, which is the separation of the three chromophores (oxyhemoglobin, deoxyhemoglobin and melanin), by comparing them to the ones obtained with the classical method [27]. It is important to note that the classical method is the unique method possible for this comparison, as it is the unique one that allows to estimate the three chromophores separately. In the second step, we evaluate the performance of our new identification criterion compared to the classical criterion used in [14,26]. These two types of performance are discussed in Sects. 3.1 and 3.2, respectively.

3.1   Performance of Chromophores Separation

We recall that the method proposed in [27] consists of exploiting the absorption coefficients of the three chromophores as well as the optical path lengths to estimate the coefficients of the mixing matrix $\mathbf{M}$. These coefficients are actually the result of multiplying the absorption coefficients by the optical path lengths that are specific to each camera. Indeed, knowing that the camera used for the acquisition of these RGB images [8] is a Nikon D3 camera, the wavelengths involved are $\lambda_1 = 473$ nm, $\lambda_2 = 532$ nm and $\lambda_3 = 620$ nm corresponding respectively to the Blue, Green and Red bands[4]. By exploiting these wavelengths, we can estimate the absorption coefficients of the three chromophores from the tables provided in [10], as well as the optical path lengths of these three wavelengths by referring to [3,28]. Therefore, using this classical method [27], the three chromophores can be estimated directly by inverting the empirically estimated mixing matrix $\mathbf{M}$.

---

[4] According to the information provided in [21].

Denoted $\mathbf{M}_{emp}$, this empirical mixing matrix is given by:

$$\mathbf{M}_{emp} = \begin{bmatrix} 15.37 & 07.55 & 76.00 \\ 35.23 & 32.60 & 82.86 \\ 01.54 & 10.21 & 97.21 \end{bmatrix}. \tag{23}$$

Due to space constraints and since the separation performance obtained for all treated images is similar, we present here only the results obtained for a melanoma image and a nevus image. These two images are shown in Fig. 3.

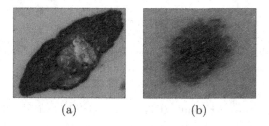

(a)   (b)

Fig. 3. Treated images: (a) Melanoma, (b) Neavus.

The results obtained in terms of estimating the distribution of chromophores for both the melanoma and the nevus, using our method and the classical method [27] are illustrated in Figs. 4 and 5, respectively.

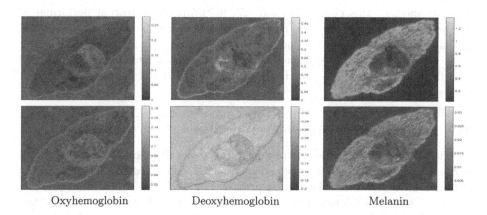

Oxyhemoglobin   Deoxyhemoglobin   Melanin

Fig. 4. Estimated distributions of oxyhemoglobin, deoxyhemoglobin and melanin using our method (in the first row) and classical method [27] (in the second row) for **melanoma**.

From the Figures 4 and 5, based on the corresponding color bar that illustrates the distribution evolution (where yellow indicates high concentration and

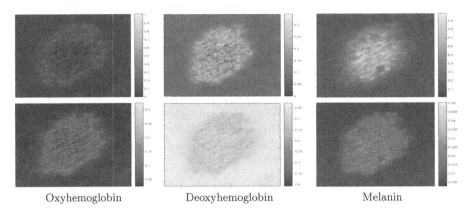

Oxyhemoglobin   Deoxyhemoglobin   Melanin

Fig. 5. Estimated distributions of oxyhemoglobin, deoxyhemoglobin and melanin using our method (in the first row) and classical method [27] (in the second row) for **neavus**.

blue indicates low concentration), it is clearly visible to the naked eye that our method of estimating the chromophores provides more satisfactory results than the classical method for both types of diseases. Indeed, the distributions estimated by our method are characterized by a low distribution of oxyhemoglobin, a medium distribution of deoxyhemoglobin and a high distribution of melanin, which is consistent with physiological knowledge of these two types of diseases [19]. On the other hand, the classical method [27] shows poor performance especially for deoxyhemoglobin, where it provides negative concentration values, which is inconsistent with its real distribution, which should always be positive. The weakness of this classic method [27] in terms of separation performance can be explained mainly by the fact that its principle is based on precise a priori information on absorption coefficients and mean optical path lengths at the three wavelengths. However, this a priori information is determined with a large margin of error when using an RGB camera known for its three non-selective bands.

3.2   Performance of Disease Identification

In this section, we focus on the performance of the disease identification using our new criterion denoted $\mathscr{C}_j$, defined by Eq. (22), compared to the performance obtained using the classical criterion proposed in [14,26]. We recall that this criterion is based on calculating the mean reflectance from the skin region affected by the disease, denoted R, on each spectral band $b_i(v)$ of the RGB image. Denoted $\mathscr{B}_i$, it is defined as follows:

$$\mathscr{B}_i = \frac{1}{N_r} \sum_{v \in R} b_i(v), \quad i \in \{1, 2, 3\}. \tag{24}$$

To delimit the region R affected by the disease, we used the *Otsu segmentation method* [24]. In order to demonstrate the relevance of our key idea which consists to initialize the mixing matrix $\mathbf{A}$ with the matrix $\mathbf{M}'$ estimated by SCA (during our first step of separation), we also tested our approach with a random initialization for this matrix $\mathbf{A}$. The results obtained using our new criterion for these two scenarios on all 85 tested images are provided in the form of box-plots in Figs. 6. As for the results obtained using the classical criterion, they are provided in Fig. 7. Note that this type of representation in the form of box-plots, which is increasingly used, is most suitable for comparing between different types of skin diseases [15,18]. Each of its boxes is delimited by a rectangle extending between the first quartile (representing 25% of the data) and the third quartile (representing 75% of the data), with a horizontal line in the middle representing the median[5]. The ends of the box delimit the interquartile range (IQR) which represents 50% of the data and values outside this range are considered outliers. The bars above and below the box represent respectively the maximum and minimum values of the data.

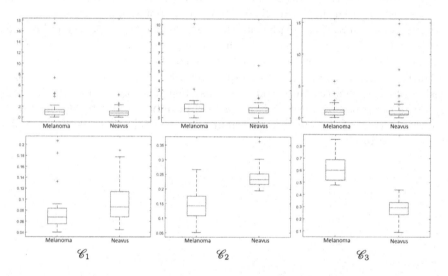

Fig. 6. Results obtained using our identification criterion with a random initialization (in the first row) and an initialization by the SCA (in the second row).

From the Fig. 6, we observe that by using our criterion, the initialization of our ALS algorithm by exploiting the sparsity of the chromophores gives much better results than random initialization. Indeed, for the latter case (corresponding to the top sub-Figures), there is a complete overlap between all the boxes corresponding to the two diseases. However, for the first case (corresponding to the bottom sub-Figures) there is no overlap at all using our criterion $\mathscr{C}_3$ related to

---

[5] The median corresponds to the value located in the middle of the data. It divides the data into two equal parts, such that 50% of the values are below the median and 50% are above it.

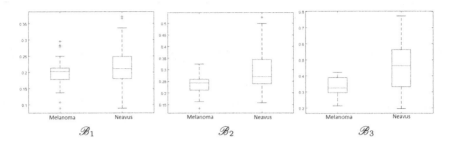

Fig. 7. Results obtained using the classic RGB criterion.

melanin. This observation also applies when comparing our criterion $\mathscr{C}_j$ related to the contributions of the chromophores $c_j(v)$ to the criterion $\mathscr{B}_i$ related to the spectral bands $b_i(v)$ according to the Fig. 7. Indeed, the latter criterion also leads to a complete overlap between all the boxes corresponding to the two diseases. Thus, our new criterion $\mathscr{C}_j$ exploiting the contributions of the chromophores in the three spectral bands is more relevant for the distinction between melanoma and naevus, particularly with respect to the contributions of melanin.

On the other hand, in order to prove the relevance of our key idea which consists of exploiting the three contributions of each chromophore $c_j(v)$ to constitute its corresponding criterion $\mathscr{C}_j$ instead of only one, we recalculated this criterion on each of the melanin contributions separately. The results obtained are presented in Fig. 8.

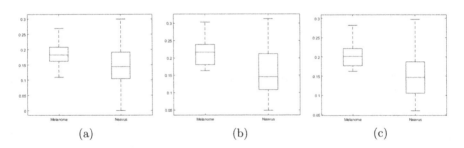

Fig. 8. Results obtained using the contribution of melanin in each of the three spectral bands: (a) Blue band (b) Green band (c) Red band. (Color figure online)

From the Fig. 8, we observe that a single contribution of melanin in one of the three spectral bands alone does not allow differentiation between the two diseases. Indeed, from this Figure, we can see that there is a complete overlap between the different boxes corresponding to the two diseases, regardless of the contribution.

4   Conclusion

In this article, we propose a new approach to identify different skin diseases from RGB dermatological images. Based on *Blind Source Separation*, our approach proceeds in two steps. We begin with a chromophore separation step based on *NMF* using the *ALS* algorithm, but instead of randomly initializing the mixing matrix, we propose an initialization exploiting the sparsity of the chromophores. In a second step, we use these chromophores to identify the type of disease. In this sense, we propose a new numerical criterion that exploits the three contributions of each chromophore on the three spectral bands. Our tests on 51 nevus images and 34 melanoma images [8] showed good performance for our approach in terms of chromophores separation, compared to the classical method proposed in [27], as well as in terms of disease identification, compared to identification based on the classical RGB criterion [14,18,26]. We have indeed found that our new numerical criterion associated with melanin allows for a clear distinction between melanoma and nevus. However, a much larger performance study on different databases and different skin diseases is desirable to confirm this result. Finally, we are fully aware of the need to improve our identification criterion and adapt it for classification purposes.

References

1. Abrard, F., Deville, Y.: A time-frequency blind signal separation method applicable to underdetermined mixtures of dependent sources. Signal Process. **85**(7), 1389–1403 (2005)
2. Adam, A., Rivlin, E., Shimshoni, I.: Robust fragments-based tracking using the integral histogram. In: 2006 IEEE Computer Society Conference on Computer Vision and Pattern Recognition (CVPR'06), vol. 1, pp. 798–805. IEEE (2006)
3. Anderson, R.R., Parrish, J.A.: The optics of human skin. J. Investig. Dermatol. **77**(1), 13–19 (1981)
4. Besse, P.: https://www.math.univ-toulouse.fr/$\sim$$besse/Wikistat/pdf/st-m-explo-nmf.pdf. Accessed 28 Mar 2023
5. Boutsidis, C., Gallopoulos, E.: Svd based initialization: a head start for nonnegative matrix factorization. Pattern Recogn. **41**(4), 1350–1362 (2008)
6. Comon, P., Jutten, C.: Handbook of Blind Source Separation: Independent component analysis and applications. Academic press (2010)
7. Févotte, C., Idier, J.: Algorithms for nonnegative matrix factorization with the β-divergence. Neural Comput. **23**(9), 2421–2456 (2011)
8. Giotis, I., Molders, N., Land, S., Biehl, M., Jonkman, M.F., Petkov, N.: Med-node: a computer-assisted melanoma diagnosis system using non-dermoscopic images. Expert Syst. Appl. **42**(19), 6578–6585 (2015)
9. Gong, H., Desvignes, M.: Hemoglobin and melanin quantification on skin images. In: International Conference Image Analysis and Recognition, pp. 198–205. Springer (2012)
10. Jacques, S.L.: https://omlc.org/~jacquess/library.html. Accessed 28 Mar 2023
11. Jacques, S.L., Samatham, R., Choudhury, N.: Rapid spectral analysis for spectral imaging. Biomed. Opt. Express **1**(1), 157–164 (2010)

12. Koprowski, R.: Processing of hyperspectral medical images (2017)
13. Kuzmina, I., Diebele, I., Asare, L., Kempele, A., Abelite, A., Jakovels, D., Spigulis, J.: Multispectral imaging of pigmented and vascular cutaneous malformations: the influence of laser treatment. In: Laser Applications in Life Sciences, vol. 7376, pp. 144–149. SPIE (2010)
14. Kuzmina, I., et al.: Towards noncontact skin melanoma selection by multispectral imaging analysis. J. Biomed. Opt. **16**(6), 060502 (2011)
15. Kuzmina, I., et al.: Skin chromophore mapping by smartphone rgb camera under spectral band and spectral line illumination. J. Biomed. Opt. **27**(2), 026004 (2022)
16. Lee, D., Seung, H.: Algorithms for non-negative matrix factorization. Adv. Neural Inform. Process. Syst. **13**, February 2001
17. Lee, D.H.: A new edge-based intra-field interpolation method for deinterlacing using locally adaptive-thresholded binary image. IEEE Trans. Consum. Electron. **54**(1), 110–115 (2008)
18. Lihachev, A., Lihacova, I., Plorina, E.V., Lange, M., Derjabo, A., Spigulis, J.: Differentiation of seborrheic keratosis from basal cell carcinoma, nevi and melanoma by rgb autofluorescence imaging. Biomed. Opt. Express **9**(4), 1852–1858 (2018)
19. Lihacova, I.: Evaluation of skin oncologic pathologies by multispectral imaging methods. Ph.D. thesis, July 2015. https://doi.org/10.13140/RG.2.2.12585.70242
20. Lin, C.J.: Projected gradient methods for nonnegative matrix factorization. Neural Comput. **19**(10), 2756–2779 (2007)
21. LLC, L.: https://www.maxmax.com/nikon_d3x.htm. Accessed 28 Mar 2023
22. Naik, G.R., Wang, W., et al.: Blind source separation. Berlin: Springer **10**, 978–3 (2014)
23. Ojima, N., Akazaki, S., Hori, K., Tsumura, N., Miyake, Y.: Application of image-based skin chromophore analysis to cosmetics. J. Imaging Sci. Technol. **48**(3), 222–226 (2004)
24. Otsu, N.: A threshold selection method from gray-level histograms. IEEE Trans. Syst. Man Cybern. **9**(1), 62–66 (1979)
25. Paatero, P., Tapper, U.: Positive matrix factorization: a non-negative factor model with optimal utilization of error estimates of data values. Environmetrics **5**(2), 111–126 (1994)
26. Saknite, I., Jakovels, D., Spigulis, J.: Diffuse reflectance and fluorescence multispectral imaging system for assessment of skin. In: Biophotonics: Photonic Solutions for Better Health Care IV, vol. 9129, pp. 593–598. SPIE (2014)
27. Spigulis, J., Oshina, I.: 3× 3 technique for rgb snapshot mapping of skin chromophores. In: Optics and the Brain, pp. JT3A-39. Optica Publishing Group (2015)
28. Spigulis, J., Oshina, I., Berzina, A., Bykov, A.: Smartphone snapshot mapping of skin chromophores under triple-wavelength laser illumination. J. Biomed. Opt. **22**(9), 091508 (2017)
29. Van Gemert, M., Jacques, S.L., Sterenborg, H., Star, W.: Skin optics. IEEE Trans. Biomed. Eng. **36**(12), 1146–1154 (1989)
30. Yamamoto, T., Takiwaki, H., Arase, S., Ohshima, H.: Derivation and clinical application of special imaging by means of digital cameras and image j freeware for quantification of erythema and pigmentation. Skin Res. Technol. **14**(1), 26–34 (2008)

Pseudo-SPR Map Generation from MRI Using U-Net Architecture for Ion Beam Therapy Application

Ama Katseena Yawson[1,2,3](✉) , Katharina Maria Paul[2,4], Cedric Beyer[2,5],
Stefan Dorsch[1,2], Sebastian Klüter[2,4] , Thomas Welzel[2,4],
Katharina Seidensaal[2,4], Jürgen Debus[2,4,5] , Oliver Jäkel[1,2,5] ,
and Kristina Giske[1,2]

[1] German Cancer Research Center (DKFZ), Division of Medical Physics in Radiation
Oncology, Heidelberg, Germany
[2] Heidelberg Institute for Radiation Oncology (HIRO), National Center for
Radiation Research in Oncology (NCRO), Heidelberg, Germany
[3] Heidelberg University, Medical Faculty, Heidelberg, Germany
a.yawson@dkfz-heidelberg.de
[4] University Clinics Heidelberg, Department of Radiation Oncology,
Heidelberg, Germany
[5] Heidelberg Ion Therapy Centre (HIT), Heidelberg, Germany

Abstract. Stopping power ratio (SPR) maps are needed for dose deposition calculations and are typically estimated from single energy CT (SECT) in clinical routine. SECT-based SPR conversion leads to large variability due to the one-to-one relationship assumed by the conversion method. Dual-energy CT (DECT) involving the acquisition of two energy spectra captures both material-specific information and tissue characterization which is essential for an accurate SPR map conversion. The goal of this study is to train a U-Net architecture to generate pseudo-SPR map from MRI (Dixon) using a DECT-converted SPR map. The model performance was validated using Head & Neck cohort of 16 patients with paired MRI and SPR maps. The proposed solution achieved a mean absolute error (MAE) and peak-signal-to-noise-ratio (PSNR) of 19.41 ± 8.67 HU and 58.76 ± 2.17 dB respectively for all test cases. From observation, the sequential incorporation of different Dixon MRI images such as fat-suppressed and water-suppressed yielded an accurate pseudo-SPR map which is comparable to its corresponding target SPR map. Furthermore, bone delineation integrated as additional channel to Dixon MRI sequence demonstrated an enhanced bone identification on predicted pseudo-SPR map. As future direction, we would like to extend this approach to a clinical SPR map which will enable dosimetric analysis of clinical target volume (CTV) to be possible in treatment planning application for ion beam therapy.

Keywords: Ion beam therapy · MRI-only treatment planning · Dual Energy CT · Stopping Power Ratio · U-Net · Head & Neck

© The Author(s), under exclusive license to Springer Nature Switzerland AG 2024
G. Waiter et al. (Eds.): MIUA 2023, LNCS 14122, pp. 257–267, 2024.
https://doi.org/10.1007/978-3-031-48593-0_19

1   Introduction

Computed tomography (CT) is fundamental for radiation treatment planning (RTP) today, as it provides a unique Hounsfield unit (HU) which can be mapped to the stopping power ratio (SPR) needed for dose calculation in ion therapy application [23]. Nonetheless, the poor soft tissue contrast inherent in CT images interferes with the treatment planning in terms of differentiating tumour volumes and healthy tissue. Thus, incorporating MRI images, which have high soft tissue contrast, promise to resolve this limitation. Due to the separate acquisition of CT and MRI images and the patient deformation between the different scans, image registration is required between the RTP phase to transfer MRI structural information to that of the planning CT. Image registration introduces a level of complexity and uncertainty that can create a spatial systematic error in the order of 2–5 mm [5]. The dosimetric impact of this systematic spatial error jeopardizes dose deposition precision aimed at small structures or when the target beam is close to an OAR. Thus, it would be practical to eliminate image registration for RTP and focus on MRI-only RTP.

With the expansion of the MR-guided linear accelerator (MR-linac), MRI-only RTP workflow has become highly desirable because of its appealing advantages of bypassing registration, reducing unnecessary imaging dose (since CT scan is eliminated) and enhancing fast adaptation of dose in online scenarios [22,24]. Despite the numerous advantages, MRI signals do not provide necessary physical material properties for dose calculation in contrast to CT scans which hinders its direct application in RTP. Therefore, the established solution to tackle this limitation is to generate pseudo-CT from MRI which serves as a surrogate for RTP. Over the last few years, various approaches with different scientific conventions, terminology and endpoints have been reported in the literature [3,4,8,10,13–16,18–20,26,27,30]. Amongst these established approaches, deep learning methods eliminate deformable image registration and can theoretically handle any random MRI sequence as an input and generate accurate pseudo-CT in very short times, thus satisfying the time requirements for online MRI-only workflows.

For proton therapy applications, studies have shown that generated pseudo-CT can be converted to stopping power ratios (SPR) using a calibration method based on the bethe-bloch formula [20,21]. SPR is a physical quantity that is used to estimate dose distribution and ion range, which is needed for accurate dose calculation [31]. SPR distribution is typically estimated from single energy CT (SECT) in a clinical routine based on a predetermined linear relationship between HU numbers and SPRs. SECT-based SPR conversion leads to large variability since the conversion method assumes a one-to-one relationship between HU and SPR for biological tissues despite different combinations of atomic composition and electron density [11]. To overcome this underlying problem, dual-energy CT (DECT) involving the acquisition of 2 energy spectra captures both material-specific information and tissue characterization by differentiating the energy dependence of different materials [6].

In this scope of the study, [19] published the first promising step towards MRI-only(T1-weighted) SPR generation from DECT using a label-GAN. However, this approach requires further validation in terms of image quality by exploring a different network architecture as well as the input MRI. Therefore, the underlying objective of this study is to generate a pseudo-SPR from MRI (Dixon) using a DECT-converted SPR map for Head & Neck cancer patients. The selected network architecture for this investigation is the standard U-Net [1] trained in a supervised manner. The U-Net architecture offers reduced complexity compared to the generative adversarial network (GAN) [22]. Additionally, this paper addresses the impact of different combinations of Dixon MRI images on the resulting pseudo-SPR images and proposes further improvement by incorporation of bone delineation as an additional channel to the U-Net.

2   Materials and Methods

2.1   Datasets Description

The dataset cohort consisted of 16 patients who were treated for Head & Neck cancer by a standard photon beam radiation therapy. For additional MRI data acquisition, an axial spoiled 3D-T1w gradient echo was obtained using the 1.5T MAGNETOM Sola scanner by Siemens Healthcare, Erlangen, Germany. The Dixon fat-suppressed and water-suppressed maps were produced by the scanner's reconstruction software. The following MRI sequences parameter were utilized for the acquisition: Field of view = $500 \times 360 \times 216 \, mm^3$, Voxel spacing = $1.3 \times 1.3 \times 3.0 \, mm^3$, Flip angle = $10°$, Total acquisition time = $3.51 \, min$, Repitition time = $6.9 \, ms$ and Echo time = $2.39/4.77 \, ms$. DECT images were acquired at a high energy and low energy of $140 \, kVp$ and $80 \, kVp$ respectively. DECT scanning was performed using Siemens Somatom Confidence with the following acquisition parameters (image size: 512×512; pixel spacing: $0.9776 \, mm$; slice thickness: $3.0 \, mm$). SPR maps were automatically generated using the Siemens DirectSPR module. Figure 1 is an illustration of a DECT with its corresponding SPR map.

2.2   Image Preprocessing

To correct MR scans for intensity inhomogeneity, N4 bias field correction was applied to all input MRI images [29]. Additionally, the level of noise in each image was reduced by performing anisotropic diffusion on all images [25]. Despite the lack of standardization present in MRI images, MRI scans were acquired with the same scanner; therefore, the intensity variations in-between patients are minimal. Thus, no normalization was adopted in this approach. Deformable image registration with the ability to preserve bony structures was utilized to align SPR images to their corresponding MRI images [17]. Subsequently, body mask for each patient was generated using morphological operations based on MR images to remove treatment couch and immobilization mask on SPR maps as shown in Fig. 2.

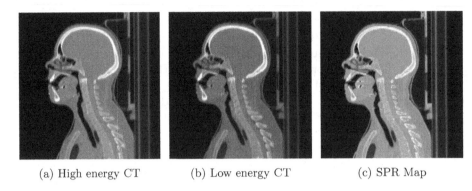

(a) High energy CT   (b) Low energy CT   (c) SPR Map

Fig. 1. An illustration of dual-energy (DECT) image with its corresponding SPR map for an exemplary Head & Neck datasets.

Using an in-house trained U-Net model which predicts individual bones on planning CT images, the model was used to automatically predict bones on the SPR images. From the predicted individual bones, all bones are merged to generate a bone mask. The transformation file of the MRI-SPR registration of each patient data was used to transform the bone mask of the SPR map to align with that of the MRI images. Next, these deformed bone masks were applied to the fat-suppressed Dixon MRI images to extract bone intensity images which were incorporated as an additional channel for the network training for later investigations. A sample case of the extraction process on fat-suppressed Dixon MRI is summarized in Fig. 3. Bone intensity images were performed on the fat-suppressed Dixon MRI because it highlights the best spatial anatomy.

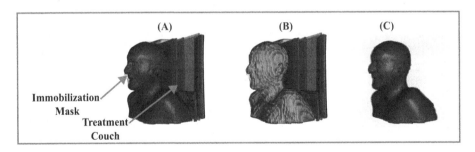

Fig. 2. Volume rendering showing body mask generation for a sample SPR map. (A) SPR map with immobilization mask and treatment couch (B) Estimated body mask (in green) after morphological operation overlaid on SPR map (C) Body mask applied on SPR map to remove immobilization mask and treatment couch. (Color figure online)

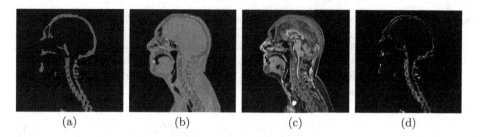

(a)   (b)   (c)   (d)

Fig. 3. Extraction of bone intensity image on the fat-suppressed Dixon MRI. (a) represents the transformed bone mask (blue) using MRI-SPR registration parameters (b) represents the bone mask overlaid on its corresponding deformed SPR map (c) represents bone mask overlaid on the fat-suppressed Dixon MRI and (d) represents bone intensity image on the fat-suppressed Dixon MRI. (Color figure online)

2.3 Model Generation

Of these 16 patient datasets, 12 cases were randomly selected as training sets and the remaining were held out as validation (1) and test (3) sets. All models were implemented using TensorFlow and trained on a double NVIDIA GeForce RTX 2080Ti GPU card. The selected U-Net architecture was trained from scratch using isotropic patch size of 64^3 by sliding window with its neighbouring patches with an overlap of 32^3. This overlap ensures that a continuous whole-label output can be obtained and allows for increased training data for the network [7]. Subsequently, standard augmentation techniques based on affine geometric transformations were explored to increase the generalization capability of the generated model. Following the patch extraction, the number of background patches was drastically reduced by randomly selecting a few of these patches as part of the training samples. The purpose of this step is to reduce the high-class imbalance skews towards background patches. The Adaptive Moment Estimation (Adam) [12] was the optimizing algorithm adopted with a learning rate of 2e-4. The batch size was limited to 4 with a fixed epoch of 200 epochs. To quantify the deviation of the predicted image patch from the target image patch, a combined loss of L1 and L2 was minimized as the objective function. Once a trained model was generated, the model was tested using the held-out test datasets. From the predicted patches, a whole 3D volume label was obtained via patch fusion of the predicted patches. The different MRI combinations used as input for generating pseudo-SPR are as follows:

- **Experiment 1:** Input MRI = T1-W only
- **Experiment 2:** Input MRI = Fat-suppressed only
- **Experiment 3:** Input MRI = Fat-suppressed + Water-suppressed
- **Experiment 4:** Input MRI = Fat-suppressed + Water-suppressed + Bone

2.4 Evaluation

The prediction accuracy of the trained model for the pseudo-SPR and its corresponding target SPR maps were evaluated both qualitatively and quantitatively.

Quantitative measures were carried out using: mean absolute error (MAE) and peak-signal-to-noise ratio (PSNR).

1. **Mean absolute error (MAE)** [28] metric assesses the overall image quality by taking the difference between the target SPR and the predicted pseudo-SPR in HU values.

$$MAE = \frac{1}{N} \sum_{i=1}^{N} |SPR_{i=1} - pSPR_{i=1}| \tag{1}$$

where N is the the total number of voxels within the entire volume; SPR_i is the ith voxel in the target SPR; $pSPR_i$ is the ith voxel in a pseudo-SPR

2. **Peak-signal-to-noise-ratio (PSNR)** [9] metric captures the maximum voxel intensity and mean squared error (MSE) of image difference in decibels (dB) by validating if the pseudo-SPR is evenly or sparsely distributed.

$$MSE = \frac{1}{N} \sum_{i=1}^{N} (SPR_{i=1} - pSPR_{i=1})^2 \tag{2}$$

where N is the the total number of voxels within the entire volume; SPR_i is the ith voxel in the target SPR; $pSPR_i$ is the ith voxel in a pseudo-SPR

$$PSNR = 10log_{10}(\frac{Q_I^2}{MSE}) \tag{3}$$

where Q is the maximal HU value between target SPR and predicted pseudo-SPR; MSE is the mean absolute as indicated in Eq. 2.

For these selected metrics, a desirable pseudo-SPR map has a low MAE (ideally, 0 HU) and high PSNR (ideally, the maximum value in dB). Evaluation metrics were performed on a voxel-wise basis on the entire predicted volume. For quantifying the quality of SPR map generated with our approach against the first published work by [19], besides the MAE (Eq. 1) and PSNR (Eq. 3), normalized mean absolute error (NMAE) was also measured which is a ratio of the MAE and the mean of the HU values of the entire volume.

3   Results and Discussion

With regards to the qualitative analysis, the detailed investigation performed on the predicted pseudo-SPR map in comparison to its target SPR map are displayed in Fig. 4 and 5. Figure 4 represents the line profile analysis while Fig. 5 elaborates the deviations observed in the predicted pseudo-SPR map against its target SPR map. Each of the experimental cases resulted in an accurate quality SPR map with significant differences observed in the bony structures. Considering the single input MRI such as Experiment 1 and 2, it can be observed that using T1-W as the input MRI resulted in more accurate SPR than Experiment 2. The difference observed in these experimental sets (Experiment 1 & 2) can be attributed to the fat information which is present in T1-W sequence. Experiment 3 illustrating the addition of the fatty signal (water-suppressed) to the fat-suppressed of the Dixon sequence triggered an improved results than using only T1-W as the input MRI. Thus, it implies that both the fat-suppressed and water-suppressed signal of the Dixon MRI images are essential information needed by the U-Net architecture to generate an accurate predicted pseudo-SPR map. Furthermore, bone delineation incorporated as an additional input MRI channel (Experiment 4) to the U-Net enhanced the bone identification on predicted pseudo-SPR map which yielded a more comparable SPR map to the target SPR map.

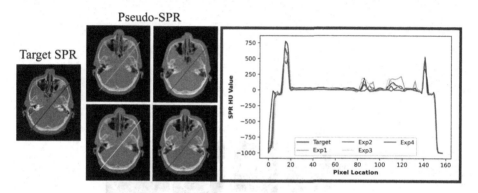

Fig. 4. Line profile analysis of target SPR map vs. predicted pseudo-SPR map for a random test case. Exp1 - Experiment 1, Exp2 - Experiment 2, Exp3 - Experiment 3 and Exp4 - Experiment 4 as defined in Subsect. 2.3.

Quantitative analysis using MAE (Eq. 1) and PSNR (Eq. 3) is presented in Table 1. The deviations discovered in the qualitative evaluation are in agreement with the quantitative results. In our study, the sequential addition of different images at the different channels (Experiments 3 & 4) increased the variability and features in the training resulting in less chances that the network overfit [2]. From Experiment 4, the model highlights the significance of bone delineation provided as an additional information to enable the differentiation of bone and air windows although they both appear dark on MRI images. As for the computational efficiency of this proposed approach, the average generation time including 3D volume reconstruction of a pseudo-SPR map was ≈ 30 sec per test case using GPU acceleration. This computation speed facilitates the smooth incorporation of such approach to online treatment settings where time is of essence. To assess our proposed approach to an already established approaches in literature, we compared to [19] as illustrated in Table 2 using the results reported in their metric table. Our approach significantly outperformed their approach for both T1-w and Dixon MRI sequences although trained with less dataset. This discovery further calls our attention to the fact that accurate deep learning models should not only be dependent on big data but rather on the quality of the dataset which will always be true in radiation therapy [32].

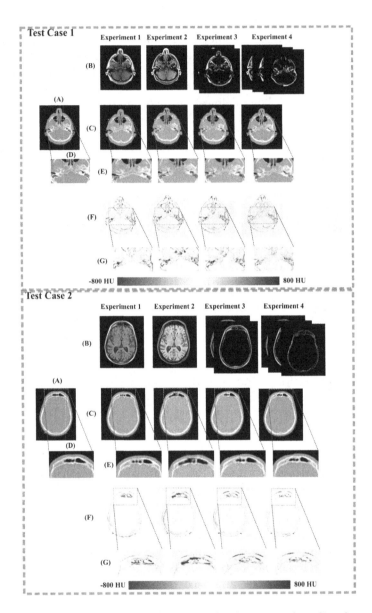

Fig. 5. Two exemplary test cases highlighting the deviation of predicted pseudo-SPR vs. target SPR slice: (A) Target SPR (B) Input MRI using experimental scenarios as explained in Subsect. 2.3 (C) Predicted pseudo-SPR (D) Random zoomed crop of target SPR (E) Random zoomed crop of predicted pseudo-SPR (F) HU-Difference map of target SPR and pseudo-SPR (G) Random zoomed crop of HU-Difference map of target SPR and pseudo-SPR.

Table 1. Numerical comparison of the MAE (HU) and PSNR (dB) evaluation metric for all experimental scenarios as explained in subsection 2.3 using all test samples.

Patient ID	Experiment 1		Experiment 2		Experiment 3		Experiment 4	
	MAE	PSNR	MAE	PSNR	MAE	PSNR	MAE	PSNR
Patient 1	32.08	55.74	35.45	55.28	31.81	55.65	**29.39**	**56.27**
Patient 2	17.60	59.06	20.72	58.68	17.74	59.00	**15.15**	**59.78**
Patient 3	16.17	59.31	19.29	58.89	16.58	59.27	**13.69**	**60.24**
Mean	21.95	58.04	25.15	57.61	22.04	57.62	**19.41**	**58.76**
(±)STD	8.80	2.00	8.95	2.02	8.47	2.02	**8.67**	**2.17**

Table 2. Numerical comparison of the mean NMAE and PSNR (dB) evaluation metric to published literature.

Approach	MRI-sequence	Number of Datasets	NMAE (%)	PSNR (dB)
CycleGAN [19]	T1-W	57	8.02 ±1.56	19.00 ± 1.03
Label-GAN [19]	T1-W	57	5.22 ±1.23	26.51 ± 1.50
Proposed	T1-W	16	2.73 ± 1.30	58.04 ± 2.00
	Dixon	16	2.41 ± 1.26	58.76 ± 2.17

4   Conclusion

In this study, we report on a U-Net-based pseudo-SPR map generation from Dixon MRI sequence and investigated the impact of different combinations of Dixon MRI images on the resulting pseudo-SPR maps. Furthermore, bone delineation incorporated as an additional channel to network was explored. The proposed solution proves that the sequential addition of different Dixon MRI sequence (i.e.: fat-suppressed and water-suppressed) significantly increased the variability and features in the training phase; thus, resulting in an accurate pseudo-SPR map rather than a single MRI input such as T1-W or fat-suppressed only. Moreover, bone delineation integrated as an additional channel to Dixon MR sequence demonstrated an enhanced bone identification on the predicted pseudo-SPR map which yielded a more comparable pseudo-SPR map to the target SPR map. Our work was compared to an already published work in literature and it achieved a promising pseudo-SPR map which outperformed their approach even on limited dataset. As future direction, we would like to extend this approach to a clinical SPR map which will enable dosimetric analysis of clinical target volume (CTV) to be possible in treatment planning application for ion beam therapy.

Funding. This study was mainly funded by German Federal Ministry of Education and Research within the track "Bildgeführte Diagnostik und Therapie - Neue Wege in der Intervention" in ARTEMIS project (13GW0436).

References

1. Çiçek, Ö., Abdulkadir, A., Lienkamp, S.S., Brox, T., Ronneberger, O.: 3D U-Net: learning dense volumetric segmentation from sparse annotation. In: Ourselin, S., Joskowicz, L., Sabuncu, M.R., Unal, G., Wells, W. (eds.) MICCAI 2016. LNCS, vol. 9901, pp. 424–432. Springer, Cham (2016). https://doi.org/10.1007/978-3-319-46723-8_49

2. Dietterich, T.: Overfitting and undercomputing in machine learning. ACM Comput. Surv. (CSUR) **27**(3), 326–327 (1995)

3. Dowling, J.A., et al.: An atlas-based electron density mapping method for magnetic resonance imaging (mri)-alone treatment planning and adaptive mri-based prostate radiation therapy. Int. J. Radiation Oncol. Biol. Phys. **83**(1), e5–e11 (2012)

4. Edmund, J.M., Kjer, H.M., Van Leemput, K., Hansen, R.H., Andersen, J.A., Andreasen, D.: A voxel-based investigation for mri-only radiotherapy of the brain using ultra short echo times. Phys. Med. Biol. **59**(23), 7501 (2014)

5. Edmund, J.M., Nyholm, T.: A review of substitute ct generation for mri-only radiation therapy. Radiat. Oncol. **12**, 1–15 (2017)

6. Forghani, R., De Man, B., Gupta, R.: Dual-energy computed tomography: physical principles, approaches to scanning, usage, and implementation: part 1. Neuroimaging Clin. **27**(3), 371–384 (2017)

7. Fu, Y., et al.: Pelvic multi-organ segmentation on cone-beam ct for prostate adaptive radiotherapy. Med. Phys. **47**(8), 3415–3422 (2020)

8. Hoffmann, A., et al.: Mr-guided proton therapy: a review and a preview. Radiat. Oncol. **15**(1), 1–13 (2020)

9. Hore, A., Ziou, D.: Image quality metrics: Psnr vs. ssim. In: 2010 20th International Conference on Pattern Recognition, pp. 2366–2369. IEEE (2010)

10. Hsu, S.H., Cao, Y., Huang, K., Feng, M., Balter, J.M.: Investigation of a method for generating synthetic CT models from MRI scans of the head and neck for radiation therapy. Phys. Med. Biol. **58**(23), 8419 (2013)

11. Hudobivnik, N., et al.: Comparison of proton therapy treatment planning for head tumors with a pencil beam algorithm on dual and single energy CT images. Med. Phys. **43**(1), 495–504 (2016)

12. Jais, I.K.M., Ismail, A.R., Nisa, S.Q.: Adam optimization algorithm for wide and deep neural network. Knowl. Eng. Data Sci. **2**(1), 41–46 (2019)

13. Johansson, A., Karlsson, M., Nyholm, T.: Ct substitute derived from MRI sequences with ultrashort echo time. Med. Phys. **38**(5), 2708–2714 (2011)

14. Karotki, A., Mah, K., Meijer, G., Meltsner, M.: Comparison of bulk electron density and voxel-based electron density treatment planning. J. Appl. Clin. Med. Phys. **12**(4), 97–104 (2011)

15. Kristensen, B.H., Laursen, F.J., Løgager, V., Geertsen, P.F., Krarup-Hansen, A.: Dosimetric and geometric evaluation of an open low-field magnetic resonance simulator for radiotherapy treatment planning of brain tumours. Radiother. Oncol. **87**(1), 100–109 (2008)

16. Largent, A., et al.: Pseudo-CT generation for MRI-only radiation therapy treatment planning: comparison among patch-based, atlas-based, and bulk density methods. Int. J. Radiation Oncol. Biol. Phys. **103**(2), 479–490 (2019)

17. Leibfarth, S., et al.: A strategy for multimodal deformable image registration to integrate pet/MR into radiotherapy treatment planning. Acta Oncol. **52**(7), 1353–1359 (2013)

18. Leu, S.C., Huang, Z., Lin, Z.: Generation of pseudo-CT using high-degree polynomial regression on dual-contrast pelvic MRI data. Sci. Rep. **10**(1), 1–11 (2020)
19. Liu, R., et al.: Synthetic dual-energy CT for MRI-only based proton therapy treatment planning using label-gan. Phys. Med. Biol. **66**(6), 065014 (2021)
20. Liu, Y., et al.: MRI-based treatment planning for liver stereotactic body radiotherapy: validation of a deep learning-based synthetic ct generation method. Br. J. Radiol. **92**(1100), 20190067 (2019)
21. Liu, Y., et al.: Evaluation of a deep learning-based pelvic synthetic CT generation technique for MRI-based prostate proton treatment planning. Phys. Med. Biol. **64**(20), 205022 (2019)
22. Ma, X., Chen, X., Li, J., Wang, Y., Men, K., Dai, J.: Mri-only radiotherapy planning for nasopharyngeal carcinoma using deep learning. Front. Oncol. **11**, 713617 (2021)
23. Minogue, S., Gillham, C., Kearney, M., Mullaney, L.: Intravenous contrast media in radiation therapy planning computed tomography scans-current practice in Ireland. Techn. Innov. Patient Support Radiation Oncol. **12**, 3–15 (2019)
24. Owrangi, A.M., Greer, P.B., Glide-Hurst, C.K.: Mri-only treatment planning: benefits and challenges. Phys. Med. Biol. **63**(5), 05TR01 (2018)
25. Perona, P., Shiota, T., Malik, J.: Anisotropic diffusion. Geometry-driven diffusion in computer vision, pp. 73–92 (1994)
26. Sjölund, J., Forsberg, D., Andersson, M., Knutsson, H.: Generating patient specific pseudo-CT of the head from MR using atlas-based regression. Phys. Med. Biol. **60**(2), 825 (2015)
27. Spadea, M.F., et al.: Deep convolution neural network (DCNN) multiplane approach to synthetic CT generation from MR images–application in brain proton therapy. Int. J. Radiation Oncol. Biol. Phys. **105**(3), 495–503 (2019)
28. Tang, H., Cahill, L.: A new criterion for the evaluation of image restoration quality. In: TENCON'92-Technology Enabling Tomorrow, pp. 573–577. IEEE (1992)
29. Tustison, N.J., et al.: N4itk: improved n3 bias correction. IEEE Trans. Med. Imaging **29**(6), 1310–1320 (2010). https://doi.org/10.1109/TMI.2010.2046908
30. Uh, J., Merchant, T.E., Li, Y., Li, X., Hua, C.: MRI-based treatment planning with pseudo CT generated through atlas registration. Med. Phys. **41**(5), 051711 (2014)
31. Yang, M., et al.: Comprehensive analysis of proton range uncertainties related to patient stopping-power-ratio estimation using the stoichiometric calibration. Phys. Med. Biol. **57**(13), 4095 (2012)
32. Zhou, L., Pan, S., Wang, J., Vasilakos, A.V.: Machine learning on big data: opportunities and challenges. Neurocomputing **237**, 350–361 (2017)

Generalised 3D Medical Image Registration with Learned Shape Encodings

Christoph Großbröhmer$^{(\boxtimes)}$ (ID) and Mattias P. Heinrich (ID)

University of Lübeck, Lübeck, Germany
{c.grossbroehmer,heinrich}@uni-luebeck.de

Abstract. Due to the high variability in medical images, task-specific solutions have often prevailed in the field of Deep Learning Image Registration (DLIR). Contrary to classical approaches, these are hardly transferable to other tasks or datasets. To overcome these limitations we propose to exploit easy-to-acquire anatomical segmentations, which provide strong semantic features as a common representation across different datasets and image modalities. We, therefore, present a new module for medical image registration based on generalised label encodings that can be applied to arbitrary label maps. We demonstrate the generalisability of our approach by employing the module trained on skeletal and pulmonary vascular structures to the registration of abdominal segmentations, yielding an average increase in Dice scores of 24% and compare their encoding qualities to classical Euclidean Distance Maps. Furthermore, we leverage the modular design to assess the generalisability of convolutional and vision-transformer based registration architectures. Source code and trained models are released on GitHub.

Keywords: Gerneralisability · Image Registration · Semantic Segmentation

1 Introduction

In recent years, Deep Learning Image Registration (DLIR) has become increasingly popular as it performs comparably in terms of accuracy to well-tuned optimisation-based registration methods while being much faster in inference. However, one drawback of DLIR methods is that their performance is highly dependent on the training data used, resulting in very task-specific solutions that do not generalise well to unseen domains. In contrast, conventional registration algorithms such as Elastix [11] or NiftyReg [13] can be applied to a variety of different tasks, although suitable hyperparameters must be found for a good solution. One of the reasons why such generalised methods have not yet entered the DLIR domain is the high variability of medical image data for registration problems (inter- and intra-patient, single- and multimodal, anatomical scope),

G. Waiter et al. (Eds.): MIUA 2023, LNCS 14122, pp. 268–280, 2024.
https://doi.org/10.1007/978-3-031-48593-0_20

which results in a small number of relevant datasets and the development of specialized solutions. Furthermore, this heterogeneity leads to the difficult choice of image features, metrics and losses suitable for registration.

In this work, we propose a new approach to generalised medical image registration. Instead of dealing with the high variability of registration-related tasks and data, we explore the possibility of decoupling semantic feature generation and learning-based image registration. Our work is mainly motivated by two aspects: First, the fact that the use of semantic segmentations together with regularisation can increase the accuracy of DLIR methods, especially for inter-patient registration tasks [6]. For example, in the multi-task registration challenge Learn2Reg, best results were obtained when semantic information has been included during training, either directly as additional input features or indirectly through the application of segmentation-dependent metrics, such as the Dice loss [6]. Second, advances in deep learning methods in recent years for image segmentation greatly facilitate the synthesis of anatomical labels. Robust and easy-to-use pre-trained tools, such as the TotalSegmentator [19], a variety of nnUNet [9] models or models from the MONAI Model Zoo [2] enable the simple, fast and straightforward generation of segmentations for a wide range of anatomical structures from a variety of image domains. Consequently, the question arises whether semantic segmentations can serve as a key building block to developing generalised registration methods beyond single dataset boundaries, effectively transforming the task of image registration into the task of shape matching.

Contrary to the common approach for registration algorithms based on joint learning of suitable features and deformation modules, we exploit the precomputed segmentations as strong semantic features for the prediction of deformation fields. Directly registering label maps as opposed to grey-valued image data results in great flexibility: Registration across domain gaps, such as various medical image systems, is possible without any adaptation to the method. Other advantages include the effortless incorporation of advances in segmentation algorithms and the fact that, since our method does not rely directly on easily identifiable image data, it is privacy-preserving by design.

Related Work

Shape matching and image registration are two closely related and integral tasks in computer vision and image processing and have been addressed using both classical iterative and deep learning methods. One approach is the minimization of Distance Functions (DF) to measure the alignment of geometric shapes [15], including Signed Distance Maps and Euclidean Distance Maps (EDMs). Segmentations of organs or other anatomical structures are one of the most common semantic annotations in medical image data. In addition to their direct clinical use for volume, shape, and location analysis, they also offer the possibility of introducing semantic knowledge into medical image processing methods. In Machine Learning (ML), this includes the employment of label-based metrics such as the Dice Loss. On the other hand, anatomical information can also guide the learning process. In the domain of image registration, Hu et al. [8] leverage a variety of annotations, including segmentations, as weak supervision to overcome the domain gap between ultrasound and MRI images. However, this non-generalised method is restricted to anatomies and image modalities previously seen in training. Other approaches preprocess medical information to facilitate their usage in learning-based methods, including the encoding of anatomical shapes with distance transforms. Popular approaches for semantic segmentations include loss functions relying on EDMs or the introduction of auxiliary tasks to boost model performances [12].

EDMs have also been employed sporadically in the field of image registration. Chen et al. use distance transforms in combination with mutual information to improve the registration in areas with homogeneous intensities in brain MRIs [3] and Canalini et al. leverage the euclidean distance transform for interoperative registration of 3D ultrasound [1]. While these works highlight the principle suitability of EDMs for image registration, both rely on conventional registration algorithms.

Augmentation and synthesis of training data is an integral part of Deep Learning methods. This is even more true in the field of medical image analysis, where training data remains scarce due to privacy concerns and high annotation costs. However, there are also approaches where the synthesis of training data exceeds the purpose of extending a real dataset. One example of this in the field of image registration is SynthMorph [7], which uses synthesized images with arbitrary shapes to develop a contrast-agnostic registration network. A Voxelmorph architecture is trained using both strongly augmented dense label maps as well as unrealistic shapes, enforcing the matching of structures based on shapes rather than image features. In terms of leveraging segmentations to achieve generalisability, we follow a similar strategy. However, instead of inferring on real grey-valued image data, we seek to predict motion completely relying on segmentation maps.

Contributions

In this work, we aim to investigate the possibility of using automatically generated semantic segmentations to develop a generalisable deep learning registration method. This approach requires two components: 1) a common, robust encoding of arbitrary label maps that is independent of particular datasets and anatomical regions and 2) a module to register the encoded label maps to each other.

We propose a novel learning-based adaptation layer for label map generalisation and evaluate its performance against Multilabel Euclidean Distance Maps. Furthermore, we investigate the effect of two spatial transformer registration modules, including a Swin Transformer architecture. To the best of our knowledge, this is both the first approach to develop a generalised deep learning registration method and the first application of a registration model fully trained with CT data on MRI scans.

In addition, source code and trained models have been released on GitHub[1].

2   Methods

The main concept of our method is a decoupling between the individual instances of a segmentation map and subsequent label-agnostic deformable image registration. The method can be trained on arbitrary pairs of segmentations as long as each pair of fixed and moving images have congruent label definitions, which allows the use of data from a variety of different datasets with manually annotated or automatically generated segmentations. Consequently, segmentations derived from arbitrary image data, anatomical regions and various segmentation models can also be used during inference, so that the trained model has no explicit domain gap with respect to unseen datasets. An overview of our pipeline is shown in Fig. 1.

Our method is composed of two building blocks: First, segmentation maps have to be encoded in a generalised manner to prevent overfitting to particular anatomies in training data. We achieve this by a) utilizing Multilabel Euclidian Distance maps and b) a novel learning-based Label Adaptation Module. Second, the encoded label maps are subject to a registration method. We employ both a simple convolution neural network-based registration module and a variant utilizing a hierarchical shifted-window vision transformer.

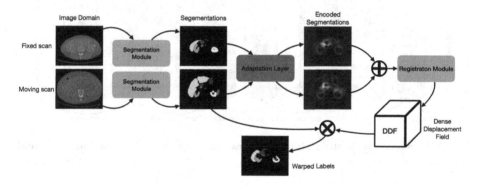

Fig. 1. Generalized Registration Pipeline. One or more suitable segmentation methods predict congruent sets of anatomical labels on input fixed and moving scans. These are then fed into the proposed Adaptation Layer, which extracts features independently of the type and number of generated semantic structures. Subsequently, the generalised single-channel encodings are concatenated and subject to a two-step registration module, predicting a dense displacement field to be used to warp input segmentations or image.

---

[1] https://github.com/multimodallearning/GenRegShapes/.

2.1 Encoded Label Maps

Multilabel Euclidean Distance Maps. Given the binarized label image $I \in \{0,1\}^n$, we elect to calculate a Euclidean Distance Map (EDM). For every label image element x_i we assign

$$y_i = \begin{cases} \sqrt{\sum_i^n (x_i - b_i)^2} & \text{if } x_i = 1 \\ 0, & \text{otherwise} \end{cases},$$

where b_i represents the background element ($i = 0$) with the smallest euclidean distance to input point x_i.

We expand this technique to multilabel segmentations S with labels $l = \{1, ..., L\}$ so that

$$\text{MEDM}(S) = \sum_j^L \frac{1}{\lambda} \left(\lambda - ELU\left(\text{EDM}(\lambda - S_j)\right) \right).$$

Here, ELU denotes the element-wise Exponential Linear Unit [4] and λ denotes the distance to structure boundaries affected by the ELU. We design the MLEMs to emphasize borders to facilitate their use for registration methods by decreasing the values of voxels inside of structures.

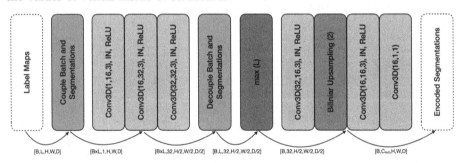

Fig. 2. Proposed Adaptation Module. Operations with dimension manipulations and convolutional layers are highlighted in green and ocher respectively. (Color figure online)

Adaptation Module. The purpose of the proposed Adaptation Module is to enable feature learning explicitly on individual segmentations. Therefore, we follow three objectives: 1) Decoupling of segmentation instances, 2) Composing agnostic feature learning and 3) Recoupling of features. First, batch and channel dimensions of one-hot-encoded segmentation maps $S \in \{0,1\}$ with dimensions [B×L×H×W×D] are joined into one dimension, resulting in a tensor of shape [B·L×1×H×W×D]. Therefore, subsequent convolutions are performed on a single channel and cannot establish a connection between segmentation instances, effectively separating feature learning for individual anatomical structures. This is necessary to prevent the learning of specific labels from the training dataset and to enable the application to previously unseen structures. We employ 3 blocks composed of $3 \times 3 \times 3$ convolutions (of which the first one has a stride of 2) with output channel dimensions of (16,32,32), Instance Normalisation and a ReLU activation function. The initial separation of batch (cases) and channels (segmentations) is reversed by splitting the initial batch and channels back into their two respective dimensions. By collecting the maximum feature map response along segmentation instance dimension, the tensor is condensed to the shape $[B \times 1 \times \frac{H}{2} \times \frac{W}{2} \times \frac{D}{2}]$. Subsequent convolutions and bilinear upsampling restore spatial input shapes. Since the second convolutional block can access the convolved and pooled segmentations of all labels, it can in principle leverage the spatial relationship of encoded individual labels. Finally, we employ a $1 \times 1 \times 1$ convolution to change output channel dimensions, yielding the final encoded segmentations of shape $[B \times 1 \times \frac{H}{2} \times \frac{W}{2} \times \frac{D}{2}]$. An overview of the proposed Adaptation Module is shown in Fig. 2.

2.2 Registration Models

Since our proposed method can be seen as a feature encoder, we are not restricted to a specific registration model. We, therefore, employ two variants of registration networks composed of a two-stream feature encoder and a spatial transformation module. To account for fine-grained registration results, both networks can be employed in a two-stage manner.

CNN-Reg. We design CNN-Reg by employing a two-stream feature-encoding architecture with shared weights for both input tensors consisting of six encoder blocks, where each block is composed of a 3D convolution, Instance Normalisation and ReLU. The final feature dimension is successively built up to a total of 64 channels, whereas the spatial dimension is halved. In the registration part of the model, fixed and moving features are concatenated and subject to two convolution blocks with decreasing channel dimensions. The resulting 3-channel deformation field is subject to a hyperbolic tangent and multiplied with a factor of 0.25, which corresponds to displacements for 1/4 of initial image size. Finally, we employ B-Spline regularisation (as a combination of trilinear upsampling and average pooling with kernels of base sizes 3 and 5) twice. The two-step architecture allows the prediction of deformations with increased accuracy by accounting for the residual, uncompensated motion. If this setting is chosen, the moving input features are warped with the predicted displacements and again subject to stage one feature extraction layers. We then employ a second registration block with B-Spline regularisation for which we choose the previous parameters.

UNETR-Reg. We leverage the interoperability of the proposed generalisation module by including SwinUNETR [5] modules provided by MONAI[2]. We build

---

[2] https://docs.monai.io/en/stable/_modules/monai/networks/nets/swin_unetr.html.

upon the two-step CNN-Reg and simply replace the convolutional registration layers with SwinUNETR blocks. Since we train the networks with variable spatial image dimensions, we set the SwinUNETR input image size to $64 \times 64 \times 64$ and resample all encoded label maps to $192 \times 192 \times 192$ voxels before feeding them to the registration module.

3   Experiments

3.1   Datasets

We aim to train our models with varying anatomical shapes. Therefore, we construct two datasets with semi- or fully automated labels (TS-Skeletal and VIS-Vascular), leveraging the broad range of predictable classes from the popular TotalSegmentator tool and dataset. For evaluation, we utilize two publicly available datasets (AbdomenCTCT [6], AMOS [10]) from established challenges, which both only share two labels (aorta and vena cava inferior) with one of the training datasets. 3D renders of used data are shown in Fig. 3.

TS-Skeletal. We use semi-automatic segmentations for a total of 28 skeletal structures from the TotalSegmentator dataset [19]. First, we extract all images containing segmentations for ribs (bilateral, rib 2–11) and vertebrae (T2 - L3). After visual inspection and discarding of scans with implausible or overlapping segmentations, we select 33 remaining cases for dataset construction. Based on the proposed segmentations, label maps have been cropped to $256 \times 160 \times 256$ voxels. We utilize Convert3D command line tool[3] (`c3d -align-landmarks`) for rigid alignment of label maps.

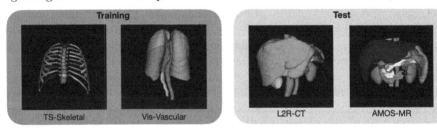

Fig. 3. 3D visualisations of training and test datasets. Both L2R-CT and AMOS-MR feature disjoint sets compared to the training datasets, with the exception of the Vena Cava Inferior and Aorta.

VIS-Vascular. We use the TotalSegmentator tool [19] to predict segmentations on 51 contrast-enhanced CT scans of the Anatomy 3 Visceral Silver Corpus Dataset [18] and choose 9 cardiovascular and lung structures which share no overlap with TS-Skeletal (Aorta, Vena Cava Inferior, Pulmonary Artery, Heart, Lung lobes 1–5), emphasizing functional and structural differences as well as spatial distances. Analogous to TS-Skeletal, segmentation maps are cropped to a uniform shape of $256 \times 192 \times 288$ voxels and aligned rigidly using Convert3D.

---

[3] http://www.itksnap.org/pmwiki/pmwiki.php?n=Convert3D.Convert3D.

L2R-CT We choose the Learn2Reg Challenge AbdomenCTCT dataset[4] to evaluate the registration accuracy of the proposed method. The data includes a total of 30 affine pre-registered CT scans ($192 \times 160 \times 256$) and manual segmentations for 13 labels (■ aorta, ■ inferior vena cava, ■ spleen, ■ left / ■ right kidney, ■ gallbladder, ■ esophagus, ■ liver, ■ pancreas, ■ stomach, ■ left/■ right adrenal glands and ■ portal and splenic vein). We apply the proposed validation and registration split (20 training, 10 validation, 45 registration pairs) and train a nnUNet for 300 epochs on all training scans and predict segmentations for all validation cases.

AMOS-MR. Furthermore, we evaluate the proposed method on publicly available MRI scans of the AMOS 2022 challenge[5] [10], highlighting the multimodal applicability of the proposed segmentation registration approach. The selected subset comprises 40 affine pre-registered MRI scans with spatial dimensions of $256 \times 160 \times 256$ and abdominal segmentations (■ aorta, ■ inferior vena cava, ■ spleen, ■ left/■ right kidney, ■ gallbladder, ■ esophagus, ■ liver, ■ stomach, ■ pancreas, ■ left / ■ right adrenal glands and ■ duodenum). Following the approach of L2R-CT, we train another nnUNet for 300 epochs on 32 training cases for the generation of predicted segmentations and choose the remaining 8 cases for testing, yielding a total of 28 registration pairs. We assess the quality of nnUNet predictions and achieve an average Dice overlap of 82% regarding all labels, with large structures (such as liver, spleen, stomach and kidneys) reaching Dice scores above 90% and small structures (such as adrenal glands) around 60%.

Table 1. Registration results for each combination of model and label encoding on the **L2R-CT** dataset shown in dice overlap percentages, 95th percentile of Hausdorff distance (HD95) and deformity of registration as standard deviation of Jacobian determinants (JD, lower is better). We report dice averages over all labels and shared anatomic labels with one training set (■,■) separately. Color Encodings for individual labels are described in Sect. 3.1.

Model	Enc	■	■	■	■	■	■	■	■	■	■	■	■	■	avg(2)	avg(all)	HD95	JD
initial		33	36	42	34	35	2	23	62	24	15	8	8	5	34.5	25.1±13	20.1	–
CNN	MEDM	39	38	67	52	56	4	23	83	46	23	11	10	12	38.1	35.6±10	16.0	0.006
	Adapt	40	46	70	68	73	11	33	79	44	33	19	23	13	42.9	**42.4±12**	14.7	0.007
UNETR	MEDM	38	39	64	46	60	4	21	80	49	18	9	7	9	38.4	34.2±11	21.2	0.031
	Adapt	46	49	66	72	79	17	43	81	47	33	24	29	24	47.5	**47.0±14**	23.3	0.022

---

[4] https://learn2reg.grand-challenge.org/Datasets/.
[5] https://amos22.grand-challenge.org/.

3.2   Training and Evaluation

We train all networks in 3 stages of each 2000 iterations each using Adam with a learning rate of 0.001 and mixed precision to accelerate training. In the second half of each stage, we employ the registration network in a two-step manner for higher accuracy predictions and increase the strength of random affine augmentations throughout all stages. In each iteration, one (i.e. batch size = 1) one-hot encoded label map from either training dataset is drawn randomly, downsampled by a factor of 2 using average pooling, encoded by MEDM ($\lambda = 7$) or the proposed Adaptation Module and processed by the registration network. We employ Soft Dice loss on warped input labels. To facilitate the convergence of the Adaptation Module, we use the EDM-pretrained registration network as a starting point.

We evaluate our method for each combination of architecture (CNN-Reg, UNETR-Reg) and label generalisation method (MEDM, learned Adaptation Module) on both test datasets (L2R-CT, AMOS-MR) independently using a 32 GB Tesla V100-SXM3 GPU. For each test case, a corresponding label map is predicted and generalised. Moving ground-truth annotations are warped with the deformation field output by the registration module, which takes about 190 ms (CNN-Reg) or 450 ms (UNETR-Reg) respectively. Inference of the Adaptation Layer takes about 50 ms. The quality of registration is measured in terms of volumetric alignment, surface distances and deformity of registration by calculation of Dice coefficients, the 95th percentile of Hausdorff Distances and the standard deviation of logarithmized Jacobian determinants respectively. We chose to include vascular labels ■ aorta and ■ inferior vena cava, which exist both in VisVascular and testing datasets, to investigate a possible bias towards previously seen anatomic shapes.

4   Results and Discussion

Quantitative results (Table 1 and Table 2) show that all combinations of models and label encoders succeed in aligning anatomical shapes in out-of-training-domain datasets. Best Dice scores of 47% (L2R-CT) and 54.2% (AMOS) are achieved by employing UNETR and the proposed Adaptation Layer. In comparison between Multilabel Euclidean Distance Maps, the Adaptation Module holds the advantage throughout nearly all individual labels, leading to an average increase of 8 (CNN-Reg) and 14 (UNETR-Reg) percentage points. Especially the alignment of small structures (■, ■, ■) improves considerably by Dice scores of 8% to 22%. MEDMs do not seem to be able to encode anatomies with small diameters well, as those may not benefit as much from the proposed boundary emphasis technique. Another explanation is that the learned Adaptation layer is to some extent able to compensate for the lower quality of predicted segmentations of small structures for registration. For large structures such as the liver and spleen, which are already well aligned by rigid pre-registration, similar results are obtained for both coding variants.

Table 2. Registration results for each combination of model and label encoding on the **AMOS-MR** dataset shown in dice overlap percentages, 95th percentile of Hausdorff distance (HD95) and deformity of registration as the standard deviation of Jacobian determinants (JD, lower is better). We report dice averages over all labels and shared anatomic labels with one training set (■,■) separately. Colour Encodings for individual labels are described in Sec. 3.1.

Model	Enc.	■	■	■	■	■	■	■	■	■	■	■	■	■	avg(2)	avg(all)	HD95	JD
initial		21	30	50	52	51	12	8	71	37	38	10	11	20	25.7	31.6±11	19.0	–
CNN	MEDM	34	37	74	69	69	23	15	89	52	42	9	8	27	35.7	42.2±09	15.2	0.006
	Adapt	46	54	79	78	75	41	33	88	58	56	19	18	32	49.9	**52.1±11**	12.9	0.006
UNETR	MEDM	33	40	69	68	65	16	11	84	50	35	3	8	22	36.3	38.6±10	21.7	0.032
	Adapt	50	56	83	75	81	35	35	84	68	59	19	26	34	53.1	**54.2±14**	27.3	0.026

A possible label bias of (■,■) is not reflected in the corresponding dice scores, as they do not show excessive improvements compared to the other labels in either data set, which strengthens the assumption of strong generalisation.

When employing the learned adaptation layer, better dice scores can be obtained by the Swin Transformer registration model UNETR-Reg (7%-16%), while the average dice score even deteriorates when using MEDM. In this setting, Vision Transformers benefit even more strongly from the learnable Adaptation Layer. In both variants, however, the warped labels show significantly higher Hausdorff distances, which partly exceed the initial values. This is accompanied by a decrease in deformation smoothness by a factor of 3–5, suggesting the need for a stronger regularization.

We further explore the possibility of refining the deformation field with Adam instance optimisation (similar to [17]). This boosts average dice scores to 56% and 61% for L2R-CT and AMOS-MR respectively, without leveraging any image-domain data, highlighting the possibility to employ our method as a coarse pre-registration.

Since we are following the L2R Challenge data split we are able to compare our results with the SOTA. Our models succeed to outperform both supervised and unsupervised Voxelmorph variants, which achieve average Dice scores of 35.5% and 43.85% respectively [16]. When using instance optimisation, our method trails only by 10% behind the upper baseline of 66.2% set by the label-supervised LapIRN[6] [14], which leveraged the target domain for training the registration network. Considering both the generalised design of our method and the fact that our models were trained not using any of the L2R data, the competitiveness of our results is very promising.

Figure 4 shows example visual results of predicted labels, encoded segmentations, and registration results. Compared to the MEDMs, the features generated by the proposed Adaptation Module do not show a uniform response at the segmentation edges, but a locally different weighting of nearby structures. Although

---

[6] https://learn2reg.grand-challenge.org/evaluation/54c99fe5-9afb-48df-8b75-6e758248be6a/.

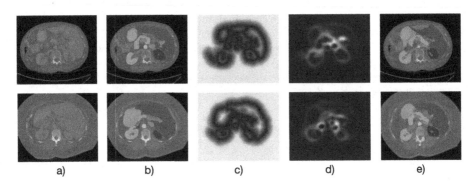

Fig. 4. Exemplary visual results. Randomly chosen a) L2R-CT scans with b) nnUNet predicted labels are encoded with c) Multilabel Euclidean Distance Maps or d) the trained Adaptation Module. Rows show the fixed and moving input images respectively. Column e) shows the overlaid warped moving labels on the fixed input scan in top and warped labels and moving scan after subsequent instance optimisation finetuning in bottom position.

the warped moving labels align reasonably well with the anatomy of the fixed scans it can further be improved by subsequent instance optimisation finetuning. Since label adaptation and registration can only access label information, the spatial extent of predicted deformation is restricted. Image contents outside these boundaries, such as skin and fat tissue, cannot be registered with our method by design. However, since in most scenarios of medical image registration, an alignment of target structures is desired, this drawback can be mitigated by selecting appropriate and high-quality input segmentations.

5   Conclusion and Outlook

The novel approach to generalised deformable Deep Learning Image Registration successfully aligns previously unseen label structures, by transforming the registration task into a shape-matching task and leveraging segmentations as strong semantic features without the constraints of being restricted to specific training datasets. A proposed Label Adaptation Module outperforms classical Euclidean Distance encodings by a large margin, especially for small structures. We assess its versatility and effectiveness with both a convolutional and a Vision Transformer registration module and subsequent instance optimisation refinement on inter-patient abdominal datasets. Future work could address its incorporation into further existing deformable image pipelines, training with large amounts of synthetical data and investigating the registration accuracy with regard to variable extends of anatomical segmentations. We hope that this work can promote research in the area of generalised image registration, which has been scarcely explored so far.

References

1. Canalini, L., Klein, J., Miller, D., Kikinis, R.: Registration of ultrasound volumes based on euclidean distance transform. In: Zhou, L., et al. (eds.) LABELS/HAL-MICCAI/CuRIOUS -2019. LNCS, vol. 11851, pp. 127–135. Springer, Cham (2019). https://doi.org/10.1007/978-3-030-33642-4_14

2. Cardoso, M.J., et al.: Monai: an open-source framework for deep learning in healthcare. arXiv preprint arXiv:2211.02701 (2022)

3. Chen, M., Carass, A., Bogovic, J., Bazin, P.L., Prince, J.L.: Distance transforms in multichannel MR image registration. In: Medical Imaging 2011: Image Processing, vol. 7962, pp. 430–436. SPIE (2011)

4. Clevert, D.A., Unterthiner, T., Hochreiter, S.: Fast and accurate deep network learning by exponential linear units (elus). arXiv preprint arXiv:1511.07289 (2015)

5. Hatamizadeh, A., Nath, V., Tang, Y., Yang, D., Roth, H.R., Xu, D.: Swin UNETR: swin transformers for semantic segmentation of brain tumors in MRI images. In: Crimi, A., Bakas, S. (eds.) International MICCAI Brainlesion Workshop, vol. 12962, pp. 272–284. Springer, Heidelberg (2021). https://doi.org/10.1007/978-3-031-08999-2_22

6. Hering, A., et al.: Learn2reg: comprehensive multi-task medical image registration challenge, dataset and evaluation in the era of deep learning. IEEE Trans. Med. Imaging **42**, 697–712 (2022)

7. Hoffmann, M., Billot, B., Greve, D.N., Iglesias, J.E., Fischl, B., Dalca, A.V.: Synthmorph: learning contrast-invariant registration without acquired images. IEEE Trans. Med. Imaging **41**(3), 543–558 (2021)

8. Hu, Y., et al.: Weakly-supervised convolutional neural networks for multimodal image registration. Med. Image Anal. **49**, 1–13 (2018)

9. Isensee, F., Jaeger, P.F., Kohl, S.A., Petersen, J., Maier-Hein, K.H.: NNU-NET: a self-configuring method for deep learning-based biomedical image segmentation. Nat. Methods **18**(2), 203–211 (2021)

10. Ji, Y., Bai, H., Ge, C., Yang, J., Zhu, Y., Zhang, R., Li, Z., Zhanng, L., Ma, W., Wan, X., et al.: AMOS: a large-scale abdominal multi-organ benchmark for versatile medical image segmentation. Adv. Neural. Inf. Process. Syst. **35**, 36722–36732 (2022)

11. Klein, S., Staring, M., Murphy, K., Viergever, M.A., Pluim, J.P.: Elastix: a toolbox for intensity-based medical image registration. IEEE Trans. Med. Imaging **29**(1), 196–205 (2009)

12. Ma, J., et al.: How distance transform maps boost segmentation CNNs: an empirical study. In: Medical Imaging with Deep Learning, pp. 479–492. PMLR (2020)

13. Modat, M., et al.: Fast free-form deformation using graphics processing units. Comput. Methods Programs Biomed. **98**(3), 278–284 (2010)

14. Mok, T.C.W., Chung, A.C.S.: Large deformation diffeomorphic image registration with Laplacian pyramid networks. In: Martel, A.L., et al. (eds.) MICCAI 2020. LNCS, vol. 12263, pp. 211–221. Springer, Cham (2020). https://doi.org/10.1007/978-3-030-59716-0_21

15. Paragios, N., Rousson, M., Ramesh, V.: Non-rigid registration using distance functions. Comput. Vis. Image Underst. **89**(2–3), 142–165 (2003)

16. Siebert, H., Hansen, L., Heinrich, M.P.: Evaluating design choices for deep learning registration networks. In: Bildverarbeitung für die Medizin 2021. I, pp. 111–116. Springer, Wiesbaden (2021). https://doi.org/10.1007/978-3-658-33198-6_26

17. Siebert, H., Hansen, L., Heinrich, M.P.: Fast 3D registration with accurate optimisation and little learning for Learn2Reg 2021. In: Aubreville, M., Zimmerer, D., Heinrich, M. (eds.) MICCAI 2021. LNCS, vol. 13166, pp. 174–179. Springer, Cham (2022). https://doi.org/10.1007/978-3-030-97281-3_25

18. Jimenez-del Toro, O., et al.: Cloud-based evaluation of anatomical structure segmentation and landmark detection algorithms: visceral anatomy benchmarks. IEEE Trans. Med. Imaging **35**(11), 2459–2475 (2016)

19. Wasserthal, J., Meyer, M., Breit, H.C., Cyriac, J., Yang, S., Segeroth, M.: Totalsegmentator: robust segmentation of 104 anatomical structures in CT images. arXiv preprint arXiv:2208.05868 (2022)

ToFi-ML: Retinal Image Screening with Topological Machine Learning

Faisal Ahmed⬤ and Baris Coskunuzer⁽✉⁾⬤

University of Texas at Dallas, Richardson, TX 75080, USA
{faisal.ahmed,coskunuz}@utdallas.edu

Abstract. The analysis of fundus images for the early screening of eye diseases is of great clinical importance. Traditional methods for such analysis are time-consuming and expensive as it requires a trained clinician. Therefore, the need for a comprehensive and automated method of retinal image screening to diagnose and grade retinal diseases has long been recognized. In the past decade, with the substantial developments in computer vision and deep learning, machine learning methods became highly effective in this field as clinical-decision support methods. However, most of these algorithms face challenges like computational feasibility, reliability, and interpretability.

In this paper, we develop a novel approach to this crucial task in retinal image screening. By employing topological data analysis tools, for the most common retinal diseases, i.e., Diabetic Retinopathy (DR), Glaucoma, and Age-related Macular Degeneration (AMD), we observe different types of topological patterns between normal and abnormal classes. These patterns enable us to produce topological fingerprints of fundus images, and we use them as feature vectors with standard ML methods. Our computationally efficient model ToFi-ML outperforms or gives highly competitive accuracy results with state-of-the-art deep learning methods. Furthermore, our topological fingerprints are both explainable and interpretable, and can easily be integrated with future early-screening ML models.

Keywords: Retinal Image Analysis · Topological Data Analysis · Persistent Homology · Glaucoma · Diabetic Retinopathy · Age-related Macular Degeneration

1   Introduction

The World Health Organization reports that as of 2019, more than 400 million people worldwide suffer from Glaucoma, diabetic retinopathy (DR), age-related macular degeneration (AMD), or other serious eye diseases [55]. As most patients with eye diseases are not aware of the aggravation of these conditions, early screening and treatment of eye diseases are quite important. Currently, detecting these conditions is a time-consuming and manual process that requires a trained

G. Waiter et al. (Eds.): MIUA 2023, LNCS 14122, pp. 281–297, 2024.
https://doi.org/10.1007/978-3-031-48593-0_21

clinician to examine and evaluate digital color fundus images of the retina, which can result in delayed treatment. Therefore, the need for clinical decision-support methods has long been recognized.

With this motivation, machine learning (ML) methods have been widely employed in retinal image analysis in the past decade [35,53]. These efforts have made substantial progress in the field by using image classification and pattern recognition [35,45]. Especially after the success of convolutional neural networks (CNN) in image classification, ML tools proved to be quite effective in retinal image analysis [52]. However, these methods are not either computationally efficient to work in large datasets or interpretable to provide insights to ophthalmologists for disease diagnosis.

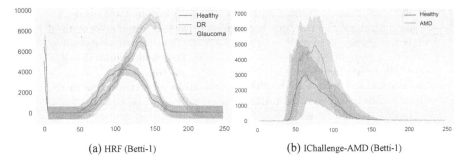

(a) HRF (Betti-1)   (b) IChallenge-AMD (Betti-1)

Fig. 1. We give the median curves and 40% confidence bands of our topological fingerprints (Betti functions). x-axis represents color values and y-axis represents count of components (Betti-0) or count of loops (Betti-1).

In this paper, we bring a novel approach to retinal disease diagnosis by introducing topological data analysis (TDA) methods to fundus imaging. In the past years, TDA tools have been quite successful in medical image analysis in various domains by capturing hidden shape patterns in the images (Sect. 2). In particular, the retinal images develop very different patterns for different color values, e.g. abundant small loops, large loops of different sizes, etc. A key method in TDA, persistent homology, recognizes these patterns and produces reliable fingerprints (feature vectors) of fundus images (Sect. 3). By combining these topological fingerprints with suitable ML methods, we obtain a high-performing, interpretable, and robust model (ToFi-ML) to detect and grade retinal diseases. In this paper, we studied the diagnosis of the three most common retinal diseases, Diabetic Retinopathy (DR), Glaucoma, and Age-related Macular Degeneration (AMD). Our topological ML model produces very competitive accuracy results with state-of-the-art deep learning models on benchmark datasets (Sect. 5.3).

Our Contributions

– We bring a new perspective to retinal image analysis by introducing TDA methods to the field.

- We produce topological fingerprints of fundus images for most common retinal diseases (DR, Glaucoma, and AMD) and observe easily detectable topological pattern differences in certain color values (Figs. 1a and 5).
- By using our fingerprints as feature vectors for standard ML techniques (RF, XGBoost, kNN), we produce a computationally very efficient ToFi-ML model which gives highly competitive results with the latest deep learning models in benchmark datasets (Sect. 5.3).
- Unlike most ML models, our fingerprints are both explainable and interpretable (Sect. 4.2). The interpretation of different topological patterns captured by our fingerprints can help ophthalmologists to better understand the subtlety of these diseases and assist the ML community to develop faster and more accurate models to address this crucial need.

2   Related Work

ML in Retinal Image Analysis. In the past decade, ML tools have been widely employed in retinal image analysis [35,52]. There are two mainstream applications of ML tools for retinal image analysis. The first is the diagnosis and grading of diseases which can be considered a classification problem for image data [43]. The second mainstream application is lesion detection/segmentation [35]. In this paper, we focus only on the diagnosis and grading of retinal diseases by using TDA methods.

Especially after the success of convolutional neural networks (CNN) in image classification in general, ML tools proved to be quite effective in medical image analysis [27] and retinal image analysis [40]. For more on ML methods in ophthalmology, see the surveys [35,45,48,53].

TDA in Image Processing. Persistent homology, the main tool in TDA, has been quite effective for pattern recognition in image and shape analysis. There have been several works in various fields in the past two decades, e.g., analysis of images of hepatic lesions [1], human and monkey fibrin images [5], tumor classification [15], fingerprint classification [23], analysis of 3D shapes [47], neuronal morphology [32], brain artery trees [4], fMRI data [44,49], genomic data [8]. See also the excellent survey [46] for a thorough review of TDA methods in biomedicine. Note that TDA Applications Library [25] presents hundreds of interesting applications of TDA in various fields.

3   Background

In this paper, we use persistent homology (PH) as a powerful feature extraction tool for retinal images. PH is one of the key approaches in topological data analysis (TDA), allowing us to systematically assess the evolution of various hidden patterns in the data as we vary a scale parameter [13,54]. The extracted patterns, or homological features, along with information on how long such features persist throughout the considered filtration of a scale parameter, convey a critical insight into salient data characteristics and data organization. In this section, we give a basic introduction to PH in image data settings. For a more thorough background and PH process for other data types (e.g., point clouds, networks), see [10,16].

Persistent Homology for Image Data. PH is a 3-step process. The first step is the *filtration* step, where one induces a sequence of simplicial complexes from the data. This is the key step, where one can integrate the domain information into the process. The second step is the *persistence diagrams*, where the machinery records the evolution of topological features (birth/death times) in the filtration, sequence of the simplicial complexes. The final step is the *vectorization* (fingerprinting) where one can convert these records to a function or vector to be used in suitable ML models. For more details on how to apply PH in image analysis, check out the references given in Sect. 2.

i. Constructing Filtrations: As PH is basically the machinery to keep track of the evolution of topological features in a sequence of simplicial complexes, the most important step is the construction of this sequence. In the case of image analysis, the most common method is to create a nested sequence of binary images (aka cubical complexes). For a given image $\mathcal{X}$ (say $r \times s$ resolution), to create such sequence, one can use grayscale (or other color channels) values γ_{ij} of each pixel $\Delta_{ij} \subset \mathcal{X}$. In particular, for a sequence of grayscale values ($t_1 < t_2 < \cdots < t_N$), one obtains nested sequence of binary images $\mathcal{X}_1 \subset \mathcal{X}_2 \subset \cdots \subset \mathcal{X}_N$ such that $\mathcal{X}_n = \{\Delta_{ij} \subset \mathcal{X} \mid \gamma_{ij} \leq t_n\}$ (See Fig. 2 and 3). In other words, we start with empty $r \times s$ image and start activating (coloring black) pixels when their grayscale value reaches the given threshold. This is called *sublevel filtration* for $\mathcal{X}$ with respect to a given function (grayscale in this case). One can also go in decreasing order to activate the pixels, which is called *superlevel filtration*.

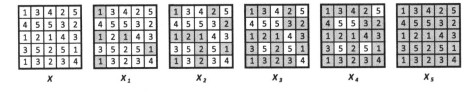

Fig. 2. Sublevel filtration. The leftmost figure represents an image of 5×5 size with the given pixel values. Then, the sublevel filtration is the sequence of binary images $\mathcal{X}_1 \subset \mathcal{X}_2 \subset \mathcal{X}_3 \subset \mathcal{X}_4 \subset \mathcal{X}_5$.

The choice of thresholds is also very crucial in this construction as it indicates how fine we want our PH machinery to detect the topological patterns. One can choose $N = 255$ which makes the filtration too fine, and most outputs would be trivial in the fingerprinting process. If we choose N very small, then one can miss many topological features. While in most cases, the thresholds are chosen evenly distributed, depending on the image dataset, using quantiles of the grayscale values is also common, e.g., when grayscale values in the dataset are concentrated around specific numbers. In our experiments, we use $N = 100$ as we obtain very good results with this choice, and increasing N did not improve the performance significantly.

Note that sublevel and superlevel filtration produces completely different information for most data types. However, in the image data setting (cubical complexes), in the complementary dimensions, sublevel and superlevel filtration produce very similar information thanks to a celebrated result in algebraic topology, Alexander Duality [29]. In other words, the choice of sublevel and superlevel filtration is not important in image data as long as one uses all possible dimensions (in this setting, $k = 0, 1$).

ii. Persistence Diagrams: The second step in PH process is to obtain persistence diagrams (PD) for the filtration $\mathcal{X}_1 \subset \mathcal{X}_2 \subset \cdots \subset \mathcal{X}_N$, i.e., the sequence of cubical complexes (binary images). PDs are formal summaries of the evolution of topological features in the filtration sequence. PDs are collection of 2-tuples, $\{(b_\sigma, d_\sigma)\}$, marking the birth and death times of the topological features appearing in the filtration. In other words, if a topological feature σ appears for the first time at $\mathcal{X}_{i_0}$, we mark the birth time $b_\sigma = i_0$. Then, if the topological feature σ disappears at $\mathcal{X}_{j_0}$, we mark the death time $d_\sigma = j_0$. i.e., $\mathrm{PD}_k(\mathcal{X}) = \{(b_\sigma, d_\sigma) \mid \sigma \in H_k(\widehat{\mathcal{X}}_i) \text{ for } b_\sigma \leq i < d_\sigma\}$. Here, $H_k(\widehat{\mathcal{X}}_i)$ represent k^{th} homology group of $\widehat{\mathcal{X}}_i$, representing k-dimensional topological features in cubical complex $\widehat{\mathcal{X}}_i$ [29]. By construction, for $2D$ image analysis, only meaningful dimensions to use are $k = 0, 1$, i.e., $\mathrm{PD}_0(\mathcal{X})$ and $\mathrm{PD}_1(\mathcal{X})$.

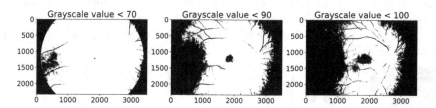

Fig. 3. Binary images $\mathcal{X}_{70}, \mathcal{X}_{90}, \mathcal{X}_{110}$ obtained from a Retinal image for threshold values $70, 90, 110$.

In layman's terms, 0-dimensional features are connected components and 1-dimensional features are the holes (loops). In our case, if a loop τ first appears at the binary image $\mathcal{X}_3$ and it gets filled in the binary image $\mathcal{X}_7$, we add 2-tuple $(3, 7)$ in the persistence diagram $\mathrm{PD}_1(\mathcal{X})$. Similarly, if a new connected component appears in the binary image $\mathcal{X}_5$ and it merges to the other components in the binary image $\mathcal{X}_8$, we add $(5, 8)$ to $\mathrm{PD}_0(\mathcal{X})$. In Fig. 2, we have $\mathrm{PD}_0(\mathcal{X}) = \{(1, \infty), (1, 2), (1, 3), (1, 3), (1, 4), (2, 3)\}$ and $\mathrm{PD}_1(\mathcal{X}) = \{(3, 5), (3, 5), (4, 5)\}$ This step is pretty standard and there are various software libraries for this task. For image data with cubical complexes, see [28]. For other types of data and filtrations, see [41].

iii. Vectorization (Fingerprinting): PDs being a collection of 2-tuples are not very practical to be used with ML tools. Instead, a common way is to convert PD information into a vector or a function, which is called *vectorization* [16]. A common function for this purpose is the *Betti function*, which basically keeps track of the number of "alive" topological features at the given threshold. In particular, the Betti function is a step function with $\beta_0(t_n)$ the count of connected components in the binary image $\mathcal{X}_n$, and $\beta_1(t_n)$ the number of holes (loops) in $\mathcal{X}_n$. In ML applications, Betti functions are usually taken as a vector β_k of size N with entries $\beta_k(t_n)$ for $1 \leq n \leq N$.

$$\overrightarrow{\beta_k}(\mathcal{X}) = [\beta_k(t_1) \ \beta_k(t_2) \ \beta_k(t_3) \ \dots \ \beta_k(t_N)]$$

For example, for the image $\mathcal{X}$ in Fig. 2, we have $\overrightarrow{\beta_0}(\mathcal{X}) = [5\ 4\ 2\ 1\ 1]$ and $\overrightarrow{\beta_1}(\mathcal{X}) = [0\ 0\ 2\ 3\ 0]$, e.g., $\beta_0(1) = 5$ is the count of components in $\mathcal{X}_1$ and $\beta_1(4) = 3$ is the count of holes (loops) in $\mathcal{X}_4$. There are other vectorization methods like Persistence Images, Persistence Landscapes or Silhouettes [16], but to keep our model interpretable, we use Betti functions in this study.

Color Channels for Retinal Images. Retinal image quality assessment is essential for controlling the quality of retinal imaging and guaranteeing the reliability of diagnoses by ophthalmologists or automated analysis systems. The three main families of conventional color spaces are primary spaces, luminance-chrominance spaces, and perceptual spaces. Appropriate color spaces can help simplify some color computations that occur during the generation of images. In 1978, Joblov et al. [31] described the significance of different color spaces in computer graphics and the feature extraction process. In our persistent homology approach, the way we construct the filtration out of the given fundus image is the key step (Sect. 3), and the different color channels induce completely different filtrations and produce different topological patterns.

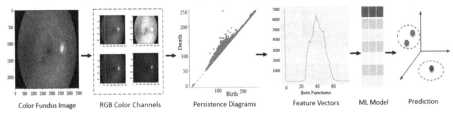

Color Fundus Image   RGB Color Channels   Persistence Diagrams   Feature Vectors   ML Model   Prediction

Fig. 4. Flowchart of our model: For any fundus image, we first get their RGB and Gray color spaces. Then, we get their persistence diagrams by using color values. Then, we obtain our topological feature vectors (Betti functions, Silhouette functions) out of these persistence diagrams. We feed these vectors to our ML models (RF, XGBoost, kNN, etc.) which give highly accurate classification results.

4   Retinal Image Screening with TDA

In this part, we first describe our topological fingerprinting machine learning model, ToFi-ML. Then, we elaborate on the explainability and interpretability of our model.

4.1 ToFi-ML Model for Retinal Images

In the flowchart (Fig. 4), we summarized our ToFi-ML model. For a given fundus image $\mathcal{X}$ (say at $r \times s$ resolution), we first get its RGB and Grayscale images (Fig. 4-Step 2). In other words, we produce 4 color functions $\mathbf{g}(i,j)$ (grayscale=average(RGB)), $\mathbf{R}(i,j)$ (red), $\mathbf{G}(i,j)$ (green), and $\mathbf{B}(i,j)$ (blue) where $f(i,j)$ assigns every pixel $\Delta_{ij} \subset \mathcal{X}$ to its assigned color value for $1 \leq i \leq r$ and $1 \leq j \leq s$. Note that all colors have a range $[0, 255]$. Then, we extract all topological features for color channel values by constructing a sublevel filtration with respect to the corresponding color function. While grayscale values vary from 0 (black) to 255 (white), we chose the number of thresholds as $N = 100$ in our filtration step, as further increasing the threshold steps did not increase the performance of our model. In other words, we renormalized $[0, 255]$ grayscale interval to $[0, 100]$. After defining the filtration $\mathcal{X}_1 \subset \mathcal{X}_2 \subset \cdots \subset \mathcal{X}_{100}$, we obtain the persistent diagrams $PD_k(\mathcal{X})$ of each fundus image $\mathcal{X}$ for dimensions $k = 0, 1$ (Sect. 3). In layman's terms, the filtration produces a sequence of binary (black-white) images where the dark points represent the pixels value less than the given threshold (Fig. 3). Then, $PD_0(\mathcal{X})$ records connected components in these binary images, and $PD_1(\mathcal{X})$ records holes in the binary images.

After getting persistent diagrams, we convert them into feature vectors as explained in Sect. 3. In this vectorization step, one can use several choices like Betti functions, Silhouettes, or Persistence Images. Since most of the topological features have short life spans, Betti and Silhouette (with $p \leq 1$) functions were the natural choice as they give the count of topological features at a given threshold. Our experiments verified this intuition and we obtained almost all the best results with Betti functions. Hence, to keep our model simple, we only use Betti functions as vectorizations (Fig. 4-Step 4).

After obtaining our topological feature vectors, the final step is to apply ML tools to these topological fingerprints. To keep our model computationally feasible, we applied standard ML methods like Random Forest, XGBoost, and kNN to our extracted topological features. Note that our topological feature vectors can easily be integrated with various deep learning models, too. We give the details of our ML steps in the experiments section. Note that topological features are invariant under rotation, flipping, and translation. This makes our model highly robust against noisy data. Furthermore, our model does not need any data augmentation and data pre-processing which makes our model computationally very feasible. Our experiments show that our model is highly successful in small datasets as well as large ones.

4.2 Explainability and Interpretability of Topological Fingerprints

As mentioned in the introduction, one of the main advantages of our model is explainability and interpretability. In Fig. 1 and Fig. 5, we illustrate the topological patterns created by normal and abnormal classes for DR, Glaucoma, and AMD diseases. In these figures, we give median curves and 40% confidence bands of each class for the corresponding dataset. The details of these non-parametric confidence bands and median curves can be found at [24].

For explainability of our model, the different behaviors observed in Fig. 5 shows the distinguishing power of our topological feature vectors. From ML perspective, these figures prove how strong our feature vectors are. For any image, we get 100-dimensional Betti-0 and 100-dimensional Betti-1 vectors. Hence, each image is embedded into the latent space $\mathbb{R}^{100}$ via these Betti functions. In the latent space, each image $\mathcal{X}$ is represented as a point $\beta(\mathcal{X}) \in \mathbb{R}^{100}$, and the median curves can be considered as the centers of the clusters corresponding to each class. The separation of the clusters can be considered as the distance of these points from each other. In our figures, we give a highly compactified summary of the latent space. While the curves might close to each other, note that each entry $\beta_k(t_i)$ in the vector $\beta_k(\mathcal{X}) = [\beta_k(0)\ \beta_k(1)\ \ldots \beta_k(100)]$ represent the ith-coordinate in $\mathbb{R}^{100}$ for an image. Hence even a little separation between the Betti curves results in a huge separation of the corresponding points in the latent space. This separation between the clusters is easily identified by the ML model (RF, XGBoost) in training, and gives a very robust model with high accuracy results.

For the interpretability of the model, we need to give some background on topological feature vectors. While different vectorizations of persistence diagrams can give very powerful feature vectors, Betti curves are the best in terms of interpretation. This is one of the reasons why we preferred Betti curves as vectorization in our model. The Betti curves keep track of the count of components and the loops while we increase the color value from 0 to 255. In other words, for any color value $t_0 \in [0, 255]$, $\beta_0(t_0)$ is the total number of components in the binary image $\mathcal{X}_{t_0}$ and $\beta_1(t_0)$ is the total number of loops (holes) in the binary image $\mathcal{X}_{t_0}$ (Sect. 3). The counts in y-axis of the figures are the real counts of components and loops, not normalized values. By keeping this property in mind, we interpret our topological fingerprints for each disease as follows.

In Fig. 5a, we see Betti-1 curves of the two classes (DR vs. normal) in the APTOS 2019 dataset. Our curves indicate that when the grayscale value is between 100 and 150, the count of loops in the DR class is almost double the ones in the normal class. In other words, if $\mathcal{X}$ is a normal fundus image and $\mathcal{V}$ is a DR fundus image in the APTOS dataset, the binary image $\mathcal{X}_{125}$ has around 1000 loops (holes), while the binary image $\mathcal{V}_{125}$ has around 2200 loops. One way to interpret this is that in a normal class, there are about 1000 light spots in normal classes, while there are about 2200 light spots spread out DR classes as holes in $\mathcal{X}_{125}$ and $\mathcal{V}_{125}$ are white regions (color value > 125) in these binary images. Recall that the color value 0 represents black, and the color value 255 represents white. Therefore, our results show that in DR fundus images, lighter spots are much more abundant and spread out than in normal fundus images.

Similarly, in Fig. 5b, we have Betti-0 curves of the two classes (Glaucoma vs. normal) in the ORIGA dataset. In this case, our curves indicate that when the grayscale value is between 80 and 130, the count of components in the Glaucoma class is almost half of the count of components in the normal class. In other

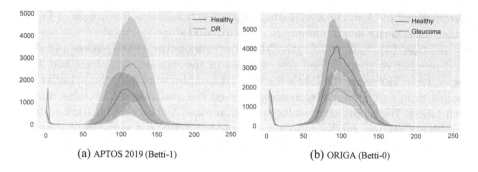

(a) APTOS 2019 (Betti-1)   (b) ORIGA (Betti-0)

Fig. 5. Median curves and 40% confidence bands of our topological fingerprints (Betti functions). x-axis represents color values and y-axis represents count of components (Betti-0) or count of loops (Betti-1).

words, if $\mathcal{X}$ is a normal fundus image and $\mathcal{V}$ is a Glaucoma fundus image in the Origa dataset, the binary image $\mathcal{X}_{90}$ has around 4000 components, while the binary image $\mathcal{V}_{90}$ has around 2000 component. One can interpret this as follows: In a normal classes, there are about 4000 dark spots in normal classes, while there are about 2000 dark spots spread out in Glaucoma classes. Less number of components means that Glaucoma images get darker faster than the normal classes. i.e., dark regions get more connected in earlier thresholds.

5   Experiments

5.1   Datasets

To see the performance of our ToFi-ML model for Glaucoma, DR, and AMD screening, we did several experiments on well-known benchmark datasets. We give the basic details of these datasets in Table 1. Further details (resolution, camera, etc.) for all the datasets can be found in [35].

IChallenge-AMD dataset is designed for the Automatic Detection challenge on Age-related Macular degeneration (ADAM Challenge) which was held as a satellite event of the ISBI 2020 conference [18,22]. There are two different resolutions of images, i.e., 2124×2056 pixels (824 images) and 1444×1444 (376 images). While the dataset has 1200 images, only 400 of them are available with labels. Like most other references, we used these 400 images in our experiments Table 2.

ORIGA dataset contains 650 high resolution (3072×2048) retinal images for Glaucoma annotated by trained professionals from Singapore Eye Research Institute [58].

APTOS 2019 dataset was used for a Kaggle competition on DR diagnosis [3]. The images have varying resolutions, ranging from 474×358 to 3388×2588. APTOS stands for Asia Pacific Tele-Ophthalmology Society, and the dataset was provided by Aravind Eye Hospital in India. It contains five stages of DR to detect the severity levels, namely No DR (0), mild stage (1), moderate stage (2), severe stage (3), and proliferative diabetic retinopathy (PDR) stage (4). The total number of training and test samples in the dataset were 3662 and 1928 respectively. However, the labels for the test samples were not released after the competition, so like other references, we used the available 3662 fundus images with labels. We report our results on a binary and 5-class classification setting. In a binary setting, fundus images with grades 1,2,3, and 4 are identified as DR group, and grade 0 images as the normal group.

5.2   Experimental Setup

Training:Test Split: Unfortunately, none of the datasets have a predefined *train:test* split, and therefore, many models used their own split. In our experiments, we used an 80:20 split in all datasets, which is the closest to the common settings in the previous works for all datasets. Because of the discrepancy between the experimental setups of different methods, we give the train:test splits of all models in our accuracy tables to facilitate a fair comparison (Table 2, 3 and 4).

No Data Augmentation: Note that as our datasets are quite small and imbalanced compared to other image classification tasks for deep learning models, hence all CNN and other deep learning methods need to use serious data augmentation (sometimes 50–100 times) to train their model and avoid overfitting [26]. Our ToFi-ML model are using topological feature vectors, and our feature extraction method is invariant under rotation, flipping and other common data augmentation techniques. Hence, we do not use any type of data augmentation or pre-processing to increase the size of training data, except SMOTE (a minority oversampling method [21]) for imbalanced data in IChallenge-AMD. This makes our model computationally very efficient, and highly robust against small alterations and the noise in the image. To avoid the effect of imbalanced data to the performance of ML model, we did downsampling for majority class for ORIGA dataset. For imbalanced datasets IChallenge-AMD and ORIGA, we report two results. One result is in the original setting, while the second result is in the balanced setting. For APTOS, we only used the original setting as the dataset is already balanced.

Table 1. Benchmark datasets for fundus images.

Dataset	Disease	Total	Normal	Abnormal
APTOS 2019	DR	3662	1805	1857
Origa	Gl	650	482	168
IChallenge-AMD	AMD	400	311	89
HRF	DR, Gl	45	15	30

ML Model: To increase the performance of our model in terms of accuracy and computational efficiency, we performed parametric tuning and feature selection methods. We extracted 800 features (Gray and RGB color spaces) from the datasets by using Betti 0 and Betti 1. To improve the performance and avoid collinearity, we used dimension reduction by choosing the most important features. For feature selection, we used *SelectFromModel* from scikit-learn. We first assign importance to each feature and sorted them in descending order according to threshold parameter. The features are considered unimportant and removed if the corresponding importance of the feature values is below the provided threshold parameter. Random Forest, XGBoost, and KNN models are trained on all of the datasets. We used default parameters as parametric tuning for XGBoost. After feature selection and fine-tuning, XGBoost gave the best results for APTOS (159 features) and IC-AMD (58 features) datasets while KNN produced the best performance for ORIGA. As our ablation study (Table 6) indicates, feature selection improved both performance and computational time.

Computational Complexity and Implementation: While for high dimensional data PH calculation is computationally expensive [41], for image data, it is highly efficient. For 2D images, PH has time complexity of $\mathcal{O}(|\mathcal{P}|^r)$ where $r \sim 2.37$ and $|\mathcal{P}|$ is the total number of pixels [39]. In other words, PH computation increases almost quadratic with the resolution. The remaining processes (vectorization, RF) are negligible compared to PH step. We used Giotto-TDA [51] to obtain persistence diagrams, and Betti and Silhouette functions. We used Jupyter notebook as an IDE for writing the code in Python 3. Our code is available at the following link[1].

Runtime: We did all our experiments on a personal laptop with a processor Intel(R), Core(TM) i7-8565U, CPU 1.80 GHz, and RAM 16 GB. For the largest dataset, APTOS (3662 images with high resolution), the end-to-end process (extracting topological features, training, and getting accuracy results) took 43.7 h. Runtime for small datasets and less resolution to extract topological features takes much less time.

5.3   Results

Here, we present the performance of our ToFi-ML model along with SOTA models on benchmark datasets. We give the accuracy results of our ToFi-ML model for Glaucoma, AMD, and DR diseases. In Tables 2, 3 and 4, we give *Normal:Abnormal* split, *Train:Test* split, and *# Classes* to describe the experimental setting used for each model.

In these tables, we report all available performance metrics (AUC, Accuracy, Precision, Recall, F-1, etc.) given in these references. For details of these performance metrics, see [9]. We used the training data to build the model only. We used no data augmentation. All our predictions and performance are based

---

[1] https://github.com/FaisalAhmed77/Topo-Ret.

on the test (unseen) data. Scikit-Learn [42] library is used for performance metric calculations. For multiclass classification (APTOS 2019), we used a weighted average of AUC values with a One-vs-Rest configuration. This method computes the AUC values of each class against the rest [19]. In all tables, the best result for each column is given in bold, and the second-best result is underlined. For missing data in the table from reference papers, we used "-". We reported the performance of our model (ToFi-ML) at the bottom of the table.

AMD Detection Results: We give our results for AMD diagnosis on IChallenge-AMD dataset in Table 2. The accuracy results for methods in rows 1–4 are taken from [36, Table 1]. While these papers did not use directly IChallenge-AMD (IC-AMD) dataset, [36] adapted these methods with the same setting to IC-AMD, and reported these results. Since the dataset is imbalanced, we developed two models. In the first model, we did not use any data processing and used the original setting. In the second model, we used SMOTE (minority oversampling method [21]) to address the imbalance issue. While our model with pre-processing gives very competitive results with state-of-the-art models, our SMOTE model outperformed all SOTA models on this dataset.

Table 2. Accuracy results for AMD diagnosis on IChallenge-AMD dataset.

IChallenge-AMD Dataset (AMD)								
Method	Nor:Abn	Train:Test	Class	Prec	Recall	Acc	AUC	F_1
MemoryBank [56]	311:89	5 fold CV	2	74.63	66.49	82.02	66.49	68.69
Invariant [57]	311:89	5 fold CV	2	83.20	74.53	86.58	71.58	77.33
Decouple [20]	311:89	5 fold CV	2	79.72	69.19	83.80	69.19	71.62
Contrastive [14]	311:89	5 fold CV	2	73.48	68.06	82.45	68.06	69.84
Rotation S [36]	311:89	5 fold CV	2	84.51	77.19	87.60	77.19	_79.71_
CycleGAN [59]	933:267	5 fold CV	2	77.10	**80.60**	87.30	86.20	73.20
DCNN [11]	311:89	10 fold CV	2	_84.87_	75.99	_91.69_	88.00	79.64
ToFi-ML	311:89	80:20	2	**88.89**	57.14	91.25	_90.58_	69.56
ToFi-ML SMOTE	311:89	80:20	2	84.62	_78.57_	**93.75**	**91.34**	**81.48**

Table 3. Accuracy results for DR diagnosis on APTOS dataset.

APTOS 2019 Dataset (DR)							
Method	Nor:Abn	Train:Test	Class	Prec	Recall	Acc	AUC
D-Net121 [12]	1805:1857	85:15	2	86.0	87.0	94.4	–
ConvNet [6]	1805:1857	80:20	2	–	–	96.1	–
DRISTI [33]	1805:1857	85:15	2	–	–	97.1	–
C-DNN [7]	1805:1857	85:15	2	_98.0_	_98.0_	_97.8_	–
LBCNN [38]	1805:1857	80:20	2	–	–	96.6	_98.7_
SCL [30]	1805:1857	85:15	2	**98.4**	**98.4**	**98.4**	**98.9**
ToFi-ML	1805:1857	80:20	2	95.5	94.0	94.5	97.9

DR Detection Results: We give our results for DR diagnosis on APTOS 2019 dataset in Table 3. We got the best accuracy result with the XGBoost ML model. While the main classification setting for APTOS is binary, but we also tried our model for multi-class setting. The results for 5-class classification are given in Appendix (Table 5). Note that because of the mixed resolution of the images, this is a very challenging dataset from ML perspective. In spite of this fact, our model proved to be very robust and can tackle mixed-resolution problems. Our accuracy results are close to the latest deep learning models with no data augmentation or preprocessing (Table 3).

Glaucoma Detection Results: We give our results for Glaucoma diagnosis on ORIGA dataset in Table 4. In this dataset, we obtained the best performance with Betti-0,1 feature vectors combined with kNN. Note that the high-performing DL models ODGNet and CNN-SVM in the table used other datasets in addition to Origa to improve their performances by providing more training data to their model. Since other deep learning models use serious data augmentation and balance the training data, to have a fair comparison, we used a balanced setting in our experiments by downsampling the majority class. In the balanced setting, in spite of the smallness of the dataset, our model outperformed the SOTA models. However, in the imbalanced setting, our model did not perform as well as the low sensitivity score shows.

Table 4. Accuracy results for Glaucoma diagnosis on Origa dataset.

ORIGA Dataset (Glaucoma)							
Method	Nor:Abn	Train:Test	Class	Sen	Spec	Acc	AUC
EAMNet [37]	482:168	2 fold CV	2	–	–	–	88.0
SVM-SMOTE [60]	482:168	10-fold CV	2	87.6	77.9	82.8	88.9
18-CNN [17]	482:168	70:30	2	58.1	92.4	78.3	–
NasNet [50]	482:168	70:30	2	78.7	91.1	87.9	–
CNN-SVM [2]	660:453	70:30	2	89.5	**100**	95.6	–
ODGNet [34]	482:168	pretrained	2	<u>94.8</u>	94.9	<u>95.8</u>	<u>97.9</u>
ToFi-ML-balanced	168:168	80:20	2	**100**	<u>97.1</u>	**98.5**	**99.7**
ToFi-ML	482:168	80:20	2	50.0	92.7	81.5	77.3

Table 5. Accuracy results for multiclass (5 labels) classification on APTOS dataset.

APTOS 2019 Dataset (DR)							
Method	Normal:Abnormal	Train:Test	Class	Prec	Rec	Acc	AUC
DRISTI [33]	1805:1857	85:15	5	59.40	54.60	75.50	–
C-DNN [7]	1805:1857	85:15	5	–	–	<u>80.96</u>	–
SCL [30]	1805:1857	85:15	5	**73.8**	<u>70.5</u>	**84.6**	**93.8**
ToFi-ML	1805:1857	80:20	5	<u>73.51</u>	**75.58**	75.58	<u>88.29</u>

Ablation Study. We give our ablation study in Table 6. For each fundus image, we produce 800 topological features by using different dimensions ($k = 0, 1$) and four different color channels (Gray, RGB). As ML model, we used XGBoost on APTOS and IChallenge datasets, and kNN on the ORIGA dataset. To avoid collinearity and improve the performance of our ML model, we used a feature selection algorithm. In APTOS 2019, we select best 159 features out of 800. In IC-AMD, we select best 58 features out of 800. Since kNN is a non-parametric ML model, we did not use feature selection in ORIGA.

Table 6. Ablation Study. Accuracy results of our model with different subsets of feature vectors for the default setting for each dataset as described in Sect. 5.3.

Feature Vector	# Features	APTOS		IC-AMD		ORIGA	
		Acc	AUC	Acc	AUC	Acc	AUC
Gray (Betti-1)	100	91.95	97.01	86.25	81.17	96.97	98.48
Gray (Betti-0, 1)	200	91.95	97.42	86.25	82.14	**98.48**	**99.92**
RGB (Betti-0, 1)	600	<u>94.00</u>	<u>98.03</u>	87.50	84.63	93.94	97.57
RGB+Gray (Betti-0, 1)	800	93.86	**98.10**	<u>88.75</u>	**92.53**	98.48	<u>99.68</u>
Feature Selection	*	**94.54**	97.94	**91.25**	<u>90.58</u>	–	–

6 Discussion

In this paper, we bring a novel approach to retinal image analysis by introducing TDA techniques to the field. By using persistent homology, we produce highly effective topological fingerprints of fundus images for the most common retinal diseases, namely DR, Glaucoma, and AMD. These topological fingerprints provide powerful feature vectors to distinguish normal and abnormal images and give highly competitive accuracy results with current deep learning models. The high accuracy results from combining different ML models show our topological fingerprints are model agnostic, and indeed extract the valuable features of the diseases in fingerprinting step. Furthermore, the model is highly robust to noise and small alterations because of the stability of topological fingerprints. Furthermore, since we use fingerprinting approach, the model performs quite well in mixed-resolution datasets like APTOS (See Sect. 4.2).

Unlike most ML models, our model ToFi-ML is highly interpretable as it expresses the topological patterns appearing in different grayscale values. The interpretation of special topological patterns created by each disease can help expert ophthalmologists to better understand the subtlety of these diseases. Furthermore, being feature vectors, our topological fingerprints can naturally be integrated with any ML and DL model. Considering the significant need for automated clinical-decision support methods to help clinicians, our unique topological feature vectors can critically help any future ML and DL models to boost their performance and aid them to have more robust results.

Acknowledgements. This work was partially supported by National Science Foundation (Grant # DMS-2202584) and by Simons Foundation (Grant # 579977).

References

1. Adcock, A., Rubin, D., Carlsson, G.: Classification of hepatic lesions using the matching metric. Comput. Vis. Image Underst. **121**, 36–42 (2014)
2. Ajitha, S., Akkara, J.D., Judy, M.: Identification of glaucoma from fundus images using deep learning techniques. Indian J. Ophthalmol. **69**(10), 2702 (2021)
3. APTOS: Asia Pacific Tele-Ophthalmology Society (APTOS) 2019 Blindness Detection Dataset (2019). https://www.kaggle.com/c/aptos2019-blindness-detection
4. Bendich, P., Marron, J.S., Miller, E., Pieloch, A., Skwerer, S.: Persistent homology analysis of brain artery trees. Ann. Appl. Stat. **10**(1), 198 (2016)
5. Berry, E., Chen, Y.C., Cisewski-Kehe, J., Fasy, B.T.: Functional summaries of persistence diagrams. J. Appl. Comput. Topol. **4**(2), 211–262 (2020)
6. Bodapati, J.D., et al.: Blended multi-modal deep convnet features for diabetic retinopathy severity prediction. Electronics **9**(6), 914 (2020)
7. Bodapati, J.D., et al.: Composite deep neural network with gated-attention mechanism for DR severity classification. J. Amb. Int. Hum. Compt. **12**(10), 9825–9839 (2021)
8. Cámara, P.G., Levine, A.J., Rabadan, R.: Inference of ancestral recombination graphs through topological data analysis. PLoS Comput. Biol. **12**(8), e1005071 (2016)
9. Campbell, M.J., Machin, D., Walters, S.J.: Medical Statistics: A Textbook for the Health Sciences. Wiley, Hoboken (2010)
10. Carlsson, G., Vejdemo-Johansson, M.: Topological Data Analysis with Applications. Cambridge University Press, Cambridge (2021)
11. Chakraborty, R., Pramanik, A.: DCNN-based prediction model for detection of AMD from color fundus images. Med. Bio. Eng. Comput. **60**(5), 1431–1448 (2022)
12. Chaturvedi, S.S., Gupta, K., Ninawe, V., Prasad, P.S.: Automated diabetic retinopathy grading using deep convolutional neural network. arXiv preprint arXiv:2004.06334 (2020)
13. Chazal, F., Michel, B.: An introduction to topological data analysis: fundamental and practical aspects for data scientists. Front. Artif. Intell. **4**, 108 (2021)
14. Chen, T., Kornblith, S., Norouzi, M., Hinton, G.: A simple framework for contrastive learning of visual representations. In: ICML, pp. 1597–1607. PMLR (2020)
15. Crawford, L., et al.: Predicting clinical outcomes in glioblastoma: an application of topological and functional data analysis. J. Am. Stat. Assoc. **115**(531), 1139–1150 (2020)
16. Dey, T.K., Wang, Y.: Computational Topology for Data Analysis. Cambridge University Press, Cambridge (2022)
17. Elangovan, P., Nath, M.K.: Glaucoma assessment from color fundus images using convolutional neural network. Int. J. Imaging Syst. Technol. **31**(2), 955–971 (2021)
18. Fang, H., et al.: ADAM challenge: detecting age-related macular degeneration from fundus images. IEEE Trans. Med. Imaging **41**, 2828–2847 (2022)
19. Fawcett, T.: An introduction to ROC analysis. Pattern Rec. Lett. **27**(8), 861–874 (2006)
20. Feng, Z., Xu, C., Tao, D.: Self-supervised representation learning by rotation feature decoupling. In: CVPR, pp. 10364–10374 (2019)

21. Fernández, A., et al.: SMOTE for learning from imbalanced data. J. Artif. Intell. Res. **61**, 863–905 (2018)

22. Fu, H., et al.: ADAM: automatic detection challenge on AMD (2020). https://doi.org/10.21227/dt4f-rt59

23. Giansiracusa, N., Giansiracusa, R., Moon, C.: Persistent homology machine learning for fingerprint classification. In: ICMLA, pp. 1219–1226. IEEE (2019)

24. Gibbons, J.D., Chakraborti, S.: Nonparametric Statistical Inference. CRC Press, Boca Raton (2014)

25. Giunti, B.: TDA applications library (2022). https://www.zotero.org/groups/2425412/tda-applications/library

26. Goutam, B., Hashmi, M.F., Geem, Z.W., Bokde, N.D.: A comprehensive review of deep learning strategies in retinal disease diagnosis using fundus images. IEEE Access **10**, 57796–57823 (2022)

27. Greenspan, H., Van Ginneken, B., Summers, R.M.: Guest editorial deep learning in medical imaging: overview and future promise of an exciting new technique. IEEE Trans. Med. Imaging **35**(5), 1153–1159 (2016)

28. GUDHI: The GUDHI project (2020). https://gudhi.inria.fr/doc/3.3.0/

29. Hatcher, A.: Algebraic Topology. Cambridge University Press, Cambridge (2002)

30. Islam, M.R., et al.: Applying supervised contrastive learning for the detection of DR and its severity levels from fundus images. Comput. Biol. Med. **146**, 105602 (2022)

31. Joblove, G.H., Greenberg, D.: Color spaces for computer graphics. In: Proceedings of the 5th Annual Conference on Computer Graphics and Interactive Techniques, pp. 20–25 (1978)

32. Kanari, L., et al.: A topological representation of branching neuronal morphologies. Neuroinformatics **16**(1), 3–13 (2018)

33. Kumar, G., Chatterjee, S., Chattopadhyay, C.: DRISTI: a hybrid deep neural network for diabetic retinopathy diagnosis. Signal Image Video Process. **15**(8), 1679–1686 (2021)

34. Latif, J., Tu, S., Xiao, C., Ur Rehman, S., Imran, A., Latif, Y.: ODGNet: a deep learning model for automated optic disc localization and glaucoma classification using fundus images. SN Appl. Sci. **4**(4), 1–11 (2022)

35. Li, T., et al.: Applications of deep learning in fundus images: a review. Med. Image Anal. **69**, 101971 (2021)

36. Li, X., et al.: Rotation-oriented collaborative self-supervised learning for retinal disease diagnosis. IEEE Trans. Med. Imaging **40**(9), 2284–2294 (2021)

37. Liao, W., et al.: Clinical interpretable deep learning model for glaucoma diagnosis. IEEE J. Biomed. Health Inform. **24**(5), 1405–1412 (2019)

38. Macsik, P., Pavlovicova, J., Goga, J., Kajan, S.: Local binary CNN for diabetic retinopathy classification on fundus images. Acta Polytech. Hung. **19**(7), 27–45 (2022)

39. Milosavljević, N., Morozov, D., Skraba, P.: Zigzag persistent homology in matrix multiplication time. In: SoCG, pp. 216–225 (2011)

40. Orlando, J.I., et al.: An ensemble deep learning based approach for red lesion detection in fundus images. Comput. Methods Programs Biomed. **153**, 115–127 (2018)

41. Otter, N., Porter, M.A., Tillmann, U., Grindrod, P., Harrington, H.A.: A roadmap for the computation of persistent homology. EPJ Data Sci. **6**, 1–38 (2017)

42. Pedregosa, F., et al.: Scikit-learn: machine learning in python. J. Mach. Learn. Res. **12**, 2825–2830 (2011)

43. Pratt, H., Coenen, F., Broadbent, D.M., Harding, S.P., Zheng, Y.: Convolutional neural networks for diabetic retinopathy. Procedia Comput. Sci. **90**, 200–205 (2016)
44. Rieck, B., et al.: Uncovering the topology of time-varying fmri data using cubical persistence. In: NeurIPS, vol. 33, pp. 6900–6912 (2020)
45. Sarhan, M.H., et al.: Machine learning techniques for ophthalmic data processing: a review. IEEE J. Biomed. Health Inform. **24**(12), 3338–3350 (2020)
46. Skaf, Y., Laubenbacher, R.: Topological data analysis in biomedicine: a review. J. Biomed. Inform. 104082 (2022)
47. Skraba, P., Ovsjanikov, M., Chazal, F., Guibas, L.: Persistence-based segmentation of deformable shapes. In: CVPR-Workshops, pp. 45–52. IEEE (2010)
48. Srivastava, O., et al.: Artificial intelligence and machine learning in ophthalmology: a review. Indian J. Ophthalmol. **71**(1), 11–17 (2023)
49. Stolz, B.J., et al.: Topological data analysis of task-based fMRI data from experiments on schizophrenia. J. Phys.: Complexity **2**(3), 035006 (2021)
50. Taj, I.A., et al.: An ensemble framework based on deep CNNs for glaucoma classification. Math. Biosci. Eng. **18**(5), 5321–5347 (2021)
51. Tauzin, G., et al.: giotto-TDA: a TDA toolkit for machine learning and data exploration (2020)
52. Ting, D.S.W., et al.: Artificial intelligence and deep learning in ophthalmology (2019)
53. Ting, D.S., et al.: Deep learning in ophthalmology: the technical and clinical considerations. Prog. Retin. Eye Res. **72**, 100759 (2019)
54. Wasserman, L.: Topological data analysis. Annu. Rev. Stat. Appl. **5**, 501–532 (2018)
55. World Health Organization, W.: World vision report (2019). https://www.who.int/publications/i/item/9789241516570
56. Wu, Z., Xiong, Y., Yu, S.X., Lin, D.: Unsupervised feature learning via nonparametric instance discrimination. In: CVPR, pp. 3733–3742 (2018)
57. Ye, M., Zhang, X., Yuen, P.C., Chang, S.F.: Unsupervised embedding learning via invariant and spreading instance feature. In: CVPR, pp. 6210–6219 (2019)
58. Zhang, Z., et al.: ORIGA-light: an online retinal fundus image database for glaucoma analysis and research. In: 2010 Annual International Conference of the IEEE Engineering in Medicine and Biology, pp. 3065–3068. IEEE (2010)
59. Zhang, Z., Ji, Z., Chen, Q., Yuan, S., Fan, W.: Joint optimization of CycleGAN and CNN classifier for detection and localization of retinal pathologies on color fundus photographs. IEEE J. Biomed. Health Inform. **26**(1), 115–126 (2021)
60. Zhao, X., et al.: Glaucoma screening pipeline based on clinical measurements and hidden features. IET Image Process. **13**(12), 2213–2223 (2019)

Neural Network Pruning for Real-Time Polyp Segmentation

Suman Sapkota[1], Pranav Poudel[2], Sudarshan Regmi[1], Bibek Panthi[1],
and Binod Bhattarai[3(✉)]

[1] NepAl Applied Mathematics and Informatics Institute (NAAMII), Lalitpur, Nepal
[2] Fogsphere (Redev AI Ltd), London, UK
[3] University of Aberdeen, Aberdeen, UK
`binod.bhattarai@abdn.ac.uk`

Abstract. Computer-assisted treatment has emerged as a viable application of medical imaging, owing to the efficacy of deep learning models. Real-time inference speed remains a key requirement for such applications to help medical personnel. Even though there generally exists a trade-off between performance and model size, impressive efforts have been made to retain near-original performance by compromising model size. Neural network pruning has emerged as an exciting area that aims to eliminate redundant parameters to make the inference faster. In this study, we show an application of neural network pruning in polyp segmentation. We compute the importance score of convolutional filters and remove the filters having the least scores, which to some value of pruning does not degrade the performance. For computing the importance score we use the Taylor First Order (TaylorFO) approximation of the change in *network output* for the removal of certain filters. Specifically, we employ a gradient-normalized backpropagation for the computation of the importance score. Through experiments in the polyp datasets, we validate that our approach can significantly reduce the parameter count and FLOPs retaining similar performance.

Keywords: Polyp Segmentation · Real-time Colonoscopy · Neural Network Pruning

1 Introduction

Polyp segmentation [6,7,37] is a crucial research problem in the medical domain involving dense classification. The primary aim of segmenting polyps in the colonoscopy and endoscopy is to identify pathological abnormalities in body parts such as the colon, rectum, etc. Such abnormalities can potentially lead to adverse effects causing colorectal cancer, thus inviting fatal damage to health. Statistics show that between 17% and 28% of colon polyps are overlooked during normal colonoscopy screening procedures, with 39% of individuals having at least one polyp missed, according to several recent studies [22,27]. However, timely diagnosis of a polyp can lead to timely treatment. It has been calculated that

G. Waiter et al. (Eds.): MIUA 2023, LNCS 14122, pp. 298–309, 2024.
https://doi.org/10.1007/978-3-031-48593-0_22

a 1% improvement of polyp detection rate reduces colorectal cancer by 3% [3]. Realizing the tremendous upside of early polyp diagnosis, medical AI practitioners have been trying to utilize segmentation models to assist clinical personnel. However, the latency of the large segmentation model has been the prime bottleneck for successful deployment. Utilizing smaller segmentation models is an option, but doing so compromises the performance of the model. In the case of bigger models, there is a good chance model learns significant redundancies thereby leaving room for improvement in performance. In such a scenario, we can prune the parameters of the model to reduce its size for the inference stage. Neural Network Pruning has established itself as an exciting area to reduce the inference time of larger models.

Neural Network Pruning [2,10,14,26] is one of the methods to reduce the parameters, compute, and memory requirements. This method differs significantly from knowledge distillation [12,16] where a small model is trained to produce the output of a larger model. Neural Network Pruning is performed at multiple levels; (i) weight pruning [13,14,35] removes per parameter basis while (ii) neuron/channel [24,43] pruning removes per neuron or channel basis and (iii) block/group [11,25] pruning removes per a block of networks such as residual block or sub-network.

Weight pruning generally achieves a very high pruning ratio getting similar performance only with a few percentages of the parameters. This allows a high network compression and accelerates the network on specialized hardware and CPUs. However, weight pruning in a defined format such as N:M block-sparse helps in improving the performance on GPUs [29]. Pruning network at the level of neurons or channels helps reduce the parameters with similar performance, however, the pruning ratio is not that high. All these methods can be applied to the same model as well.

In this work, we are particularly interested in neuron-level pruning. Apart from the benefit of reduced parameter, memory, and computation time (or FLOPs), neuron or channel level pruning, the number of neurons in a neural network is small compared to the number of connections and can easily be pruned by measuring the global importance [15,26,28,34,44]. We focus on the global importance as it removes the need to inject bias about the number of neurons to prune in each layer. This can simplify our problem to remove less significant neurons globally, allowing us to extend it to differently organized networks such as VGG, ResNet, UNet or any other Architecture. However, in this work, we focus only on the layer-wise, block-wise and hierarchical architecture of UNet [38].

Our experiment on Kvasir Segmentation Dataset using UNet model shows that we can successfully prune ≈1K Neurons removing ≈14% of parameters and reducing FLOPs requirement to ≈0.5x the original model with approximately the same performance of the original (from 0.59 IoU to 0.58 IoU). That is half the computational requirement of previous model with negligible performance loss.

2   Related Works

2.1   Real-Time Polyp Segmentation

Convolution-based approaches [30, 38, 46] have mostly dominated the literature while recently attention-based models [6, 23] have also been gaining traction in polyp segmentation. A number of works have been done in the area of real-time settings too. One of the earliest works [39], evidencing the ability of deep learning models for real-time polyp, has shown to achieve 96% accuracy in screening colonoscopy. Another work [41] utilizing a multi-threaded system in a real-time setting, has shown the deep learning models' ability to process at 25 fps with 76.80 ± 5.60 ms latency. Specialized architectures for polyp segmentation have also been studied in the medical imaging literature accounting for real-time performance. MSNet [45] introduced a subtraction unit, performing inference on 352×352 at 70 fps, instead of the usual addition as used in many works such as UNet [38], UNet++ [46], etc. Moreover, NanoNet [21] introduced a novel architecture tailor-made for real-time polyp segmentation primarily relying on a lightweight model hence compromising the learning capacity. SANet [42] has been shown to achieve strong performance with an inference speed of about 72 FPS. It showed samples collected under different conditions show inconsistent colors, causing the feature distribution gap and overfitting issue. Another work [36] used 2D gaussian instead of binary maps to better detect flat and small polyps which have unclear boundaries.

2.2   Neural Network Pruning

Works in pruning have somewhat lagged behind in medical imaging as compared to other domains. A recent work [1] has focused its study on reducing the computational cost of model retraining after post-pruning. DNNDeepening-Pruning [8] proposed the two-stage model development algorithm to build the small model. In the first stage, the residual layers are added until the overfitting starts and in the latter stage, pruning of the model is done with some user instructions. Furthermore, [9] has demonstrated evolution strategy-based pruning in generative adversarial networks (GAN) framework for medical imaging diagnostics purposes. In biomedical image segmentation, [19] applied a pruning strategy in U-Net architecture achieving 2x speedup trading off a mere 2% loss in mIOU(mean Intersection Over Union) on PhC-U373 and DIC-HeLa dataset. STAMP [4] tackles the low data regime through online simultaneous training and pruning achieving better performance with a UNet model of smaller size as compared to the unpruned one. In histological images, the superiority of layer-wise pruning and network-wide magnitude pruning has been shown for smaller and larger compression ratios respectively [32]. For medical image localization tasks, pruning has also been used to automatically and adaptively identify hard-to-learn examples [18]. In our study, we make use of pruning to reduce the model's parameters.

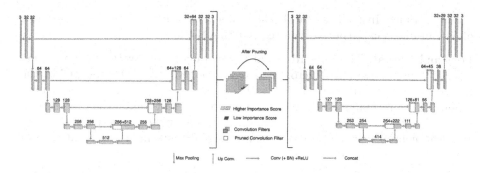

Fig. 1. Left: Unpruned UNet Model. Right: Model After Purning convolution filters with low importance score. *The exact number of pruned filters is 956, extracted from experiment shown in Fig. 2 (top).*

Previous works showed that global importance estimation can be computed using one or all of forward (activation) [17], parameter (weight) [14] or backward (gradient) [5, 31, 40] signals. Some of the previous techniques use Feature Importance propagation [44] or Gradient propagation [28] to find the neuron importance. Others use both activation and gradient information for pruning [33, 34]. Although there are methods using such signals for pruning at initialization [40], we limit our experiment to the pruning of trained models for a given number of neurons.

In this work, we use importance metric similar to Taylor First Order (Taylor-FO) approximations [33, 34] but from heuristics combining both forward and backward signals. The forward signal, namely the activation of the neuron, and the backward signal, the gradient. We use a normalized gradient signal to make the contribution of each example similar for computing the importance score.

3   Methodology

In this section, we discuss the pruning method in detail, and the application of the pruning method for polyp segmentation tasks, specifically focusing on the UNet architecture. However, it can be applied to other architecture as well. Instead of pruning all layers, we specifically target the convolutional layers for pruning. It is important to note that the term 'neurons' refers to the channels in the context of pruning convolutional layers. Furthermore, we present a method to select the pruned model that is best suited for the task at hand.

3.1   Pruning Method

Previous works on global importance-based post-training pruning of neurons focus on using forward and backward signals. Since most of these methods are based on Taylor approximation of the change in loss after removing a neuron or group of parameters, these methods require input and target value for computing the importance. Instead, we tackle the problem of pruning from the perspective of overall function output without considering the loss.

Forward Signal: The forward signal is generally given by the pre-activation (x_i). If a pre-activation is zero, then it has no impact on the output of the function, i.e. the output deviation with respect to the removal of the neuron is zero. If the incoming connection of a neuron is zero-weights, then the neuron can be removed, i.e. it has no significance. If the incoming connection is non-zero then the neuron has significance. Forward signal takes into consideration how data affects a particular neuron.

Backward Signal: The backward signal is generally given by back-propagating the loss. If the outgoing connection of the neuron is zeros, then the neuron has no significance to the function, even if it has positive activation. The gradient (δx_i) provides us with information on how the function or loss will change if the neuron is removed.

Importance Metric: Combining the forward and backward signal we can get the influence of the neuron on the loss or the function for given data. Hence, the importance metric (I_i) of each neuron (n_i) for dataset of size M is given by $I_i = \frac{1}{M} \sum_{n=1}^{M} x_i.\delta x_i$, where x_i is the pre-activation and δx_i is its gradient. It fulfills the criterion that importance should be low if incoming or outgoing connections are zeros and higher otherwise.

Problem 1: This importance metric (I_i) is similar to Taylor-FO [34]. However, the metric gives low importance when the gradient is negative, which to our application, is a problem as the function will be changed significantly, even if it lowers the loss. Hence, we calculate the square of importance metric to make it positive. The squared importance metric (I_i^s) is computed as below:

$$I_i^s = \frac{1}{M} \sum_{n=1}^{M} (x_i.\delta x_i)^2$$

Problem 2: During the computation of the gradients, some input examples produce a higher magnitude of gradient, and some input examples produce a lower magnitude of the gradient. Since the magnitude is crucial for computing the importance, different inputs contribute differently to the overall importance score. To this end, we normalize the gradient to the same magnitude of 1. Doing so makes the contribution of each data point equal for computing the importance.

Pruning Procedure: Consider that pruning is performed using dataset $\mathbf{D} \in [\mathbf{x}_0, \mathbf{x}_1, ...\mathbf{x}_N]$ of size N. We have a Convolutional Neural Network (CNN) whose output is given by: $\mathbf{y}_n = f_{CNN}(\mathbf{x}_n)$. We first compute the gradient w.r.t $\mathbf{y}_n$ for all $\mathbf{x}_n$ for given target $\mathbf{t}_n$ as:

$$\Delta \mathbf{y}_n = \frac{\delta E(\mathbf{y}_n, \mathbf{t}_n)}{\delta \mathbf{y}_n}$$

We then normalize the gradient $\Delta \mathbf{y}_n$ as:

$$\Delta \hat{\mathbf{y}}_n = \frac{\Delta \mathbf{y}_n}{\|\Delta \mathbf{y}_n\|}$$

This gradient $\Delta \hat{\mathbf{y}}_n$ is then backpropagated through the f_{CNN} network to compute the squared Importance score (I_i^s) of each convolution filter.

3.2 Pruning UNet for Polyp-Segmentation

UNet [38] is generally used for Image Segmentation Tasks. It consists of only Convolutional Layers including Upsampling and Downsampling layers organized in a hierarchical structure as shown in Fig. 1.

We compute the Importance Score for each Convolutional layer and prune the least important ones. Removing a single convolution filter removes a channel of the incoming convolution layer and the outgoing convolution channel. When used with many channels, we can get a highly pruned UNet with only a slight change in performance.

This method can be used to drastically reduce the computation and memory requirements without degrading the performance even without fine-tuning the pruned model. A single computation of Importance Score allows us to prune multiple numbers of neurons and select sparsity with the best FLOPs (or Time-taken) and IoU trade-off.

3.3 Measuring Pruning Performance

Performance metrics are crucial for measuring the effectiveness of different pruning algorithms. Some of them are listed below.

FLOPs: FLOP stands for floating point operation. Floating point operation refers to the mathematical operations performed on floating point numbers. FLOP measures model complexity, with a higher value indicating a computationally expensive model and a lower value indicating a computationally cheaper model with faster inference time. We evaluate an algorithm's efficiency by how many FLOPs it reduces.

Parameters: Parameters represent learnable weights and biases typically represented by floating point numbers. Models with many parameters need a lot of memory, while models with fewer parameters need less memory. The effectiveness of the pruning algorithm is measured by the reduction in the model's parameters.

Time-Taken: It is the actual wall-clock inference time of model. We measure time taken before and after pruning the network. Time-taken is practical but not the most reliable metric for efficiency gain as it might vary with device and with different ML frameworks.

4   Experiments

We conduct the experiment for Polyp-Segmentation model pruning using the
Kvasir Dataset [20]. We use the pretrained UNet Model for segmentation and
prune the Convolutional Filters of the Network to reduce the computational cost
as shown in Fig. 2.

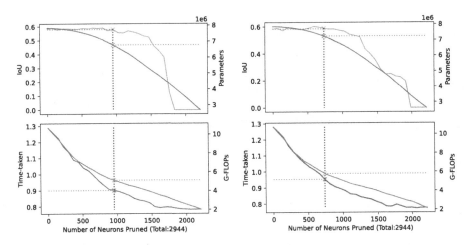

Fig. 2. (Top) row is the Number of Neurons Pruned vs IoU and Parameters plot.
(Bot) row is the Number of Neurons Pruned vs Time-taken and Giga-FLOPs plot.
Here, Time-taken is measured in seconds for the inference of 100 samples with 10
batch size. **(Left)** column shows pruning performance using 39 data samples for importance estimation. A sample pruning of 956 neurons reduces the FLOPs to 0.477× and
parameters to 0.864× while retaining performance to 0.99× the original performance
(≈0.5795 IoU). The time taken is reduced by ≈30%. **(Right)** column shows pruning
performance using 235 data samples for importance estimation. A sample pruning of
736 neurons reduces the FLOPs to 0.54× and parameters to 0.922× while retaining
the same performance (≈0.5879 IoU). Here, we manage to reduce time taken by ≈26%.

Procedure: First, we compute the importance score for each neuron/channel
on a given dataset. Secondly, we prune the P least important neurons of total
N by importance metric (I^s) given by our method. We measure the resulting
accuracy and plot the Number of neurons pruned as shown in Fig. 2. The pruning
is performed using one split of the test dataset and the IoU is measured on
another split of the test dataset.

Although the pruned models could be finetuned to see an increase in performance, we do not finetune pruned model in our case. We analyse the change in
performance (IoU), the efficiency achieved (FLOPs) and the compression (the
number of parameters) for different values of the *number-of-neurons-pruned* in
the UNet Model.

Observation: The experiments show that model generally consists of redundant and less important convolutional channels, which can be pruned with little to no effect on the output of the model. We see in Figure (2 left) that about 50% ($\approx$1500 out of 2944) of the neurons can be pruned before the IoU starts to decrease drastically. Furthermore, this result is observed for a varying numbers of data points, which suggests that pruning under different settings creates different pruned architectures, while still following the same pattern of performance retention after pruning of an increasing number of neurons (up to some point).

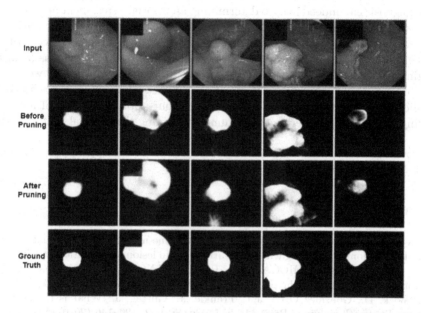

Fig. 3. Qualitative comparison of Polyp Segmentation before and after pruning of the UNet model. The pruned model samples are generated from experiment in Fig. (2 *left* with 956 neurons pruned).

The qualitative evaluation (see Fig. 3) of the Pruned UNet Model on the Polyp-Segmentation Dataset shows that the pruned model makes slight changes in the output of the unpruned model while preserving most of the important characteristics. We find that these slight changes can be improvements or degradation to the original model outputs but without significantly distorting the model output.

5   Conclusion

In this work, we propose to use Neuron Level Pruning in the application of the polyp segmentation task for the first time. The benefit of proposed channels or filter pruning can be realized immediately with parallel hardware like GPUs to significantly reduce the computation cost to less than 50% without degrading the performance. Such a reduction in computational cost automatically leads to the potential application in a real-time setting. Computer-assisted treatment of patients, especially during medical tasks like colonoscopy, requires low latency with satisfactory performance such that the pace of treatment is not hindered. Since the polyp's nature can exhibit significant variability during colonoscopy, real-time polyp segmentation models can indeed provide medical personnel useful insights into locating the abnormal growths in the colon thereby assisting in early diagnosis. Moreover, the advanced visualizations aided through real-time diagnosis can indeed lead to determining appropriate treatment approaches. Moreover, it also allows safe, methodical, and consistent diagnosis of patients. Our work paves the path for off-the-shelf models to be significantly accelerated through neural network pruning in tasks requiring fast inference such as medical imaging, reducing the inference and storage cost. To sum up, in this work, we explore a promising research direction of neural network pruning demonstrating its efficacy in polyp segmentation. We validate our approach of neural network pruning with various experiments by almost retaining the original performance.

Acknowledgments. This work is partly funded by the EndoMapper project by Horizon 2020 FET (GA 863146).

References

1. Bayasi, N., Hamarneh, G., Garbi, R.: Culprit-prune-net: efficient continual sequential multi-domain learning with application to skin lesion classification. In: de Bruijne, M., et al. (eds.) MICCAI 2021, Part VII. LNCS, vol. 12907, pp. 165–175. Springer, Cham (2021). https://doi.org/10.1007/978-3-030-87234-2_16
2. Blalock, D., Gonzalez Ortiz, J.J., Frankle, J., Guttag, J.: What is the state of neural network pruning? Proc. Mach. Learn. Syst. **2**, 129–146 (2020)
3. Corley, D.A., et al.: Adenoma detection rate and risk of colorectal cancer and death. N. Engl. J. Med. **370**(14), 1298–1306 (2014)
4. Dinsdale, N.K., Jenkinson, M., Namburete, A.I.: Stamp: simultaneous training and model pruning for low data regimes in medical image segmentation. Med. Image Anal. **81**, 102583 (2022)
5. Evci, U., Ioannou, Y., Keskin, C., Dauphin, Y.: Gradient flow in sparse neural networks and how lottery tickets win. In: Proceedings of the AAAI Conference on Artificial Intelligence, vol. 36, pp. 6577–6586 (2022)
6. Fan, D.-P., et al.: PraNet: parallel reverse attention network for polyp segmentation. In: Martel, A.L., et al. (eds.) MICCAI 2020. LNCS, vol. 12266, pp. 263–273. Springer, Cham (2020). https://doi.org/10.1007/978-3-030-59725-2_26
7. Fang, Y., Chen, C., Yuan, Y., Tong, K.: Selective feature aggregation network with area-boundary constraints for polyp segmentation. In: Shen, D., et al. (eds.) MICCAI 2019, Part I. LNCS, vol. 11764, pp. 302–310. Springer, Cham (2019). https://doi.org/10.1007/978-3-030-32239-7_34
8. Fernandes, F.E., Yen, G.G.: Automatic searching and pruning of deep neural networks for medical imaging diagnostic. IEEE Trans. Neural Netw. Learn. Syst. **32**(12), 5664–5674 (2020)

9. Fernandes, F.E., Jr., Yen, G.G.: Pruning of generative adversarial neural networks for medical imaging diagnostics with evolution strategy. Inf. Sci. **558**, 91–102 (2021)

10. Gale, T., Elsen, E., Hooker, S.: The state of sparsity in deep neural networks. arXiv preprint arXiv:1902.09574 (2019)

11. Gordon, A., et al.: MorphNet: fast & simple resource-constrained structure learning of deep networks. In: Proceedings of the IEEE Conference on Computer Vision and Pattern Recognition, pp. 1586–1595 (2018)

12. Gou, J., Yu, B., Maybank, S.J., Tao, D.: Knowledge distillation: a survey. Int. J. Comput. Vision **129**, 1789–1819 (2021)

13. Han, S., Mao, H., Dally, W.J.: Deep compression: compressing deep neural networks with pruning, trained quantization and Huffman coding. arXiv preprint arXiv:1510.00149 (2015)

14. Han, S., Pool, J., Tran, J., Dally, W.: Learning both weights and connections for efficient neural network. In: Advances in Neural Information Processing Systems, vol. 28 (2015)

15. Hassibi, B., Stork, D., Wolff, G.: Optimal brain surgeon: extensions and performance comparisons. In: Advances in Neural Information Processing Systems, vol. 6 (1993)

16. Hinton, G., Vinyals, O., Dean, J.: Distilling the knowledge in a neural network. arXiv preprint arXiv:1503.02531 (2015)

17. Hu, H., Peng, R., Tai, Y.W., Tang, C.K.: Network trimming: a data-driven neuron pruning approach towards efficient deep architectures. arXiv preprint arXiv:1607.03250 (2016)

18. Jaiswal, A., Chen, T., Rousseau, J.F., Peng, Y., Ding, Y., Wang, Z.: Attend who is weak: pruning-assisted medical image localization under sophisticated and implicit imbalances. In: Proceedings of the IEEE/CVF Winter Conference on Applications of Computer Vision, pp. 4987–4996 (2023)

19. Jeong, T., Bollavaram, M., Delaye, E., Sirasao, A.: Neural network pruning for biomedical image segmentation. In: Medical Imaging 2021: Image-Guided Procedures, Robotic Interventions, and Modeling, vol. 11598, pp. 415–425. SPIE (2021)

20. Jha, D., et al.: Kvasir-SEG: a segmented polyp dataset. In: YM, R., et al. (eds.) MMM 2020, Part II. LNCS, vol. 11962, pp. 451–462. Springer, Cham (2020). https://doi.org/10.1007/978-3-030-37734-2_37

21. Jha, D., et al.: NanoNet: real-time polyp segmentation in video capsule endoscopy and colonoscopy. In: 2021 IEEE 34th International Symposium on Computer-Based Medical Systems (CBMS), pp. 37–43 (2021). https://doi.org/10.1109/CBMS52027.2021.00014

22. Kim, N.H., et al.: Miss rate of colorectal neoplastic polyps and risk factors for missed polyps in consecutive colonoscopies. Intest. Res. **15**(3), 411 (2017)

23. Kim, T., Lee, H., Kim, D.: UACANet: uncertainty augmented context attention for polyp segmentation. In: Proceedings of the 29th ACM International Conference on Multimedia, pp. 2167–2175 (2021)

24. Lebedev, V., Lempitsky, V.: Fast convnets using group-wise brain damage. In: Proceedings of the IEEE Conference on Computer Vision and Pattern Recognition, pp. 2554–2564 (2016)

25. Leclerc, G., Vartak, M., Fernandez, R.C., Kraska, T., Madden, S.: Smallify: learning network size while training. arXiv preprint arXiv:1806.03723 (2018)

26. LeCun, Y., Denker, J., Solla, S.: Optimal brain damage. In: Advances in Neural Information Processing Systems, vol. 2 (1989)

27. Lee, J., et al.: Risk factors of missed colorectal lesions after colonoscopy. Medicine **96**(27) (2017)

28. Lee, N., Ajanthan, T., Torr, P.H.: SNIP: single-shot network pruning based on connection sensitivity. arXiv preprint arXiv:1810.02340 (2018)

29. Liu, S., Wang, Z.: Ten lessons we have learned in the new "sparseland": a short handbook for sparse neural network researchers. arXiv preprint arXiv:2302.02596 (2023)

30. Long, J., Shelhamer, E., Darrell, T.: Fully convolutional networks for semantic segmentation. In: Proceedings of the IEEE Conference on Computer Vision and Pattern Recognition, pp. 3431–3440 (2015)

31. Lubana, E.S., Dick, R.P.: A gradient flow framework for analyzing network pruning. arXiv preprint arXiv:2009.11839 (2020)

32. Mahbod, A., Entezari, R., Ellinger, I., Saukh, O.: Deep neural network pruning for nuclei instance segmentation in hematoxylin and eosin-stained histological images. In: Wu, S., Shabestari, B., Xing, L. (eds.) AMAI 2022. LNCS, vol. 13540, pp. 108–117. Springer, Cham (2022). https://doi.org/10.1007/978-3-031-17721-7_12

33. Molchanov, P., Mallya, A., Tyree, S., Frosio, I., Kautz, J.: Importance estimation for neural network pruning. In: Proceedings of the IEEE/CVF Conference on Computer Vision and Pattern Recognition, pp. 11264–11272 (2019)

34. Molchanov, P., Tyree, S., Karras, T., Aila, T., Kautz, J.: Pruning convolutional neural networks for resource efficient inference. arXiv preprint arXiv:1611.06440 (2016)

35. Mozer, M.C., Smolensky, P.: Using relevance to reduce network size automatically. Connect. Sci. **1**(1), 3–16 (1989)

36. Qadir, H.A., Shin, Y., Solhusvik, J., Bergsland, J., Aabakken, L., Balasingham, I.: Toward real-time polyp detection using fully CNNs for 2D gaussian shapes prediction. Med. Image Anal. **68**, 101897 (2021)

37. Qiu, J., Hayashi, Y., Oda, M., Kitasaka, T., Mori, K.: Boundary-aware feature and prediction refinement for polyp segmentation. Comput. Methods Biomech. Biomed. Eng.: Imaging Vis. **11**, 1–10 (2022)

38. Ronneberger, O., Fischer, P., Brox, T.: U-net: convolutional networks for biomedical image segmentation. In: Navab, N., Hornegger, J., Wells, W.M., Frangi, A.F. (eds.) MICCAI 2015, Part III. LNCS, vol. 9351, pp. 234–241. Springer, Cham (2015). https://doi.org/10.1007/978-3-319-24574-4_28

39. Urban, G., et al.: Deep learning localizes and identifies polyps in real time with 96% accuracy in screening colonoscopy. Gastroenterology **155**(4), 1069–1078 (2018)

40. Wang, C., Zhang, G., Grosse, R.: Picking winning tickets before training by preserving gradient flow. arXiv preprint arXiv:2002.07376 (2020)

41. Wang, P., et al.: Development and validation of a deep-learning algorithm for the detection of polyps during colonoscopy. Nat. Biomed. Eng. **2**(10), 741–748 (2018)

42. Wei, J., Hu, Y., Zhang, R., Li, Z., Zhou, S.K., Cui, S.: Shallow attention network for polyp segmentation. In: de Bruijne, M., et al. (eds.) MICCAI 2021, Part I. LNCS, vol. 12901, pp. 699–708. Springer, Cham (2021). https://doi.org/10.1007/978-3-030-87193-2_66

43. Wen, W., Wu, C., Wang, Y., Chen, Y., Li, H.: Learning structured sparsity in deep neural networks. In: Advances in Neural Information Processing Systems, vol. 29 (2016)

44. Yu, R., et al.: NISP: pruning networks using neuron importance score propagation. In: Proceedings of the IEEE Conference on Computer Vision and Pattern Recognition, pp. 9194–9203 (2018)

45. Zhao, X., Zhang, L., Lu, H.: Automatic polyp segmentation via multi-scale subtraction network. In: de Bruijne, M., et al. (eds.) MICCAI 2021, Part I. LNCS, vol. 12901, pp. 120–130. Springer, Cham (2021). https://doi.org/10.1007/978-3-030-87193-2_12
46. Zhou, Z., Rahman Siddiquee, M.M., Tajbakhsh, N., Liang, J.: UNet++: a nested U-Net architecture for medical image segmentation. In: Stoyanov, D., et al. (eds.) DLMIA/ML-CDS -2018. LNCS, vol. 11045, pp. 3–11. Springer, Cham (2018). https://doi.org/10.1007/978-3-030-00889-5_1

A Novel Approach to Breast Cancer Segmentation Using U-Net Model with Attention Mechanisms and FedProx

Eyad Gad[1]([envelope]) [ID], Mustafa Abou Khatwa[1] [ID], Mustafa A. Elattar[2,3] [ID], and Sahar Selim[2,3]([envelope]) [ID]

[1] School of Engineering and Applied Sciences, Nile University, Sheikh Zayed, Giza, Egypt
{e.gad,m.aboukatwa}@nu.edu.eg
[2] Medical Imaging and Image Processing Research Group, Center for Informatics Science, Nile University, Sheikh Zayed, Giza, Egypt
melattar@nu.edu.eg
[3] School of Information Technology and Computer Science, Nile University, Sheikh Zayed, Giza, Egypt
sselim@nu.edu.eg

Abstract. Breast cancer is a leading cause of death among women worldwide, emphasizing the need for early detection and accurate diagnosis. As such Ultrasound Imaging, a reliable and cost-effective tool, is used for this purpose, however the sensitive nature of medical data makes it challenging to develop accurate and private artificial intelligence models. A solution is Federated Learning as it is a promising technique for distributed machine learning on sensitive medical data while preserving patient privacy. However, training on non-Independent and non-Identically Distributed (non-IID) local datasets can impact the accuracy and generalization of the trained model, which is crucial for accurate tumour boundary delineation in BC segmentation. This study aims to tackle this challenge by applying the Federated Proximal (FedProx) method to non-IID Ultrasonic Breast Cancer Imaging datasets. Moreover, we focus on enhancing tumour segmentation accuracy by incorporating a modified U-Net model with attention mechanisms. Our approach resulted in a global model with 96% accuracy, demonstrating the effectiveness of our method in enhancing tumour segmentation accuracy while preserving patient privacy. Our findings suggest that FedProx has the potential to be a promising approach for training precise machine learning models on non-IID local medical datasets.

Keywords: Ultrasound (US) · Ultrasonic Imaging (USI) · Breast Cancer (BC) · Federated Learning (FL) · Ultrasonic Breast Cancer Imaging (USBCI) · Federated Proximal (FedProx)

1 Introduction

Breast Cancer (BC) is a major public health concern worldwide where it affects millions of people, and early detection is crucial for successful treatment while

G. Waiter et al. (Eds.): MIUA 2023, LNCS 14122, pp. 310–324, 2024.
https://doi.org/10.1007/978-3-031-48593-0_23

increasing the likelihood of a patient's survival. According to The National Breast Cancer Foundation, and The Center for Disease Control and Prevention (CDC). BC is the most common cancer among women in the USA, with a rate of 129.7 new cases per 100,000 in 2019 while being the second leading cause of cancer related deaths among women, at a rate of 19.4 per 100,000 in 2019. Globally, BC is the most common cancer diagnosed among women, with an estimated 2.3 million new cases diagnosed in 2020. Preventative measures and early detection strategies, such as regular screening, can help reduce the impact of BC on people' s lives, according to the World Health Organization (WHO) [1–3]. Hence, the early detection and accurate diagnosis of BC is a critical affair for effective treatment and improved patient health. Modern diagnostic approaches utilize Ultrasonic Images (USI), where Ultrasonic (US) sensors send acoustic waves into the subject to determine distance, composition, and density. US sensors have become valuable tools for BC screening, as they can detect lesions that may not be visible on mammography, which is a radiological examination of breasts used to detect cancers [4].

Clinics and hospitals often face difficulties in accurately interpreting and diagnosing breast cancer from medical images. To address this challenge, Artificial Intelligence (AI) has been employed to enhance diagnosis and classification processes . Deep Learning (DL) algorithms have proven effective in extracting valuable information from medical images, enabling tasks such as segmenting and classifying Ultrasound Breast Cancer Images (USBCI). These AI-driven approaches have significantly improved the accuracy of diagnosis and treatment planning, as highlighted in a study by Xu et al. Additionally, several research studies have demonstrated that AI-based models for lesion detection and classification can enhance the sensitivity and specificity of breast cancer diagnosis, leading to improved outcomes [5–12].

The incorporation of AI has demonstrated promising potential in augmenting the capabilities of healthcare professionals in the field of BC diagnosis. However, as imaging technology advances and the volume and quality of imaging data increase, DL algorithms like ResNet, U-Net, and V-Net face challenges in accurately segmenting USBCI. Although the U-Net model is widely recognized for its effectiveness in cancer segmentation, it encounters difficulties in capturing intricate details and handling complex image structures. To overcome these limitations, this paper proposes the adoption of the Attention U-Net model. By incorporating attention mechanisms into the U-Net architecture, the model gains the ability to selectively focus on important regions within the image while disregarding irrelevant information. This attention-based approach enhances the U-Net's capability to capture fine-grained details, resulting in more precise and accurate segmentation outcomes, however, effectively analyzing such extensive datasets can be time-consuming and resource-intensive. Additionally, concerns regarding privacy arise when dealing with large amounts of data stored in a centralized location [5,6,8,13].

To address these challenges, FL is proposed, allowing the training of machine learning models on distributed datasets across different locations without sharing the raw data. In the medical field, FL is employed to maintain the privacy of medical data and alleviate network strain by training a predictor in a distributed manner, rather than transmitting raw data to a central server. This setup involves remote devices periodically communicating with a central server to learn a global model. In each communication round, a subset of selected edge devices conducts local training using their non-identically distributed user data and sends local updates to the server. Upon receiving the updates, the server incorporates them and returns the updated global model to another subset of devices. This iterative training process continues throughout the network until convergence, or a stopping criterion is met. This approach preserves privacy, minimizes resource consumption, and enables efficient utilization of local resources. Additionally, many researchers have studied the potential of FL for medical image analysis, including segmentation, classification, and synthesis [8,14].

However, FL, with its involvement of multiple parties and the non-independent and non-identically distributed (non-IID) nature of data generated and collected across the network, presents specific challenges, particularly in medical image tasks such as diagnosis and segmentation, including those related to USBCI. These challenges are further compounded by the significant variations in the data characteristics used for training, which stem from different imaging devices. Consequently, the datasets utilized in FL exhibit diverse properties, including variations in quality, resolution, and contrast. This data generation paradigm violates commonly used assumptions of independent and identically distributed (IID) data in distributed optimization introduces complexities, such as an increased likelihood of encountering stragglers and added intricacies in terms of modeling, analysis, and evaluation. Hence, to address the challenges of developing accurate and private AI models for BC segmentation using non-IID USBCI datasets, his study proposes the utilization of FL with the Fed-Prox method [15,16]. FedProx has the potential to be a promising approach for training precise models on non-IID medical datasets. The proposed approach is implemented in a setup involving three nodes or clients, leveraging the availability of datasets from two distinct sources, which were introduced in [17,18]. Furthermore, data augmentation techniques are applied to enhance the datasets, contributing to the overall improvement of the BC segmentation model's performance.

The main aim of the paper lies in improving the accuracy of breast tumor segmentation. To achieve this, the paper proposes an enhanced approach that involves modifying the U-Net model with attention mechanisms. This attention U-Net model serves as a segmentation DL model utilized by three clients within the FL framework. Additionally, the paper employs the FedProx method to aggregate the trained weights of the models from the clients, ensuring collaborative learning and integration of knowledge from the distributed clients while preserving privacy and security.

2   Literature Review

Over the past few years, several FL techniques have been proposed to enhance the accuracy and efficiency of medical image segmentation. This review aims to discuss recent advances in FL for medical image segmentation with a focus on breast lesion and tumor segmentation.

One recent FL approach called FedZaCt, developed by T. Yang et al. [9], combines Z-average aggregation with cross-teaching to improve image segmentation performance. The method was evaluated on the task of breast lesion segmentation and achieved superior performance compared to other FL methods. The Z-average aggregation reduces the impact of noisy updates from different devices, while cross-teaching encourages model diversity and enhances the ability of the models to generalize to new data.

The paper by Althobaiti et al. (2023) [19] presents a deep learning-based technique, SEODTL-BDC, for detecting and classifying breast cancer using photoacoustic multimodal imaging (PAMI). The SEODTL-BDC approach utilizes bilateral filtering for noise removal, LEDNet model for image segmentation, and ResNet-18 as a feature extractor. It incorporates social engineering optimization with an RNN-based classifier for assigning class labels. Experimental results on a benchmark dataset demonstrate the superior performance of SEODTL-BDC, achieving high precision, recall, accuracy, and F-score. This research contributes to the field of breast cancer detection and classification with PAMI, offering a promising approach for earlier detection and improved patient outcomes.

Furthermore, the study by Yuchao Lyu et al. [7] addresses the need for accurate segmentation of USBCI. Manual segmentation by physicians is challenging due to poor image quality and irregular target edges. To overcome this, the authors propose an improved segmentation model called AMS-PAN, which combines attention mechanism and multi-scale features. The model utilizes depthwise separable convolutions for multi-scale feature extraction and forms a feature pyramid. It incorporates a Global Attention Upsample (GAU) feature fusion technique and a Spatial and Channel Attention (SCA) module to focus on edge texture information. Experimental evaluation on the BUSI [18] and OASBUD datasets demonstrates the superior performance of AMS-PAN, outperforming traditional non-deep learning and mainstream deep learning methods in terms of Dice and IoU metrics. Additionally, AMS-PAN exhibits high computational efficiency, making it a valuable tool for assisting physicians in the detection and diagnosis of breast tumors in USI.

Additionally, Z. Yang et al. [14] developed a Robust Split FL approach for U-shaped medical image networks that aims to improve the efficiency and robustness of FL for medical image analysis. The proposed approach includes a novel split learning method that partitions the model between the clients and the server to reduce the communication overhead. The proposed method was evaluated on a dataset of Brain Tumor Images and achieved superior performance compared to existing FL methods.

Another proposed approach by J. Wicaksana et al. [10] is a mixed supervised FL method for medical image segmentation that combines supervised and unsupervised learning to improve segmentation performance. The proposed method includes a supervised learning phase where the model is trained on labeled data and an unsupervised learning phase where the model is further refined on unlabeled data. The approach was evaluated on a dataset for USBCI and achieved promising results. The unsupervised learning phase helps to overcome the limitations of supervised learning by leveraging the large amounts of unlabeled data available in medical imaging. The use of unsupervised learning has been shown to improve the generalization of the model, making it more robust to unseen data.

The use of such techniques can benefit medical professionals by enabling more accurate diagnoses and thus improving patient outcomes. FL has shown great potential in improving the accuracy and efficiency of medical image segmentation and classification. The proposed approaches aim to address the non-IID challenge in FL and improve the ability of the models to generalize to new data. These advances contribute to the standardization of medical image analysis and provide a benchmark for future studies in this field.

3   Materials and Methods

3.1   Proposed Approach

In this study, we propose FL approach for BC segmentation, incorporating the FedProx method and an attention U-Net model. The architecture, as depicted in Fig. 1, consists of a server side and a client side. On the server side, a global attention U-Net model is utilized, and the model's performance is evaluated using a testing dataset after each round. On the client side, three clients are set up, considering the limited availability of public USBCI datasets, each assigned with their respective data for training and testing, along with their attention U-Net models. Each client trains their models locally and shares the updated weights with the server. The FL Server then employs the FedProx algorithm to aggregate the updated weights. The newly aggregated weights are subsequently communicated back to the clients for further iterations. The proposed FL architecture aims to enable collaborative BC segmentation, ensuring data privacy, and improving model performance.

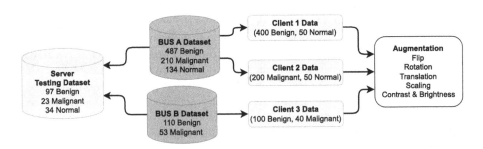

Fig. 1. The Proposed Architecture of FL for BC Segmentation using Attention U-Net Model and FedProx

3.2   Data Sources and Preparation

This study utilized two datasets for our research. The first dataset, named "Dataset of breast ultrasound images," was annotated and published by Al-Dhabyani et al. [18]. The second dataset, known as the "BUS B Dataset," was annotated and experimented with by Yap et al. [17]. For simplicity, we will refer to the "Dataset of breast ultrasound images" as the BUS A dataset, while the BUS B dataset will retain its original name.

The BUS A dataset was collected in 2018 and provides a comprehensive collection of breast ultrasound images. The dataset includes scans from 600 female patients, totalling 780 scans, each accompanied by annotated masks. The images in this dataset are categorized into three groups based on the presence of normal, benign, or malignant features. It comprises 487 scans of benign cases, 210 scans of malignant cases, and 133 scans of normal cases. This dataset has been widely utilized in various studies for the development and evaluation of breast lesion classification and segmentation models. The BUS B dataset was collected in 2012 at the UDIAT Diagnostic Centre of the Parc Tauli Company in Sabadell, Spain, using a Siemens ACUSON Sequoia C512 system with a 17L5 HD linear array transducer [17]. It consists of 163 scans of breast lesions along with their annotated masks. Among the scans, 110 belong to benign lesions, while 53 scans depict malignant tumors.

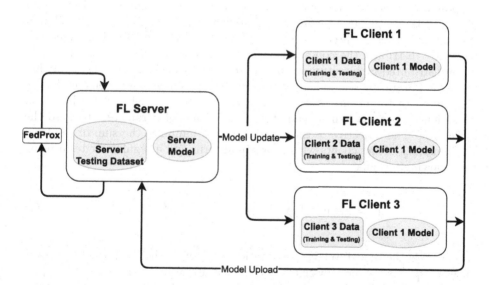

Fig. 2. Distribution of BC data across the server and client sides

Figure 2 illustrates the division of data between the client side and the server side. To ensure a comprehensive evaluation of our approach, Each client contains distinct data with different features, representing various aspects of the problem.

Client 1's data consists of 400 benign cases and 50 normal cases, while client 2's data includes 200 malignant cases and 50 normal cases, all sourced from the BUS A dataset. On the other hand, client 3 stands apart from the other clients and contains 110 benign cases and 53 malignant cases sourced from the BUS B dataset. In order to enhance the diversity and robustness of the training process, we applied various augmentation techniques to the clients' data. These techniques, including flip, rotation, translation, scaling, contrast adjustments, and brightness adjustments, were employed to increase the variability of the data samples. On the server side, a testing dataset is set up consisting of 97 benign cases, 23 malignant cases, and 34 normal cases, sourced from both datasets, enabling us to thoroughly evaluate the performance of the global model. This distribution of data across the clients and server ensures that the data is non-IID, allowing us to assess the effectiveness of our approach in handling diverse data and achieving accurate segmentation results of BC.

3.3   FedProx

The Fedprox method tackles the non-IID challenge in the data by partitioning it into non-overlapping subsets and distributing them to the clients. This ensures that each client has a representative sample of the data, which reduces the impact of distribution shift and improves the accuracy of the model. The proposed approach in this paper aims to enhance the accuracy and efficiency of the USBCI segmentation model by implementing Fedprox in a distributed system consisting of three clients and a server. The algorithm is designed to efficiently distribute the computational workload across multiple nodes, utilizing a distributed architecture to distribute resources effectively. The algorithm starts by initializing the model parameters on the server, and then the data is distributed to the clients. Each client performs local model updates based on its assigned data, and then sends the updated model parameters back to the server [16]. The objective of the algorithm is to minimize the sum of the local loss functions of the clients with a proximal term and a regularizer term using the following formula:

$$w^k = \arg\min_w \left(\sum_{j=1}^{n} \left(f_i(w) + \frac{\mu}{2} \|w - w_i^{k-1}\|_2^2 \right) \right) \tag{1}$$

where w^k is the model weights at iteration k, $f_i(w)$ is the local loss function of client i, w_i^{k-1} is the model weights of client i at iteration $k-1$, and μ is a hyper-parameter that controls the strength of the proximal term. The proximal term enforces the model parameters to be close to the previous iteration, while the regularizer term promotes sparsity in the model. The resulting aggregated model parameters are sent back to the clients for further local updates. In this study, Fedprox was utilized for the USBCI segmentation task, with carefully chosen parameter values through experimental evaluation to optimize performance for the specific task, resulting in a learning rate of 0.01, a regularization parameter of 0.001, a proximal parameter of 0.01, and a hyperparameter value of $\mu = 0.1$. Fedprox has been shown to improve the performance of the distributed deep learning model on the USBCI dataset. Specifically, it addresses the inter-client variability of the dataset by encouraging the clients to learn from their local data while sharing the model updates.

3.4 Attention U-Net

In this study, The Attention U-Net is used in this study for segmenting BC tumors. The architecture of the Attention U-Net includes a contracting path with convolutional and max pooling layers followed by an expanding path with convolutional and up-sampling layers. The key difference from a standard U-Net is that it incorporates an attention mechanism in the skip connections. The attention mechanism enhances the network's ability to focus on relevant features while discarding irrelevant ones, resulting in improved segmentation accuracy.

The graphical model of the attention mechanism is presented in Fig. 3. [20], the attention mechanism works by taking two inputs, vectors X and G as shown in the following figure. Vector G is taken from the next lowest layer of the network and has smaller dimensions and better feature representation. Vector X goes through a strided convolution, and vector G goes through a 1×1 convolution. The two vectors are then summed element-wise, and the resulting vector goes through a ReLU activation layer and a 1×1 convolution that collapses the dimensions to $1 \times 32 \times 32$. This vector then goes through a sigmoid layer, which scales the vector between the range (0–1), producing the attention coefficients (weights), where coefficients closer to 1 indicate more relevant features. The attention coefficients are unsampled to the original dimensions of vector X using trilinear interpolation. The attention coefficients are multiplied element-wise to

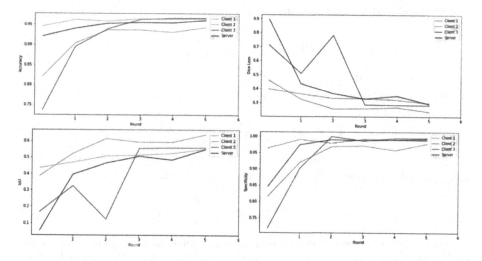

Fig. 3. The Attention Mechanism [20]

the original vector X, scaling the vector according to relevance, and then passed along in the skip connection as normal.

Attention U-Net model was implemented in the three clients with identical parameters. The training process utilized the dice loss function and the Adam optimizer with a learning rate of 0.0001 and a batch size of 16. The training was conducted over 10 epochs for each round, and evaluation was performed on a test set of 20% of the local data for each client. The performance of the model was measured using the average dice coefficient and other relevant performance metrics.

3.5 Performance Metrics

To assess the performance of our model, we employed various performance metrics, including the confusion matrix. In this study, we calculated metrics, including Dice Loss, Intersection over Union (IoU), Sensitivity, Specificity, F1 Score, and Accuracy. These metrics were evaluated during both the training and validation phases of the system.

Dice Loss measures the dissimilarity between the predicted and true positive regions. It takes into account true positives (TP), true negatives (TN), false positives (FP), and false negatives (FN) [21]. Intersection over Union (IoU) quantifies the overlap between the predicted and true positive regions and is sometimes referred to as the Jaccard index [22].

Sensitivity, also known as recall or true positive rate, calculates the proportion of actual positives correctly identified by the model. It is computed using the following formula:

$$\text{Sensitivity} = \frac{\text{TP}}{\text{TP} + \text{FN}} \tag{2}$$

Specificity measures the proportion of actual negatives correctly identified by the model and is calculated as:

$$\text{Specificity} = \frac{\text{TN}}{\text{TN} + \text{FP}} \tag{3}$$

F1 Score is a balanced measure that combines precision and recall, providing an overall evaluation of the model's performance. It is computed using the following formula:

$$\text{F1 Score} = 2 \times \frac{\text{Precision} \times \text{Recall}}{\text{Precision} + \text{Recall}} \tag{4}$$

Accuracy measures the proportion of correct predictions out of all predictions made by the model. It is calculated using the formula:

$$\text{Accuracy} = \frac{\text{TP} + \text{TN}}{\text{TP} + \text{TN} + \text{FP} + \text{FN}} \tag{5}$$

These performance metrics provide insights into the effectiveness and accuracy of the model in capturing true positive regions and distinguishing between positives and negatives.

4   Experimental Results

In our proposed model, we conducted experiments to showcase the utilization of the FedProx algorithm and attention U-Net for the server model and for the client models, resulting in the development of a simple yet effective FL model. The training process consisted of 6 rounds, where each round involved training the client models over 10 epochs and subsequently aggregating the model using the FedProx algorithm on the server. To ensure unbiased outcomes, the performance of the global model in the server was evaluated using predefined metrics, while also calculating the metrics of the individual client models. Table 1 presents a comprehensive summary of the server's (global model) performance across different training rounds. In the initial round (Round 1), the server model exhibited relatively high loss of 0.895 and low IoU of 0.0555, indicating a suboptimal level of segmentation accuracy. However, the model showed promise in specificity of 0.8452, suggesting its ability to correctly identify true negatives.

Table 1. Server (global model) results

Global	Dice Loss	IOU	Sensitivity	Specificity	F1 Score	Accuracy
Round 1	0.895	0.0555	0.1575	0.8452	0.105	0.9204
Round 2	0.4364	0.3951	0.5214	0.9742	0.5636	0.9399
Round 3	0.3664	0.4639	0.5226	0.9897	0.6336	0.9525
Round 4	0.3282	0.5064	0.5842	0.9876	0.6718	0.9549
Round 5	0.3495	0.484	0.5518	0.9895	0.6505	0.9541
Round 6	0.2924	0.5494	0.6066	0.9919	0.7076	0.9607

As training progressed to subsequent rounds, notable improvements in the server model's performance were observed. The loss values consistently decreased, indicating reduced overall error in the model. The IoU scores steadily increased, reflecting improved segmentation accuracy and better alignment with ground truth masks. Furthermore, the F1 scores, providing a balanced measure of precision and recall, showed a consistent upward trend, indicating overall performance enhancements. By the final round (Round 6), the server model demonstrated significant advancements. It achieved a substantially lower loss of 0.2924, indicating a considerable reduction in error rate. The IoU score reached 0.5494, indicating a higher degree of overlap between predicted and ground truth masks. Additionally, the model exhibited excellent specificity of 0.9919, highlighting its proficiency in accurately identifying true negatives.

The performance of the models is visually represented in Fig. 4, showcasing various metrics evaluated for the server model and three client models throughout the rounds of FL. The line Fig. 4(a) demonstrates the increasing accuracy of the FL global model as the rounds progress. In Fig. 4(b), the dice

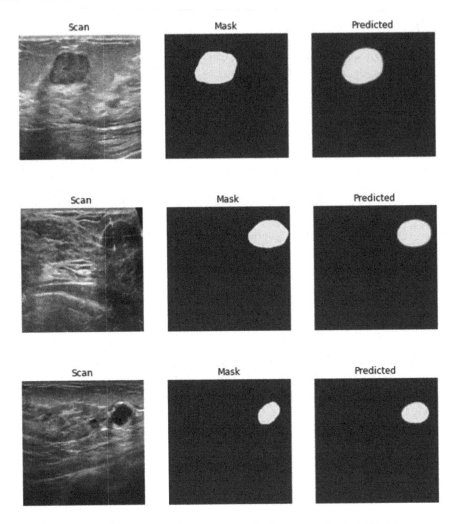

Fig. 4. Performance metrics vs. round of server and client Models

loss of the models exhibits a decreasing trend, indicating improved segmentation performance. Figure 4(c) displays the IoU scores, showing an increasing trend and enhanced overlap between predicted and ground truth segmentation masks. Lastly, Fig. 4(d) presents the sensitivity of the models, with the FL global model showcasing a progressive increase, suggesting an improved classification of positive instances.

In Fig. 5, the visualization of the segmented global test scans obtained from our proposed model are displayed. The visualization highlight the high accuracy achieved by our model in segmenting the USBCI, with minimal noise observed in the predicted images. Upon comparing the predicted images with the ground truth images or mask images, it is evident that our model successfully captures

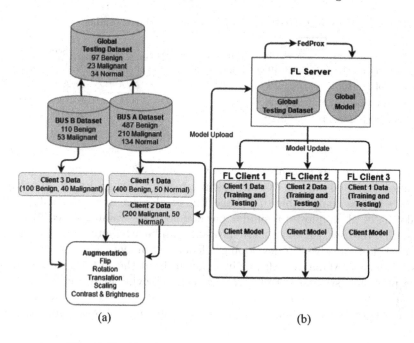

Fig. 5. Predicted outcome using the proposed model

the essential features of the lesions. This is demonstrated by the remarkable alignment between the boundaries of the predicted and ground truth images, emphasizing the model's ability to accurately identify and delineate BC lesions.

5 Discussion

The success of our proposed FedProx model in segmenting non-IID USBCI holds significant implications for medical image analysis, particularly in accurate BC diagnosis. By leveraging the Attention U-Net model, our method achieves a harmonious balance between local and global learning. The incorporation of attention mechanisms allows the model to focus on relevant features while disregarding irrelevant ones, resulting in improved segmentation accuracy.

The effectiveness of our proposed system was evaluated through a comparative analysis with other relevant studies (Table 2). Few papers have focused on decentralized learning on Ultrasound Breast Cancer Imaging (USBCI), making our research novel in this area. Maha M. Althobaiti et al. and Wicaksana et al. used centralized models but with variant datasets of USBCI. Yuchao Lyu et al. achieved outstanding performance with the AMS-PAN central model.

Table 2. FedProx comparison with related studies

Study	Dice Loss	IoU	Sensitivity	Specificity	F1 Score	Accuracy
Yuchao Lyu et al. [7]	0.1929	0.6853	0.7930	0.9854	N/A	0.9713
Wicaksana et al. [10]	0.19	N/A	N/A	N/A	N/A	N/A
Maha M. Althobaiti et al. [19]	N/A	N/A	0.9890	N/A	0.9890	0.9943
This Study(FedProx)	**0.29**	**0.55**	**0.64**	**0.99**	**0.71**	**0.96**

Our objective was to achieve accuracy within a 3% error range compared to the aforementioned papers. We successfully reached this goal with an accuracy of 96%, which is comparable to Yuchao Lyu et al.'s accuracy of 97.13%. While Wicaksana et al. employed a similar methodology by training a FedMix model on a similar dataset, the differences in results can be attributed to the challenges and constraints involved in developing a federated model.

Additionally, Maha M. Althobaiti et al. utilized the SEODTL-BDC model with a photoacoustic dataset of breast cancer, achieving high accuracy. To validate our results, we compared them to theirs using a 70-30 split, and the close proximity between the outcomes suggests the reliability of our simpler method.

6   Conclusion

Our study shows that combining USBCI with the FedProx algorithm enhances BC detection accuracy. The results indicate that FedProx is an excellent method for medical image analysis across devices, preserving privacy without compromising accuracy. The proposed model achieves 96% accuracy in image segmentation, allowing for a small error in tumor boundary detection. This high accuracy enables medical professionals to locate tumors precisely and identify regions of interest. The model also demonstrates high specificity, correctly identifying true negative cases. Overall, the proposed model offers a viable approach to medical image classification, ensuring privacy preservation and improved generalization while maintaining comparable performance to traditional models. These findings highlight the potential of FL in medical imaging and the importance of exploring novel DL techniques for healthcare challenges. Our work contributes to the research on FL in medical image analysis, particularly in improving BC detection.

References

1. National Breast Cancer Foundation 2023. www.nationalbreastcancer.org/breast-cancer-facts/
2. Centers for Disease Control and Prevention 2022. www.cdc.gov/cancer/breast/statistics/index.htm
3. World Health Organization. www.who.int/activities/preventing-cancer

4. Moghbel, M., Ooi, C.Y., Ismail, N., Hau, Y.W., Memari, N.: A review of breast boundary and pectoral muscle segmentation methods in computer-aided detection/diagnosis of Breast Mammography. Artif. Intell. Rev. **53**(3), 1873–1918 (2019). https://doi.org/10.1007/s10462-019-09721-8

5. Yuan, X., Wang, Y., Yuan, J., Cheng, Q., Wang, X., Carson, P.L.: Medical breast ultrasound image segmentation by machine learning. Ultrasonics **2**(5), 99–110 (2016)

6. Vakanski, A., Xian, M., Freer, P.E.: Attention-enriched deep learning model for breast tumor segmentation in ultrasound images. Ultrasound Med. Biol. **46**(10), 2819–2833 (2020). https://doi.org/10.1016/j.ultrasmedbio.2020.06.015

7. Lyu, Y., Yinghao, X., Jiang, X., Liu, J., Zhao, X., Zhu, X.: AMS-PAN: breast ultrasound image segmentation model combining attention mechanism and multiscale features. Biomed. Signal Process. Control **81**, 104425 (2023). https://doi.org/10.1016/j.bspc.2022.104425

8. Tedeschini, B.C., et al.: Decentralized federated learning for healthcare networks: a case study on tumor segmentation. IEEE Access **10**, 8693–8708 (2022). https://doi.org/10.1109/ACCESS.2022.3141913

9. Yang, T., Jingshuang, X., Zhu, M., An, S., Gong, M., Zhu, H.: FedZaCt: federated learning with Z average and cross-teaching on image segmentation. Electronics **11**(20), 3262 (2022). https://doi.org/10.3390/electronics11203262

10. Wicaksana, J.E.: FedMix: mixed supervised federated learning for medical image segmentation. IEEE Trans. Med. Imaging **42**, 1955–1968 (2022). https://doi.org/10.1109/TMI.2022.3233405

11. Mouhni, N., Elkalay, A., Chakraoui, M., Abdali, A., Ammoumou, A., Amalou, I.: Federated learning for medical imaging: an updated state of the art. Ingénierie des systèmes d'information **27**(1), 143–150 (2022). https://doi.org/10.18280/isi.270117

12. Jabeen, K., et al.: Breast cancer classification from ultrasound images using probability-based optimal deep learning feature fusion. Sensors **22**(3), 807 (2022). https://doi.org/10.3390/s22030807

13. Lazo, J.F., Moccia, S., Frontoni, E., De Momi, E.: Comparison of different CNNs for breast tumor classification from ultrasound images. arXiv:2012.14517 [eess.IV] (2020)

14. Yang, Z., et al.: Robust split federated learning for U-shaped medical image networks. arXiv:2212.06378 [eess.IV] (2022)

15. Roth, H.R., et al.: Federated learning for breast density classification: a real-world implementation. In: Albarqouni, S., et al. (eds.) DART/DCL -2020. LNCS, vol. 12444, pp. 181–191. Springer, Cham (2020). https://doi.org/10.1007/978-3-030-60548-3_18

16. Li, T., Sahu, A.K., Zaheer, M., Sanjabi, M., Talwalkar, A., Smith, V.: Federated optimization in heterogeneous networks. arXiv:1812.06127 [cs.LG] (2020)

17. Yap, M.H., et al.: Automated breast ultrasound lesions detection using convolutional neural networks. IEEE J. Biomed. Health Inf. **22**(4), 1218–1226 (2017). https://doi.org/10.1109/jbhi.2017.2731873

18. Al-Dhabyani, W., Gomaa, M., Khaled, H., Fahmy, A.: Dataset of breast ultrasound images. Data Brief **28**, 104863 (2020). https://doi.org/10.1016/j.dib.2019.104863

19. Althobaiti, M.M., et al.: Deep transfer learning-based breast cancer detection and classification model using photoacoustic multimodal images. Biomed. Res. Int. **2022**, 1–13 (2022). https://doi.org/10.1155/2022/3714422

20. Oktay, O., et al.: Attention U-net: learning where to look for the pancreas (2018). arXiv:1804.03999 [cs.CV]

21. Milletari, F., Navab, Ahmadi, S.A.: V-net: fully convolutional neural networks for volumetric medical image segmentation. In: 2016 Fourth International Conference on 3D Vision (3DV), pp. 565–571 (2016). https://doi.org/10.1109/3DV.2016.79
22. Everingham, M., Van Gool, L., Williams, C.K., Winn, J., Zisserman, A.: The pascal visual object classes (VOC) challenge. Int. J. Comput. Vision **88**(2), 303–338 (2009). https://doi.org/10.1007/s11263-009-0275-4

Super Images - A New 2D Perspective on 3D Medical Imaging Analysis

Ikboljon Sobirov[(✉)], Numan Saeed, and Mohammad Yaqub

Mohamed Bin Zayed University of Artificial Intelligence, Abu Dhabi, UAE
{ikboljon.sobirov,numan.saeed,mohammad.yaqub}@mbzuai.ac.ae

Abstract. In medical imaging analysis, deep learning has shown promising results. We frequently rely on volumetric data to segment medical images, necessitating the use of 3D architectures, which are commended for their capacity to capture interslice context. However, because of the 3D convolutions, max pooling, up-convolutions, and other operations utilized in these networks, these architectures are often more inefficient in terms of time and computation than their 2D equivalents. Furthermore, there are few 3D pretrained model weights, and pretraining is often tricky. We present a simple yet effective 2D method to handle 3D data while efficiently embedding the 3D knowledge during training. We propose transforming volumetric data into 2D super images and segmenting with 2D networks to solve these challenges. Our method generates a super-resolution image by stitching slices side by side in the 3D image. We expect deep neural networks to capture and learn these properties spatially despite losing depth information. This work aims to present a novel perspective when dealing with volumetric data, and we test the hypothesis using CNN and ViT networks as well as self-supervised pretraining. While attaining equal, if not superior, results to 3D networks utilizing only 2D counterparts, the model complexity is reduced by around threefold. Because volumetric data is relatively scarce, we anticipate that our approach will entice more studies, particularly in medical imaging analysis.

Keywords: Medical Image Analysis · 3D Segmentation · 2D Segmentation · Cancer Diagnosis · Self-supervised Learning · Super Images

1 Introduction

3D medical imaging modalities, such as computed tomography (CT), magnetic resonance imaging (MRI), and positron emission tomography (PET), are used extensively in clinical practice. Doctors rely on them to understand the volumetric information (depth of tumor, etc.) to perform their diagnosis more accurately. As such, a significant number of the developed data-driven techniques for volumetric medical images process the data in 3D using 3D deep neural networks. While such an approach produces promising results in the research community, its applications are limited in clinical practice. Although it may hinder the accuracy of the solution, 2D approaches have benefits such as faster and more cost-effective performance. Methods that could utilize the benefits of 2D slice-wise and 3D processing are of great importance.

The real-life medical applications for volumetric data analysis via 3D networks are still limited for several reasons, one of which is the high complexity of the models, especially when working with volumetric data. Hence, the deployment of such DL models in medical practices becomes difficult. Recent works [7,16] mainly focus on increasing the performance by a small margin, which in turn increases the model sizes in parameters and FLOPs. We argue that reducing model sizes while keeping similar performance encourages the easier deployment of these models in medical practices.

It is true that performing segmentation using 3D data directly is praised to produce better results. Several authors [1,4,10] support using 3D datasets primarily because 3D networks can capture depth information that their 2D counterpart lacks and claim that this information is crucial to model learning. Another argument is that the nature of 3D data is closer to real life, which is why 3D models ought to perform better [1,10]. On the downside, instead of 2D, 3D convolutions, max pooling, and up-convolutions are applied during model learning, thus requiring much more computation power and training/inference time [3,6].

On the other hand, those who argue that 2D should still be in heavy use reinforce their claim that utilizing 2D images is more cost- and time-effective and offers more options to apply transfer learning [6,12,15]. Transfer learning, with weights pretrained on large-scale datasets, such as ImageNet [5], can be considerably beneficial to model learning, especially when medical datasets are scarce. Another advantage of using 2D images is that 3D can be easily converted to multiple 2D slices, generating a larger scale set than relying on a limited number of 3D counterparts. Another vivid upside is that there are numerous 2D architectures available for the encoder of U-Net [4] like models [11]. This makes 2D models easier to customize and adjust to the need of the problem at hand.

Considering the abovesaid advantages of the 2D approach, this paper introduces a new perspective on using volumetric data in a 2D fashion. The intuition behind the approach is visualized in Fig. 1 (top), where a clinician is examining the scan from a bird's-eye view for slice-wise comparative analysis. We generate 2D super images (SIs) from 3D input by stacking the depth information (i.e., slices) side-by-side, and we train a 2D network for the same task. A similar notion was proposed by [6] on natural videos; unlike them, this concept is newly introduced to the medical field and volumetric data in particular. This novel approach in the segmentation task can achieve comparable results to a 3D model counterpart and reduce the model complexity by around threefold. The main contributions of our paper are:

- We introduce a new perspective on biomedical volumetric data by casting them into super images and training 2D models with them.

– We empirically show that pretraining techniques can easily boost the performance of the models that use super images.
– We validate our method on different CT, PET, and MRI datasets to show the effectiveness of our approach.
– The approach achieves comparable results to a 3D model counterpart and reduces the model complexity by around threefold.

2 Methodology

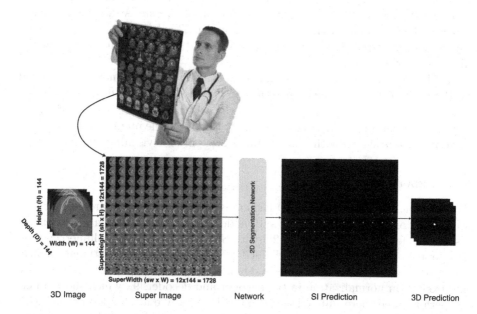

Fig. 1. (Top) The figure shows the intuition behind using super images. Similar to how clinicians examine the scans from a bird's-eye view in a slice-wise comparative approach, we generate SIs side by side such that DL models can analyze the images similarly. Image: Adobe Stock [14]. (Bottom) The figure shows the construction of super images from volumetric data. We rearrange the depth dimension by assembling the slices to generate the super image. It is then fed to a 2D segmentation network. The model yields the prediction mask, which is then rearranged back to the original shape. Note that the volumetric prediction mask shows a tumor region for visualization purposes.

The proposed approach is relatively simple to implement yet effective in training. In brief, volumetric data are converted to SIs, a 2D network of choice is trained on them, and the model outputs are cast back to the original dimensions. More details are provided below.

2.1 Super Image Generation

A 3D volume can provide features from the depth information for the model to learn since they use 3-dimensional kernels. Still, we expect these characteristics to be detectable and learnable in 2D SIs by well-designed deep neural networks. With that in mind, we generate SIs from volumetric data by taking slices and stitching them together side by side in order, as shown in Fig. 1(bottom). Given a 3D image $x_{inp} \in \mathbb{R}^{H \times W \times D \times C}$, where H is the height, W is the width, D is the depth, and C is the number of channels, the depth dimension is rearranged. The resulting image $s_{inp} \in \mathbb{R}^{\hat{H} \times \hat{W} \times C}$ is now 2D, where $\hat{H} = H \times sh$, and $\hat{W} = W \times sw$; sh and sw represent the degree by which the height and width should be rearranged respectively to generate a grid size of $sh \times sw$. As a demonstration, the size of $144 \times 144 \times 144 \times 2$ (2 for CT and PET), having 144 as the depth, can be considered with sh of 12 and sw of 12, thus generating the SI in the dimensions of $1728 \times 1728 \times 2$.

2D U-Net (or any other 2D segmentation model, for that matter) is trained on these SIs to perform the segmentation. The model output in the dimensions of $s_{out} \in \mathbb{R}^{\hat{H} \times \hat{W} \times C}$ is rearranged back to the original data dimensions of $x_{out} \in \mathbb{R}^{H \times W \times D \times C}$ as depicted in Fig. 1 (bottom). Note that the predicted mask on the final 3D data is shown with a large tumor size only for visualization purposes.

2.2 Experiments

The approach of using SIs in a 2D fashion is tested with several architectures to see the performance discrepancies and model complexities between various models. All models were compared in 2D with SIs and 3D with the volumetric data. The first model was the vanilla U-Net [4], which was the foundational medical image segmentation model. The second model was modified U-Net with a squeeze and excitation normalization in the encoder and decoder [9], which showed better performance (here dubbed as SE-norm U-Net). We also experimented with vision transformer architectures; specifically, we studied the behavior of Swin UNETR [8] both in 2D with SIs and 3D with the volumetric data. Implementing Swin UNETR in both 2D and 3D is relatively easy, and the model is praised as one of the latest and best-performing ViT networks.

On top of training from scratch, we performed SSL-based pretraining for SIs. Inpainting (i.e., masking) and jigsaw puzzle approaches were experimented with as objective tasks where the perturbed image was reconstructed during pretraining. During finetuning, the models were initialized with the pretrained weights rather than the random initialization.

2.3 Datasets and Preprocessing

To validate our new approach, we experimented with two different datasets: head and neck tumor segmentation and outcome prediction (HECKTOR) challenge [2] and atrial segmentation challenge [13] datasets. Several preprocessing techniques are performed on both datasets accordingly.

Head and Neck Tumor: HECKTOR dataset comprises 224 CT and PET scans of patients with head and neck tumors for the training set (i.e., the dataset is available online[1]). Bounding box information comes with the dataset for localization of the tumor region, which was used to crop the scans and the mask down to the size of 144×144×144 mm^3 with consistency between the scans. Sample CT and PET slices are depicted in Fig. 2 (a) and (b), respectively, where the red area corresponds to the tumor region. Since the challenge organizers provide the bounding box information, the tumor is within the cropped region, and mappings between both modalities and mask are accurate. Further pre-processing techniques were re-sampling the data to have isotropic voxel spacing (1×1×1 mm^3) and the intensity normalization of both CT and PET data. CT scans were clipped in the range of (−1024, 1024) and normalized to (−1, 1), and Z-normalization was used for PET scans.

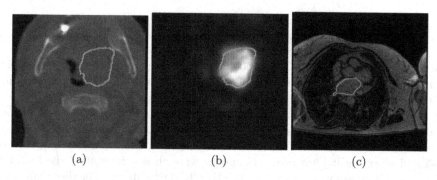

| (a) | (b) | (c) |

Fig. 2. The figure shows sample slices from the two datasets. (a) and (b) depict CT and PET scans with red regions highlighting tumors from the HECKTOR dataset, respectively; and (c) shows an MRI slice with the red region corresponding to the atrium from the Atria segmentation dataset (Color figure online)

Atria: Atrial segmentation challenge dataset includes 100 3D gadolinium contrast (GE) MRIs for the training set[2]. Figure 2 (c) shows an MRI sample with a red line delineating the atrial region. The scans are of different dimensions and thus were resized to the same size of 512 × 512 × 88^3. Similarly, intensity normalization was applied to the scans.

No further data augmentations were applied on either dataset unless reported otherwise. The testing set ground truth is inaccessible in both datasets; therefore, they are not used, and instead, k-fold cross-validation was utilized for all the experiments.

We purposely chose two datasets of different modalities (i.e., CT and PET in one and MRI in another) to test the hypothesis for generalizability. Moreover, the tasks are tumor segmentation in one dataset and atria in another. This shows that the idea is not limited to a specific tissue type with specific characteristics.

---

[1] www.aicrowd.com/challenges/miccai-2021-hecktor
[2] www.atriaseg2018.cardiacatlas.org/data

Table 1. Mean values of DSC, precision, recall, and HD95, and the number of parameters and FLOPs of different models on the validation set from 5-fold cross-validation are reported. Dim. correspond to dimensionality (either 3D volumetric or 2D SI). LK stands for the large kernel.

Models	Dim.	DSC	Precision	Recall	HD95	Params (M)	FLOPs (G)
U-Net	3D	0.728	0.734	0.783	3.035	3.61	518.61
U-Net	2D SI	0.725	0.743	0.772	4.689	1.21	319.98
SE-norm U-Net	3D	0.747	0.766	0.789	3.638	21.75	642.6
SE-norm U-Net	2D SI	0.737	0.767	0.764	5.185	8.51	493.87
SE-norm LK	2D SI	0.739	0.757	0.780	4.170	28.09	1181.84
Swin UNETR	3D	0.729	0.744	0.775	3.110	15.7	280.53
Swin UNETR	2D SI	0.732	0.753	0.774	5.176	6.3	214.86

3 Experimental Setup

We used a single NVIDIA RTX A6000 GPU for our experiments, and the implementation was done using the PyTorch library. We ran all the experiments for 100 epochs (both random initialization and finetuning). Pretraining was performed for 300 epochs. An AdamW optimizer with the initial learning rate of 0.001 and weight decay of 1e-5 was used, and a cosine annealing schedule that starts with the initial learning rate, decreasing it to the base learning rate of 1e-5 and resetting it after every 25 epochs were chosen to control the learning rate. The batch size was set to 2 for the HECKTOR dataset. The dice similarity coefficient (DSC) was chosen as the primary evaluation metric, and an additional 95% Hausdorff distance (HD95), precision, and recall were also calculated. Since we claim that the 2D approach is less costly, the number of parameters (in M) and FLOPs (in G) is also provided for all the experiments/models. k-fold cross-validation was utilized for all the experiments, and their mean values are listed in Sect. 4.

4 Results

We used two different datasets to verify this new approach to dealing with volumetric data. We used various models for the HEKCTOR dataset to see the approach's applicability. The atrial segmentation dataset was also used as additional support to showcase the generalizability of the approach.

Table 1 shows the mean results of 5-fold cross-validation for the HECKTOR dataset for different models. Dim. column indicates the dimensionality of the model used for that experiment, with 2D SI meaning a 2D model trained on SIs. The vanilla U-Net models on 3D and 2D SI achieved similar DSC of 0.728 and 0.725, respectively, whereas the number of parameters for the 2D model is three times less than that of the 3D model. Similar values can be observed for the other metrics for both approaches too.

Table 2. The table shows the results of the ablation study on the grid size of the super images. The HECKTOR dataset scans were cropped around the tumor region to form a small size of 80 × 80 × 48 for quick experimentation. The SIs were arranged in different ways, i.e., the *sw* and *sh* were chosen with different arrangements to see how they would affect the model learning. The results are the means of 5-fold cross-validation. The results indicate that the more square-like formation of SIs performs better than elongated rectangular arrangements.

Model	Image Size	sh	sw	DSC	Precision	Recall
3D U-Net	80 × 80 × 48	-	-	0.779	0.787	0.822
2D U-Net	640 × 480	8	6	0.778	0.799	0.810
2D U-Net	480 × 640	6	8	0.777	0.793	0.816
2D U-Net	960 × 320	12	4	0.770	0.809	0.801
2D U-Net	320 × 960	4	12	0.759	0.790	0.797
2D U-Net	1920 × 160	24	2	0.744	0.765	0.809
2D U-Net	160 × 1920	2	24	0.762	0.779	0.809

With the SE-norm U-Net, the performance increased across the metrics. Again the 3D training against the 2D with SIs reached similar DSC of 0.747 and 0.737, respectively. The model parameters for 3D, however, is 21.75M in contrast to 8.51M with the 2D approach. Before moving to the ViT models, we additionally experimented with a larger kernel model for the SI approach with the 2D network. We increased the kernel size to 7 (as opposed to 3 in the vanilla approach) in the training, increasing the model complexity (28.09M), but this did not seem to show too much of an improvement in DSC. Precision, recall, and HD95 values were similar in the three experiments.

Swin UNETR performance for the two approaches was also similar. The 3D network training with volumetric data achieved a DSC of 0.729, with a model complexity of 15.7M params, whereas the 2D SIs approach reached a slightly better DSC of 0.732 with only 6.3M params. Although precision and recall were similar for the two experiments, HD95 for the SI-based learning was higher with the Swin UNETR model. This is hypothesized to be caused by the nature of HD95 calculation on the 2D plane against the 3D plane.

Further ablation studies were conducted to study the behavior of the image arrangements as listed in Table 2. The HECKTOR dataset scans were cropped around the tumor region to form a small size of 80 × 80 × 48 for quick experimentation. The SIs were arranged in different ways, i.e., the *sw* and *sh* were chosen with different arrangements to see how they would affect the model learning. The results shown are the means of 5-fold cross-validation. The results indicate that the more square-like formation of SIs performs better than elongated rectangular arrangements.

Finally, SSL pretraining was applied for model initialization for the HECKTOR dataset. Masking (i.e., inpainting) and jigsaw puzzle techniques were selected as the pretraining tasks. During pretraining, the models were trained for 300 epochs for the reconstruction. During fine-tuning, they were trained for 100

Table 3. The table shows the results of vanilla 3D U-Net (comparison target) to SI-based 2D U-Net on the atrial segmentation dataset. The results are the mean of 4-fold cross-validation. PT stands for 2D U-Net pretrained on ImageNet1k and A stands for augmentations.

Model	Dim.	Image Size	sh	sw	DSC	Precision	Recall
U-Net	3D	$512 \times 512 \times 88$	–	–	0.893	0.898	0.894
U-Net	2D SI	5632×4096	11	8	0.812	0.902	0.785
U-Net	2D SI	4096×4096	8	8	0.851	0.913	0.822
PT	2D SI	4096×4096	8	8	0.895	0.872	0.878
PT&A	2D SI	4096×4096	8	8	0.901	0.919	0.890

epochs. Masking and jigsaw with the SE-norm U-Net improved DSC, reaching 0.741 and 0.738, respectively. Swin UNETR was experimented with the masking approach, ending up with a similar performance in DSC. An interesting observation here is that all the models achieved better HD95 values with the pretraining.

In the atrial segmentation task, because the dataset contains only 100 scans, we used 4-fold cross-validation, leaving more to the validation set for better generalizability. In this set of experiments listed in Table 3, there are two separate settings: (i) U-Net comparison and (ii) experimental image preprocessing for SIs. The first setting is to repeat the performance generalizability in a different task, and the second setting now focuses on how preprocessing techniques can easily be used and can boost the performance of the SI-based model.

For the first setting of the comparison of the networks, the images with the size of $512 \times 512 \times 88$ were used. The DSC of 0.893, the precision of 0.898, and the recall of 0.894 were achieved with the 3D U-Net. The SI generation with this size used the grid layout of 11×8. From ablation studies provided in Table 2, we can see that a better arrangement for SIs is having a square-like grid of SIs rather than elongated rectangles. A grid size of 11×8 was not the most favorable combination for generating SIs since its aspect ratio is high; therefore, it could reach only 0.812 DSC, 0.902 precision, and 0.785 recall values.

In the second setting, the images were preprocessed. This component was performed to show that preprocessing techniques are easily applicable and that they can help the model improve. We started with a one-on-one aspect ratio for the SIs, having 64 slices. Simple preprocessing such as this boosted the DSC score of the 2D model to 0.851. In the next step, 2D U-Net was initialized with ImageNet1k pretrained weights, pushing the DSC to 0.895, which is a 0.044 DSC jump from the base model. In the last experiment, this pretrained model was used with several sets of augmentations to see how far the model could go using a simple 2D U-Net. The augmentations were random flip, random affine, random elastic deformation, random anisotropy, and random gamma, and they are specific only for this experiment. This aggressive augmentation pushed the DSC to 0.901, which provides a substantial increase from the baseline that could reach 0.812 DSC. This shows that small preprocessing techniques are highly useful for the SI-based network to learn a better representation.

Table 4. The table shows the results of different models that are initialized with the pretrained weights and finetuned for the task. Mean values of DSC, precision, recall, and HD95 of 5-fold cross-validation are reported.

Models	SSL	DSC	Precision	Recall	HD95
SE-norm U-Net	Masking	0.741	0.764	0.773	4.708
SE-norm U-Net	Jigsaw	0.738	0.756	0.777	4.267
Swin UNETR	Masking	0.732	0.763	0.764	4.756

4.1 Qualitative Results

We studied the performance comparison in terms of qualitative analysis as well. Figure 3 shows a sample slice from the HECKTOR dataset for 3D U-Net and 2D U-Net, respectively. Note that here we performed inference on the cropped size of $80 \times 80 \times 48$. White and red represent ground truth and model prediction, respectively. The prediction from the 3D U-Net was cast to super image formation to compare the full view for both model predictions. In most of the visualized scans, the 3D network produces results that are much closer to the ground truth but are generally under-segmented; as such, beginning and ending slices (i.e., tiny tumor regions) are missed. The 2D network is more flexible with that, generally over-segmenting slightly more so as to capture even the tiny regions.

Figure 4 illustrates a sample output for the pretrained model reconstruction using the SE-norm U-Net. The reconstruction is visually reasonable, with only a few mistakes in the masked regions. Three regions are highlighted: red indicates a masked region that is poorly restored, blue shows a masked region that is well restored, and green shows a non-masked region that was reconstructed well during pretraining. We assume that the model learned useful features during pretraining for the masking task that should help the finetuning performance.

5   Discussion

The proposed concept of casting the 3D problem to 2D was validated using two datasets of different modalities and tissue types. We can see that 2D models based on SIs can achieve up-to-par or better results compared to 3D models. The head and neck tumor task is challenging, and the variability in the CT and PET image appearance is large. When visually analyzed, we found that CT scans in this dataset for underperforming models contain artifacts, generally in the teeth area. However, when we investigated the segmentation results on multiple scans, the 2D model on SIs performs well with tumor edges, generally over-segmenting, whereas the 3D model ignores these tiny regions, as is exemplified in Fig. 3. The 3D model performs better overall due to the tumor region's main areas, where it delineates the regions more accurately.

The atrial segmentation task, being completely dissimilar from the other task, did not pose as much difficulty as the first dataset. Applying basic preprocessing

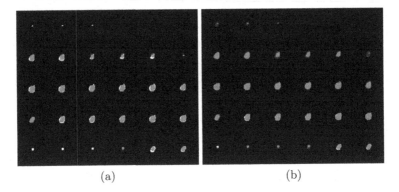

(a)   (b)

Fig. 3. The figure shows qualitative results on 3D U-Net (on volume) and 2D U-Net (on SI) segmentation results on a HECKTOR dataset sample, respectively. Note that the inference is on the cropped size of $80 \times 80 \times 48$. White is the ground truth, and red represents the prediction mask. Note that 3D U-Net results were cast to an SI form after its prediction for full-view comparison. 3D network results (left) are much stricter than the 2D results (right), whereas the 2D model allows more over-segmentation, especially in the small-size tumor regions (Color figure online)

techniques that easily push the SI-based 2D network to have an almost 9 percent increase in DSC is a good indicator of how much it can improve.

Thoroughly analyzing the problem, we put forth four main arguments as to why it is preferred to use 2D with SIs over 3D networks. First, although sometimes that extra bit of improvement in the performance is considered useful, it is imperative to have feasibly deployable models when it comes to the applications. With our approach of converting the 3D data into 2D SIs, the performance in DSC is still competitive while lowering the complexity by threefold. Second, it allows the application of pretrained weights from large-scale natural image datasets on medical images. Its effect using ImageNet1k can be seen in our results. Pretraining the model on large-scale natural image datasets and fine-tuning it on medical applications can be much easier with this approach. Third, because of natural images and a larger general computer vision community, a higher number of 2D networks are available. Such networks first come into the 2D world before moving to the medical imaging tasks. When dealing with volumetric data using SIs, employing these models is much easier. Finally, SSL on 2D datasets and 2D networks is much simpler and quicker. Plus, the availability of SSL pretrained models is higher in the 2D community. We experimented with two methods that indeed showed improvement in DSC, but more importantly, the implementation in 2D is preferred due to the simplicity and cheaper cost.

The 2D method performs equally well with the 3D approaches, even without the depth information. We believe that the underlying mechanism for its success is the "big picture" format of the SIs. In other words, discriminative features that are extracted from the depth information in 3D are still captured by the 2D restructured images. Because of the downsampling process, the field of view for the lower-level blocks of the models captures a large enough receptive field from multiple 2D images to learn efficiently and effectively.

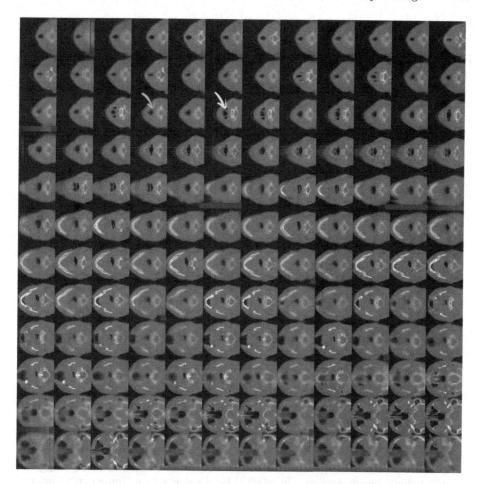

Fig. 4. The figure shows the pretraining output sample for the SE-norm U-Net model for the masking approach. The reconstruction seems visually reasonable, with a few errors in the masked regions. The red arrow shows a masked region that is poorly reconstructed, the blue arrow shows a well-reconstructed masked region, and the green arrow shows a region that was not masked. It can be assumed that the model learned relatively useful features during pretraining for the reconstruction (e.g., masking) task that should be helpful during finetuning (Color figure online)

Certain challenges may be faced using the SI formation when real-world applications are concerned, such as the variation in the number of slices in each scan (used as an essential metric in SI generation). Notably, this issue also exists in the 3D scenario. There are several techniques to address such challenges, including resizing, cropping and/or padding to bring the scans to uniform dimensions. These techniques that are used to address the potential issues already exist for 2-dimensional data in the literature.

6   Conclusion

In this work, we present a 2D DL method that can efficiently process 3D data, reducing the model complexity by around three times and still reaching similar or even better performance. In the HECKTOR dataset, simple 2D models on SIs can achieve results comparable to more powerful 3D U-Net results with much less complexity. Similarly, in the atrial segmentation dataset, the approach shows promising potential, primarily when powered by additional preprocessing techniques. We believe there is a potential for this new method of handling 3D medical data to reach the state-of-the-art with much less complexity, making it easier to deploy the models in real-world medical applications.

References

1. Ahn, B.B.: The compact 3D convolutional neural network for medical images. Standford University (2017)
2. Andrearczyk, V., Oreiller, V., Depeursinge, A.: Head and neck tumor segmentation in pet/ct. Medical Image Analysis (2021), www.aicrowd.com/challenges/miccai-2021-hecktor
3. Baumgartner, C.F., Koch, L.M., Pollefeys, M., Konukoglu, E.: An exploration of 2D and 3D deep learning techniques for cardiac MR image segmentation. In: Pop, M., Sermesant, M., Jodoin, P.-M., Lalande, A., Zhuang, X., Yang, G., Young, A., Bernard, O. (eds.) STACOM 2017. LNCS, vol. 10663, pp. 111–119. Springer, Cham (2018). https://doi.org/10.1007/978-3-319-75541-0_12
4. Çiçek, Ö., Abdulkadir, A., Lienkamp, S.S., Brox, T., Ronneberger, O.: 3D U-Net: learning dense volumetric segmentation from sparse annotation. In: Ourselin, S., Joskowicz, L., Sabuncu, M.R., Unal, G., Wells, W. (eds.) MICCAI 2016. LNCS, vol. 9901, pp. 424–432. Springer, Cham (2016). https://doi.org/10.1007/978-3-319-46723-8_49
5. Deng, J., Dong, W., Socher, R., Li, L.J., Li, K., Fei-Fei, L.: Imagenet: a large-scale hierarchical image database. In: 2009 IEEE Conference on Computer Vision and Pattern Recognition, pp. 248–255. IEEE (2009)
6. Fan, Q., Chen, C.F., Panda, R.: Can an image classifier suffice for action recognition? In: International Conference on Learning Representations (2021)
7. Feng, X., Tustison, N.J., Patel, S.H., Meyer, C.H.: Brain tumor segmentation using an ensemble of 3D u-nets and overall survival prediction using radiomic features. Front. Comput. Neurosci. 14, 25 (2020), https://doi.org/10.3389/fncom.2020.00025, https://www.frontiersin.org/article/10.3389/fncom.2020.00025
8. Hatamizadeh, A., et al.: Swin unetr: swin transformers for semantic segmentation of brain tumors in MRI images. In: International MICCAI Brainlesion Workshop. pp. 272–284. Springer (2022)
9. Iantsen, A., Visvikis, D., Hatt, M.: Squeeze-and-excitation normalization for automated delineation of head and neck primary tumors in combined PET and CT Images. In: Andrearczyk, V., Oreiller, V., Depeursinge, A. (eds.) HECKTOR 2020. LNCS, vol. 12603, pp. 37–43. Springer, Cham (2021). https://doi.org/10.1007/978-3-030-67194-5_4
10. Milletari, F., Navab, N., Ahmadi, S.A.: V-net: fully convolutional neural networks for volumetric medical image segmentation. In: 2016 Fourth International Conference on 3D Vision (3DV), pp. 565–571. IEEE (2016)

11. Patravali, J., Jain, S., Chilamkurthy, S.: 2D-3D fully convolutional neural networks for cardiac MR segmentation. In: Pop, M., Sermesant, M., Jodoin, P.-M., Lalande, A., Zhuang, X., Yang, G., Young, A., Bernard, O. (eds.) STACOM 2017. LNCS, vol. 10663, pp. 130–139. Springer, Cham (2018). https://doi.org/10.1007/978-3-319-75541-0_14

12. Roth, H.R., et al.: A New 2.5D representation for lymph node detection using random sets of deep convolutional neural network observations. In: Golland, P., Hata, N., Barillot, C., Hornegger, J., Howe, R. (eds.) Medical Image Computing and Computer-Assisted Intervention – MICCAI 2014: 17th International Conference, Boston, MA, USA, September 14-18, 2014, Proceedings, Part I, pp. 520–527. Springer International Publishing, Cham (2014). https://doi.org/10.1007/978-3-319-10404-1_65

13. Xiong, Z., et al.: A global benchmark of algorithms for segmenting the left atrium from late gadolinium-enhanced cardiac magnetic resonance imaging. Med. Image Anal. **67**, 101832 (2021)

14. Yakobchuk, V.: Smart male practitioner examining mri scan image stock photo, stock.adobe.com/images/smart-male-practitioner-examining-mri-scan-image/169 384005

15. Yang, J., et al.: Reinventing 2D convolutions for 3D images. IEEE J. Biomed. Health Inform. **25**(8), 3009–3018 (2021)

16. Zhang, Y., Liu, H., Hu, Q.: Transfuse: Fusing transformers and CNNs for medical image segmentation. CoRR abs/2102.08005 (2021)

Author Index

G. Waiter et al. (Eds.): MIUA 2023, LNCS 14122, pp. 339–340, 2024.
https://doi.org/10.1007/978-3-031-48593-0

Printed in the United States
by Baker & Taylor Publisher Services